M16A1 AND M16A2 RIFLE MARKSMANSHIP

United States Army

Fredonia Books
Amsterdam, The Netherlands

M16A1 and M16A2 Rifle Marksmanship

by
United States Army

ISBN: 1-4101-0014-6

Copyright © 2002 by Fredonia Books

Reprinted from the 1989 edition

Fredonia Books
Amsterdam, The Netherlands
http://www.fredoniabooks.com

All rights reserved, including the right to reproduce
this book, or portions thereof, in any form.

FM 23-9

CONTENTS

	Page
PREFACE	iv

CHAPTER 1. INTRODUCTION
- Training Strategy 1-1
- Combat Conditions 1-4

CHAPTER 2. OPERATION AND FUNCTIONING
- Section I. Operational Characteristics 2-2
  - M16A1 Rifle 2-2
  - M16A2 Rifle 2-2
- II. Functioning 2-3
  - Steps of Functioning 2-3
  - Semiautomatic Mode (M16A1 and M16A2) 2-8
  - Automatic Fire Mode (M16A1) 2-8
  - Burst Fire Mode (M16A2) 2-9
- III. Malfunctions and Corrections 2-10
  - Stoppage 2-10
  - Major Categories of Malfunctions 2-11
  - Other Malfunctions 2-14
- IV. Ammunition 2-16
  - Types and Characteristics 2-16
  - Care and Handling 2-17
- V. Destruction of Materiel 2-18
  - Means of Destruction 2-18
  - Field-Expedient Methods 2-18

CHAPTER 3. RIFLE MARKSMANSHIP TRAINING
- Section I. Basic Program Implementation 3-1
  - Instructor/Trainer Selection 3-2
  - Duties of the Instructor/Trainer 3-2
- II. Conduct of Training 3-3
  - Mechanical Training 3-3
  - Marksmanship Fundamentals 3-5
  - Firing Positions 3-12
- III. Dry Fire 3-18
  - Conduct of Dry-Fire Training 3-18
  - Peer Coaching 3-18
  - Checklist for the Coach 3-20
  - Position of the Coach 3-21
  - Grouping 3-21
  - Concept of Zeroing 3-21

	Page
M16A1 Standard Sights and Zeroing	3-22
M16A2 Standard Sights and Zeroing	3-25
Downrange Feedback Training	3-27
Field Fire Training	3-27
Practice Record Fire	3-27
Record Fire	3-27

CHAPTER 4. COMBAT FIRE TECHNIQUES

Section		Page
I.	Suppressive Fire	4-1
	Nature of the Target	4-1
	Point of Aim	4-1
	Rate of Fire	4-2
II.	Rapid Semiautomatic Fire	4-3
	Effectiveness of Rapid Fire	4-3
	Modifications for Rapid Fire	4-3
	Rapid-Fire Training	4-7
III.	Automatic Fire	4-7
	Effectiveness of Automatic Fire	4-8
	Modifications for trhe Automatic Fire Position	4-9
	Training of Automatic Fire Techniques	4-10
IV.	Quick Fire	4-11
	Effectiveness of Quick Fire	4-11
	Modifications for Quick-Fire Techniques	4-13
	Training for Quick-Fire Techniques	4-13
V.	MOPP Firing	4-14
	Effects of MOPP Equipment on Firing	4-14
	Effects of Aiming Modifications	4-16
	Operation and Function Modifications	4-17
	MOPP Fire Exercises	4-18
VI.	Moving Target Engagement	4-19
	Moving Target Techniques	4-19
	Moving Target Fundamentals	4-20
	Single-Lead Rule for Moving Targets	4-21
	Lead Requirements	4-22
	Multipurpose Range Complex Train-Up	4-25

CHAPTER 5. NIGHT FIRING

	Page
Considerations	5-1
Principles of Night Vision	5-2
Target Engagement Techniques	5-4
Training	5-5

APPENDIX A. YEAR-ROUND MARKSMANSHIP TRAINING A-1

APPENDIX B. TARGET DETECTION B-1

APPENDIX C. TRAINING AIDS AND DEVICES C-1

                                                                                              Page

APPENDIX D. RIFLE RANGE SAFETY BRIEFING/RANGE
                            OPERATIONS CHECKLIST D-1
APPENDIX E. SCALED SILHOUETTE TARGETS E-1
APPENDIX F. PRECISION FIRING INFORMATION F-1
APPENDIX G. LIVE-FIRE EXERCISES G-1
APPENDIX H. REPRODUCIBLE FORMS H-1
                      GLOSSARY Glossary-1
                      REFERENCES References-1
                      INDEX Index-1

Preface

This manual provides guidance for planning and executing training on the 5.56-mm M16A1 and M16A2 rifles to include the conduct of basic rifle marksmanship and advanced rifle marksmanship. It is a guide for commanders, leaders, and instructors to develop training programs, plans, and lessons that meet the objectives/intent of the United States Army rifle marksmanship program and FM 25-100.

This manual is organized to lead the trainer through the material needed to conduct training in IET and units. Preliminary subjects include discussions on mechanical training, the weapons' capabilities, and the principles and fundamentals of marksmanship. Live-fire applications are scheduled after the soldier has demonstrated preliminary skills. Initial firing will be a grouping exercise that leads to the soldier adjusting the sights on the weapon and to setting the battlesight zero.

The proponent of this publication is HQ TRADOC. Submit changes for improving this publication on DA Form 2028 (Recommended Changes to Publications and Blank Forms) and forward it to Commandant, US Army Infantry School, ATTN: ATSH-I-V-P, Fort Benning, GA 31905-5593.

> Unless otherwise stated, whenever the masculine gender is used, both men and women are included.

FM 23-9

CHAPTER 1

Introduction

The procedures and methods used in the Army rifle marksmanship program are based on the concept that soldiers must be skilled marksmen who can effectively apply their firing skills in combat. FM 25-100 stresses marksmanship as a paramount soldier skill. The basic firing skills and exercises outlined in this manual must be a part of every unit's marksmanship training program. Unit commanders must gear their advanced marksmanship training programs to their respective METLs. The proficiency attained by a soldier depends on the proper training and application of basic marksmanship fundamentals. During initial marksmanship training, emphasis is on learning the firing fundamentals, which are taught in a progressive program to prepare soldiers for combat-type exercises.

TRAINING STRATEGY

Training strategy is the overall concept for integrating resources into a program to train individual and collective skills needed to perform a unit's wartime mission.

Training strategies for rifle marksmanship are implemented in TRADOC institutions (IET, NCOES, basic and advanced officer's courses) and in units. The overall training strategy is multifaceted and is inclusive of the specific strategies used in institution and unit programs. Also included are the supporting strategies that use resources such as publications, ranges, ammunition, training aids, devices, simulators, and simulations. These strategies focus on developing critical soldier skills, and on leader skills that are required for success in combat.

Two primary components compose the training strategies: **initial training** and **sustainment training**. Both may include individual and collective skills. Initial training is critical. A task that is taught correctly and learned well is retained longer and skills can be quickly regained and sustained. Therefore, initial training must be taught correctly the first time. However, eventually an individual or unit loses skill proficiency. This learning decay depends on many factors such as the difficulty and complexity of the task. Personnel turnover is a main factor in decay of collective skills, since the loss of critical team members requires retraining to regain proficiency. If a long period elapses between initial and sustainment training sessions or training doctrine is altered, retraining may be required.

The training strategy for rifle marksmanship begins in IET and continues in the unit. An example of this overall process is illustrated in Figure 1-1 and provides a concept of the flow of unit sustainment training (Appendix A). IET provides field units with soldiers who have been trained and who have demonstrated proficiency to standard in basic marksmanship tasks. The soldier graduating from these courses has been trained to maintain the rifle and to hit a point target. He has learned target detection, application of marksmanship fundamentals, and other skills needed to engage a target. The specific tasks and programs taught in IET are explained in Appendix A, FM 21-3, and in commanders' manuals.

FM 23-9

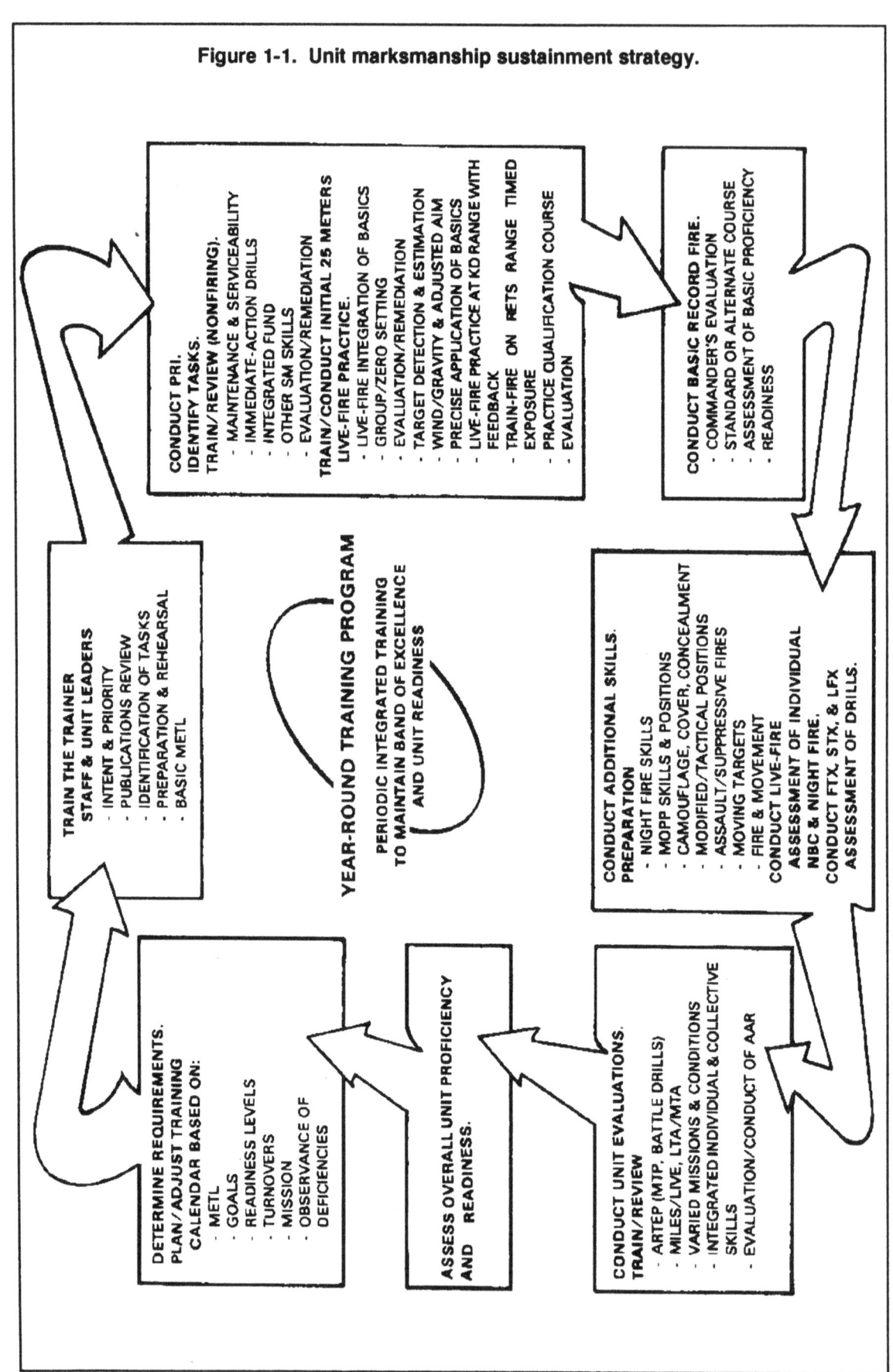

Figure 1-1. Unit marksmanship sustainment strategy.

Training continues in units on the basic skills taught in IET. Additional skills such as area fire are trained and then integrated into collective training exercises, which include platoon and squad live-fire STXs. (A year-round unit marksmanship training program is explained in Appendix A.) The strategy for sustaining the basic marksmanship skills taught in IET is periodic preliminary rifle instruction, followed by instructional and qualification range firing. However, a unit must set up a year-round program to sustain skills. Key elements include training of trainers, refresher training of nonfiring skills, and use of the Weaponeer or other devices for remedial training.

Additional skills trained in the unit include semiautomatic and automatic area fires, night fire, MOPP firing, and moving target training techniques. Related soldier skills of camouflage, cover and concealment, fire and movement, and preparation and selection of a fighting position are addressed in FM 21-3, which must be integrated into tactical training.

In the unit, individual and leader proficiency of marksmanship tasks are integrated into collective training to include squad, section, and platoon drills and STXs; and for the collective tasks in these exercises, and how they are planned and conducted, are in the MTP and battle drills books for each organization. (Force-on-force exercises using MILES are discussed in detail in TC 25-6). Based on the type organization, collective tasks are evaluated to standard and discussed during leader and trainer after-action reviews. Objective evaluations of both individual and unit proficiency provide readiness indicators and future training requirements.

A critical step in the Army's overall marksmanship training strategy is to train the trainers and leaders first. Leader courses and unit publications develop officer and NCO proficiencies necessary to plan and conduct marksmanship training and to evaluate the effectiveness of unit marksmanship programs. Training support materials are provided by the proponent schools to include field manuals, training aids, devices, simulators, and programs that are doctrinal foundations and guidance for training the force.

Once the soldier understands the weapon and has demonstrated skill in zeroing, additional live-fire training and a target acquisition exercise at various ranges are conducted. Target types and scenarios of increasing difficulty must be mastered to develop proficiency.

Initial individual training culminates in the soldier's proficiency assessment, which is conducted on the standard record fire range or approved alternates. This evaluation also provides an overview of unit proficiency and training effectiveness.

General marksmanship training knowledge and firing well are acquired skills, which perish easily. Skill practice should be conducted for short periods throughout the year. Most units have a readiness requirement that all soldiers must zero their rifles within a certain time after unit assignment. Also, soldiers must confirm the zeros of their assigned rifles before conducting a qualification firing. Units should conduct preliminary training and practice firing throughout the year due to personnel turnover. A year-round marksmanship sustainment program is needed for the unit to maintain the individual and collective firing proficiency requirements to accomplish its mission (see Appendix A).

COMBAT FACTORS

The ultimate goal of a unit rifle marksmanship program is well-trained marksmen. In order for a unit to survive and win on the battlefield, the trainer must realize that rifle qualification is not an end but a step toward reaching this combat requirement. To reach this goal, the soldier should consider some of the factors of combat conditions.

- Enemy personnel are seldom visible except when assaulting.

- Most combat fire must be directed at an area where the enemy has been detected or where he is suspected of being located but cannot be seen. Area targets consist of objects or outlines of men irregularly spaced along covered and concealed areas (ground folds, hedges, borders of woods).

- Most combat targets can be detected by smoke, flash, dust, noise, or movement and are visible only for a moment.

- Some combat targets can be engaged by using nearby objects as reference points.

- The range at which enemy soldiers can be detected and effectively engaged rarely exceeds 300 meters.

- The nature of the target and irregularities of terrain and vegetation may require a firer to use a variety of positions in addition to the prone or supported position to fire effectively on the target. In a defensive situation, the firer usually fires from a supported position.

- Choosing an aiming point in elevation is difficult due to the low contrast outline and obscurity of most combat targets.

- Time-stressed fire in combat can be divided into three types:

    - A single, fleeing target that must be engaged quickly.

    - Area targets that must be engaged with distributed fires that cover the entire area. The firer must maintain sustained fire on the sector he is assigned.

    - A surprise target that must be engaged at once with accurate, instinctive fire.

FM 23-9

CHAPTER 2

Operation and Function

The procedures and techniques described in this chapter provide commanders, planners, and trainers information on the M16A1 and M16A2 rifles. These include mechanical training, operation, functioning, preventive maintenance, and common malfunctions. Technical data are presented in a logical sequence from basic to the more complex. Additional information is provided in technical manuals for the rifle.

CLEAR the RIFLE

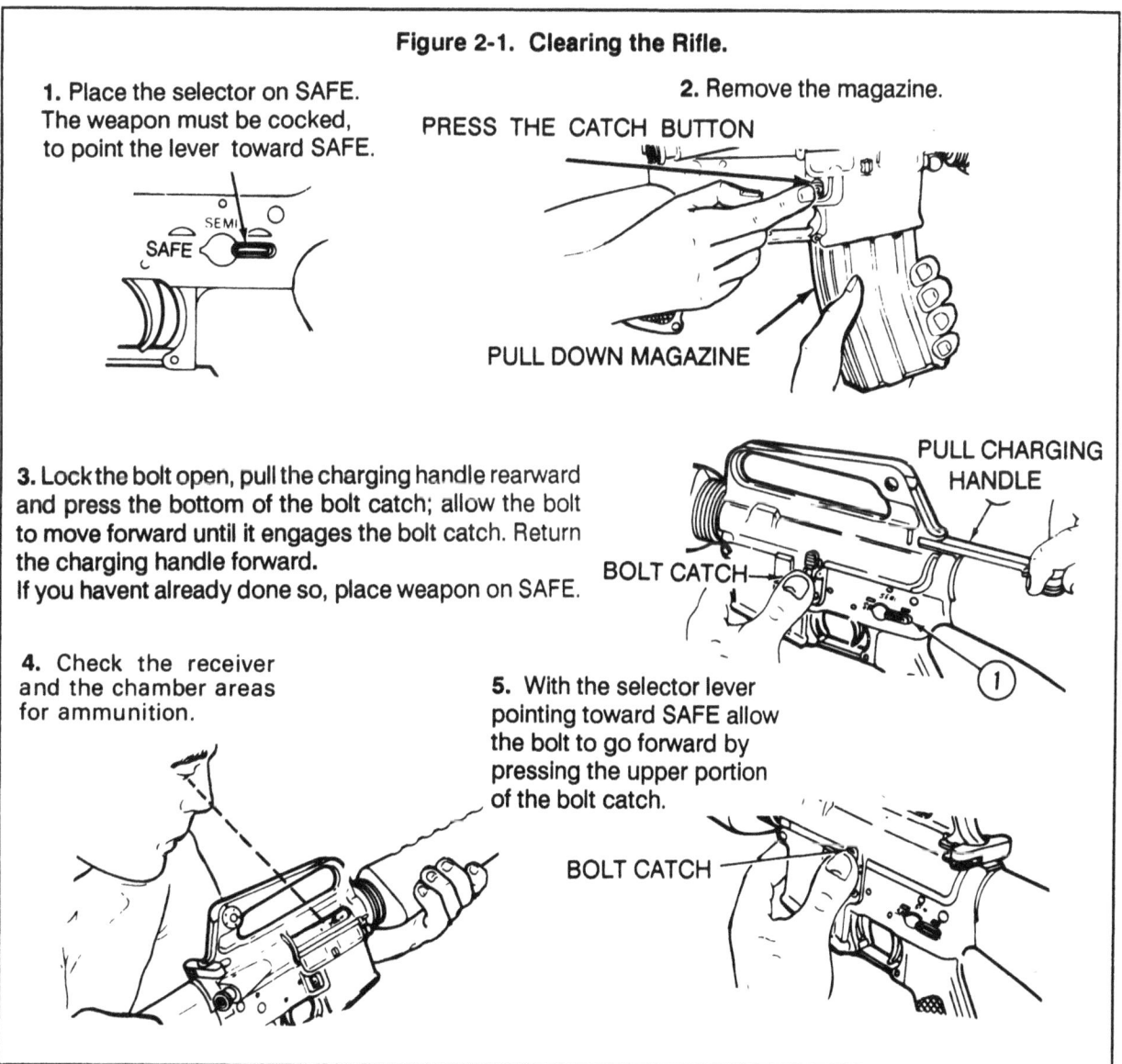

Figure 2-1. Clearing the Rifle.

1. Place the selector on SAFE. The weapon must be cocked, to point the lever toward SAFE.

2. Remove the magazine. PRESS THE CATCH BUTTON. PULL DOWN MAGAZINE.

3. Lock the bolt open, pull the charging handle rearward and press the bottom of the bolt catch; allow the bolt to move forward until it engages the bolt catch. Return the charging handle forward.
If you havent already done so, place weapon on SAFE.

4. Check the receiver and the chamber areas for ammunition.

5. With the selector lever pointing toward SAFE allow the bolt to go forward by pressing the upper portion of the bolt catch.

2-1

Section I. OPERATIONAL CHARACTERISTICS

This section describes general characteristics of the M16A1 and M16A2 rifles.

M16A1 RIFLE
The M16A1 rifle (Figure 2-2) is a 5.56-mm, magazine-fed, gas-operated, shoulder-fired weapon. It is designed for either semiautomatic or automatic fire through the use of a selector lever (SAFE, SEMI, and AUTO).

M16A2 RIFLE
The M16A2 rifle features several product improvements illustrated in this chapter and the operator's manual. The rifle (Figure 2-3) is a 5.56-mm, magazine-fed, gas-operated, shoulder-fired weapon. It is designed to fire either semiautomatic or a three-round burst through the use of a selector lever (SAFE, SEMI, and BURST).

NOTE: The procedures for disassembly, inspection, and maintenance of the M16A1 and M16A2 rifles are contained in the appropriate operator's technical manual.

Figure 2-2. Rifle, 5.56-mm, M16A1.

WEIGHT:	Kilograms	Pounds
M16A1 rifle, without cartridge magazine and sling	2.97	6.55
Firing weight with sling and loaded magazine:		
20-round	3.45	7.6
30-round	3.60	7.9
Bipod, M3	.27	.60
Bipod case	.09	.20
Bayonet knife, M7	.27	.60
Scabbard	.14	.30
Sling, M1	.18	.40
LENGTH:	**Centimeters**	**Inches**
M16A1 rifle with bayonet knife	112.40	44.25
M16A1 rifle overall with flash suppressor	99.06	39.00
Barrel with flash suppressor	53.34	21.00
Barrel without flash suppssor	41.80	20.00
AMMUNITION:		
M16A1, M193		
Complete round	179 grains	
Projectile	.55 grains	

Figure 2-3. Rifle, 5.56-mm, M16A2.

WEIGHT:	Kilograms	Pounds
M16A2 rifle, without cartridge magazine and sling	3.53	7.78
Firing weight with sling and loaded magazine:		
20-round	3.85	8.48
30-round	3.99	8.79
Bipod, M3	.27	.60
Bipod case	.09	.20
Bayonet knife, M9	.68	1.50
Scabbard	.14	.30
Sling, M1	.18	.40
LENGTH:	**Centimeters**	**Inches**
M16A2 rifle with bayonet knife,	113.99	44.88
M16A2 rifle overall with compensator	100.66	39.63
Barrel with compensator	53.34	21.00
Barrel without compensator	41.80	20.00
AMMUNITION:		
M16A2, M855		
Complet e round	190 grains	
Projectile	.62 grains	

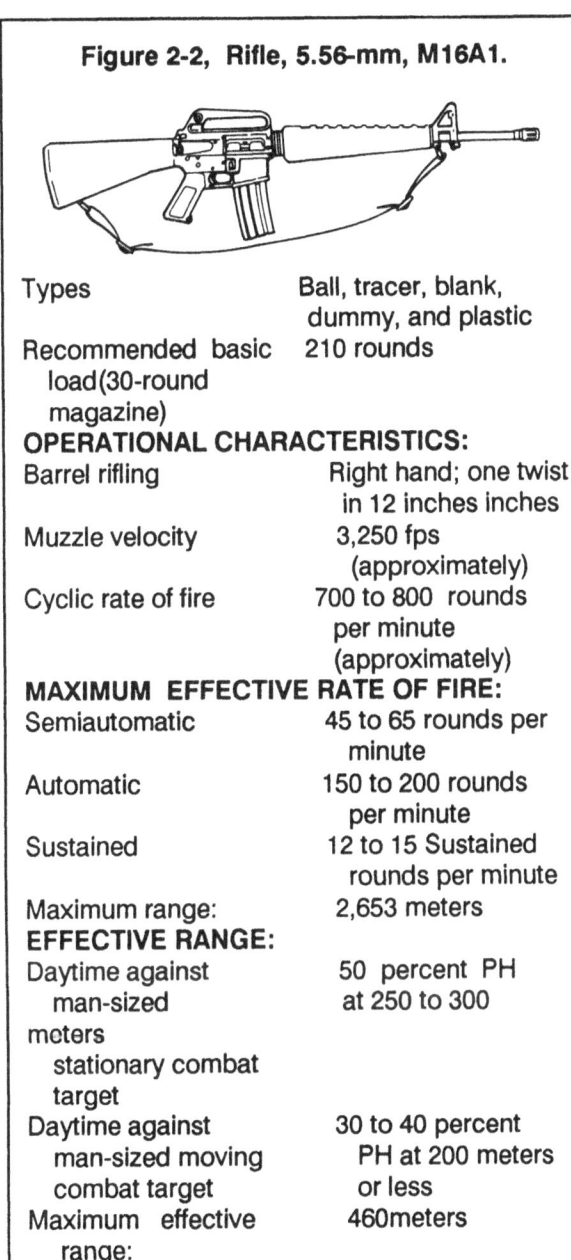

Figure 2-2. Rifle, 5.56-mm, M16A1.

Types	Ball, tracer, blank, dummy, and plastic
Recommended basic load (30-round magazine)	210 rounds
OPERATIONAL CHARACTERISTICS:	
Barrel rifling	Right hand; one twist in 12 inches inches
Muzzle velocity	3,250 fps (approximately)
Cyclic rate of fire	700 to 800 rounds per minute (approximately)
MAXIMUM EFFECTIVE RATE OF FIRE:	
Semiautomatic	45 to 65 rounds per minute
Automatic	150 to 200 rounds per minute
Sustained	12 to 15 Sustained rounds per minute
Maximum range:	2,653 meters
EFFECTIVE RANGE:	
Daytime against man-sized stationary combat target	50 percent PH at 250 to 300 meters
Daytime against man-sized moving combat target	30 to 40 percent PH at 200 meters or less
Maximum effective range:	460 meters

Figure 2-3. Rifle, 5.56-mm, M16A2.

Types	Ball, tracer, blank, dummy, and plastic
Recommended basic load (30-round magazine)	210 rounds
OPERATIONAL CHARACTERISTICS:	
Barrel rifling	Right hand; one twist in 7 inches
Muzzle velocity	3,100 fps (approximately)
Cyclic rate of fire	700 to 800 rounds per minute (approximately)
MAXIMUM EFFECTIVE RATE OF FIRE:	
Semiautomatic	45 rounds per minute
Automatic (3 round burst)	90 rounds per minute
Sustained	12 to 15 rounds per minute
Maximum range:	3,600 meters
EFFECTIVE RANGE:	
Daytime against man-sized stationary combat target	50 percent PH at 250 to 300 meters
Daytime against man-sized moving combat target	30 to 40 percent PH at 200 meters combat or less
Maximum effective range:	
Point target	550 meters
Area target	800 meters

Section II. FUNCTION

The soldier must understand the rifles' components and the mechanical sequence of events during the firing cycle. The M16A1 rifle is designed to function in either the semiautomatic or automatic mode. The M16A2 is designed to function in either the semiautomatic or three-round burst mode.

STEPS OF FUNCTIONING

The eight steps of functioning (feeding, chambering, locking, firing, unlocking, extracting, ejecting, and cocking) begin after the loaded magazine has been inserted into the weapon.

STEP 1: Feeding (Figure 2-4). As the bolt carrier group moves rearward, it engages the buffer assembly and compresses the action spring into the lower receiver extension. When the bolt carrier group clears the top of the magazine, the expansion of the magazine spring forces the follower and a new round up into the path of the forward movement of the bolt. The expansion of the action spring sends the buffer assembly and bolt carrier group forward with enough force to strip a new round from the magazine.

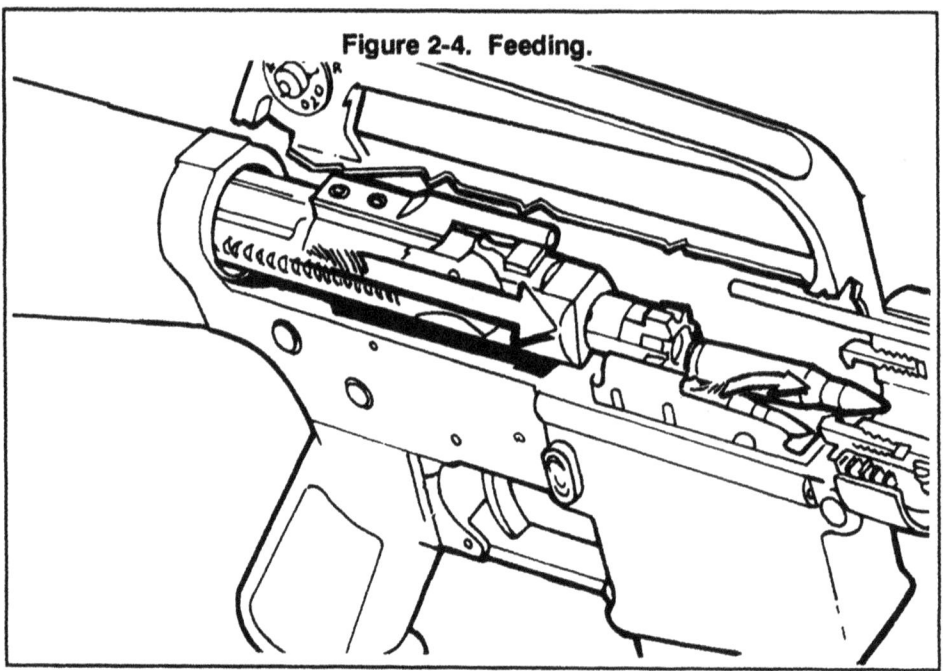

Figure 2-4. Feeding.

STEP 2: Chambering (Figure 2-5). As the bolt carrier group continues to move forward, the face of the bolt thrusts the new round into the chamber. At the same time, the extractor claw grips the rim of the cartridge, and the ejector is compressed.

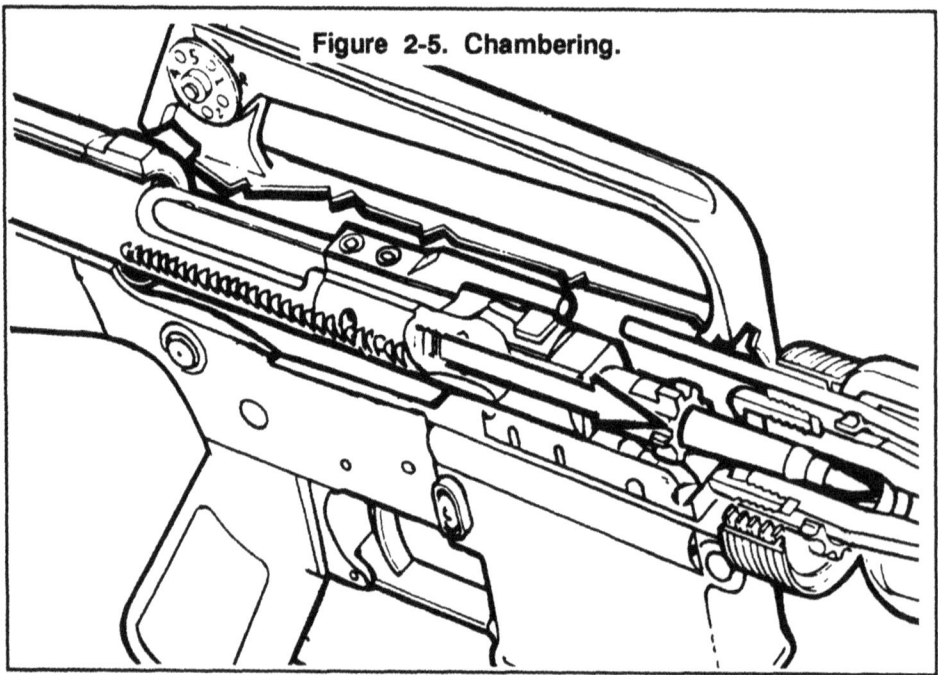

Figure 2-5. Chambering.

STEP 3: Locking (Figure 2-6). As the bolt carrier group moves forward, the bolt is kept in its most forward position by the bolt cam pin riding in the guide channel in the upper receiver. Just before the bolt locking lugs make contact with the barrel extension, the bolt cam pin emerges from the guide channel. The pressure exerted by the contact of the bolt locking lugs and barrel extension causes the bolt cam pin to move along the cam track (located in the bolt carrier) in a counterclockwise direction, rotating the bolt locking lugs in line behind the barrel extension locking lugs. The rifle is then ready to fire.

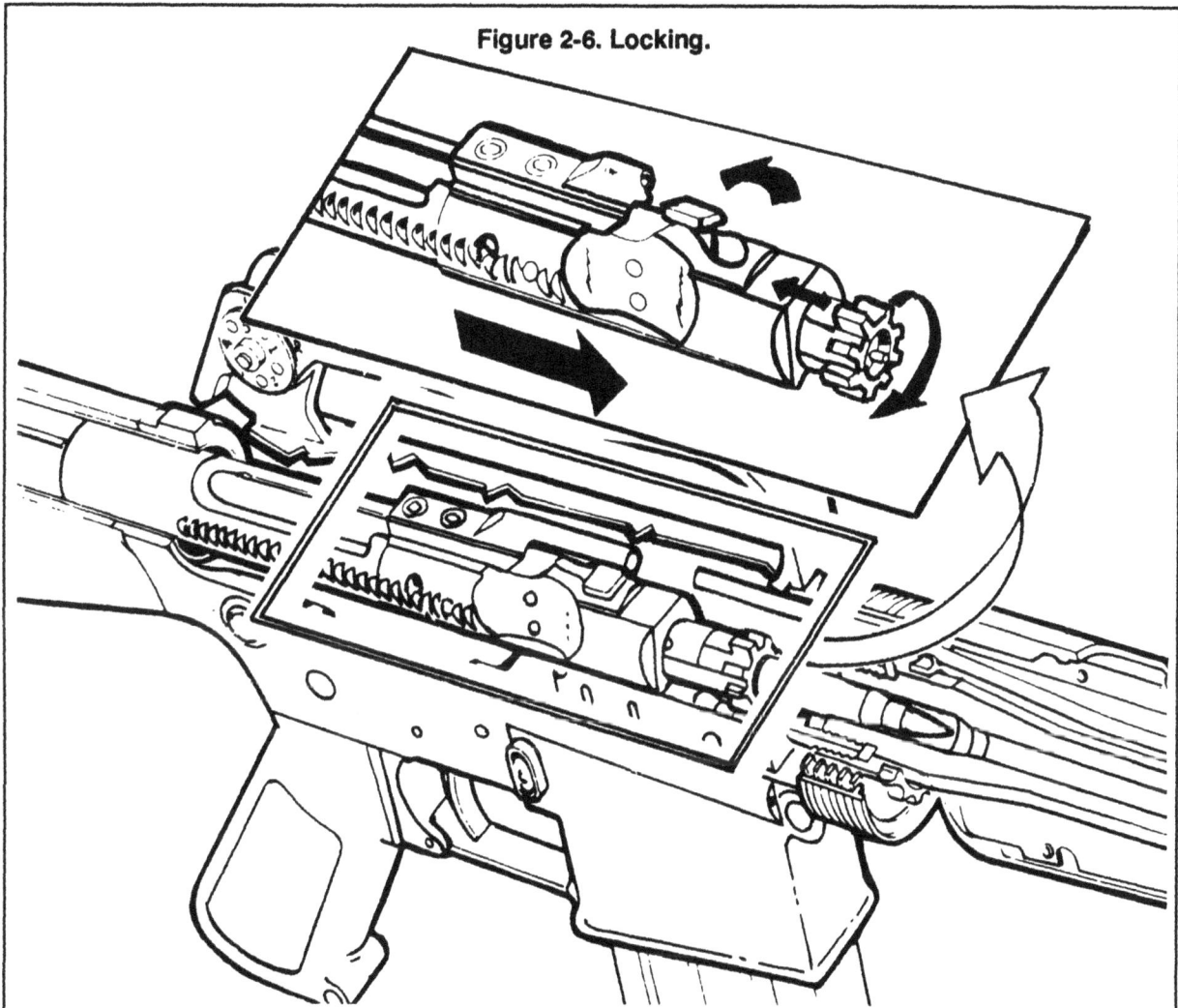

Figure 2-6. Locking.

STEP 4: Firing (Figure 2-7). With a round in the chamber, the hammer cocked, and the selector on SEMI, the firer squeezes the trigger. The trigger rotates on the trigger pin, depressing the nose of the trigger and disengaging the notch on the bottom on the hammer. The hammer spring drives the hammer forward. The hammer strikes the head of the firing pin, driving the firing pin through the bolt into the primer of the round.

When the primer is struck by the firing pin, it ignites and causes the powder in the cartridge to ignite. The gas generated by the rapid burning of the powder forces the projectile from the cartridge and propels it through the barrel. After the projectile has passed the gas port (located on the upper surface of the barrel under the front sight) and before it leaves the barrel, some gas enters the gas port and moves into the gas tube. The gas tube directs the gas into the bolt carrier key and then into the cylinder between the bolt and bolt carrier, causing the carrier to move rearward.

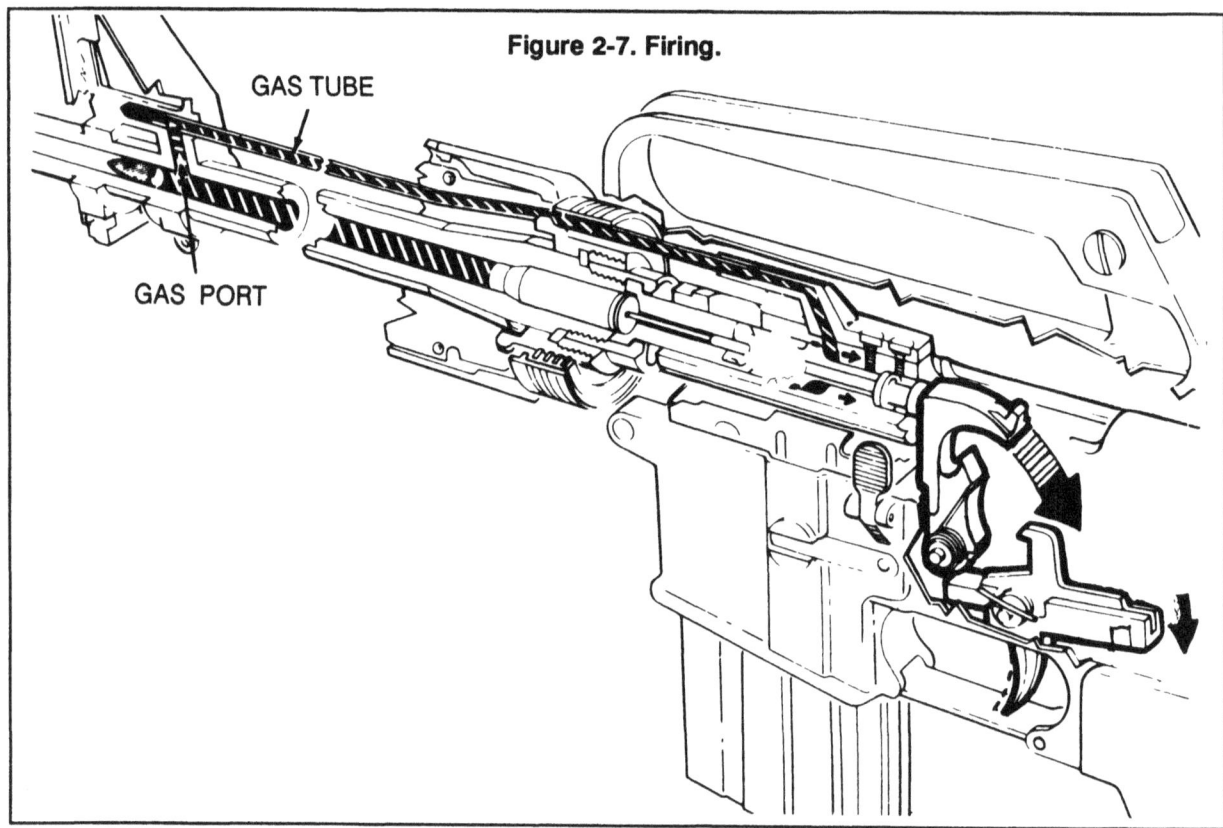

Figure 2-7. Firing.

STEP 5: Unlocking (Figure 2-8). As the bolt carrier moves to the rear, the bolt cam pin follows the path of the cam track (located in the bolt carrier). This action causes the cam pin and bolt assembly to rotate at the same time until the locking lugs of the bolt are no longer in line behind the locking lugs of the barrel extension.

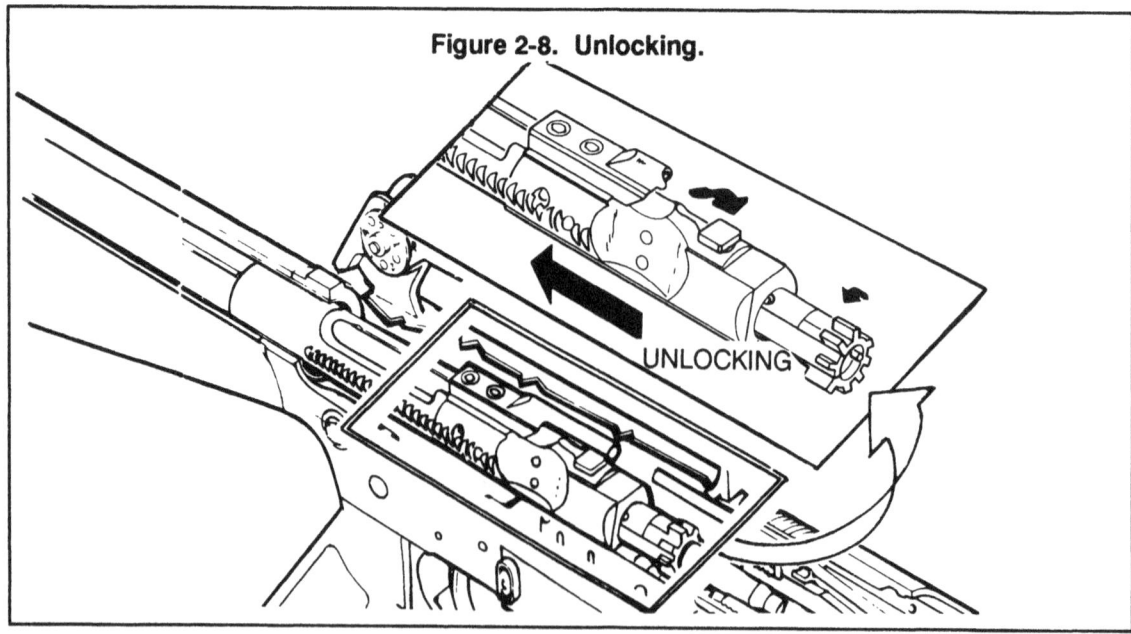

Figure 2-8. Unlocking.

STEP 6: Extracting (Figure 2-9). The bolt carrier group continues to move to the rear. The extractor (which is attached to the bolt) grips the rim of the cartridge case, holds it firmly against the face of the bolt, and withdraws the cartridge case from the chamber.

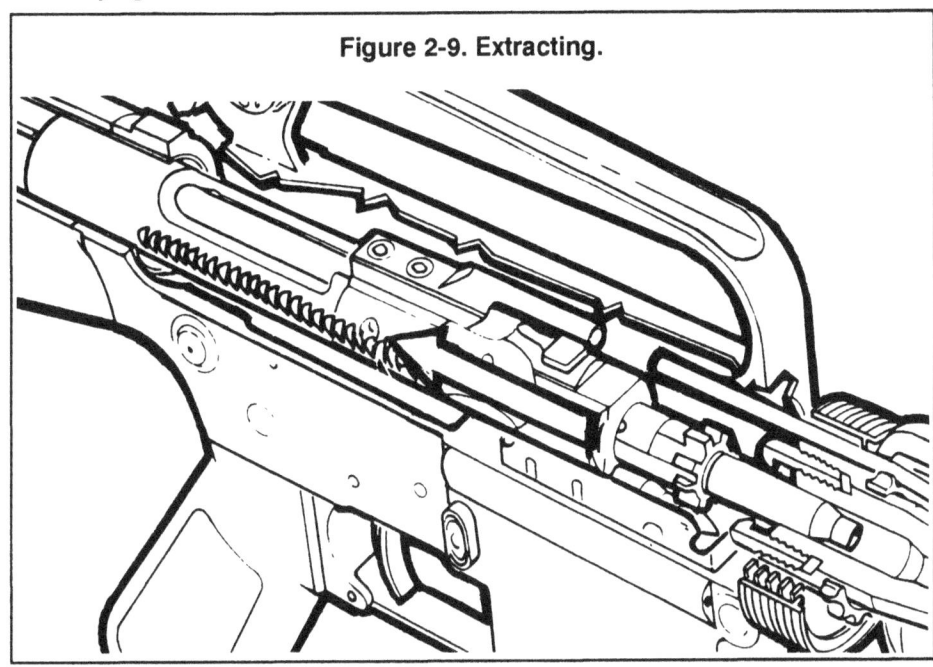

Figure 2-9. Extracting.

STEP 7: Ejecting (Figure 2-10). With the base of a cartridge case firmly against the face of the bolt, the ejector and ejector spring are compressed into the bolt body. As the rearward movement of the bolt carrier group allows the nose of the cartridge case to clear the front of the ejection port, the cartridge is pushed out by the action of the ejector and spring.

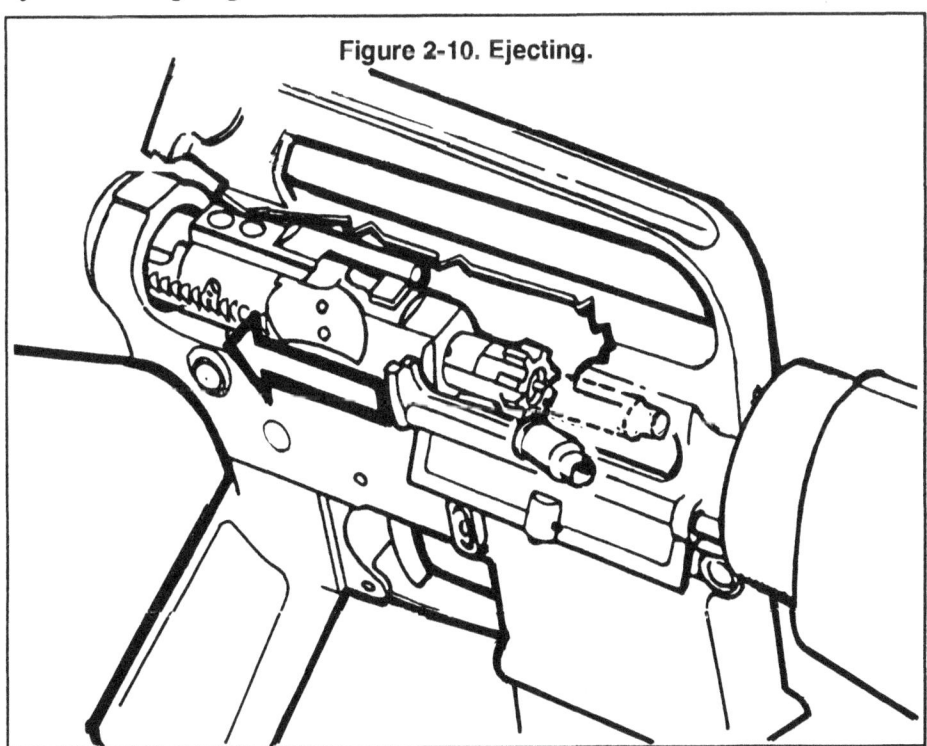

Figure 2-10. Ejecting.

STEP 8: Cocking (Figure 2-11). The rearward movement of the bolt carrier overrides the hammer, forcing it down into the receiver and compressing the hammer spring, cocking the hammer in the firing position. The action of the rifle is much faster than human reaction; therefore, the firer cannot release the trigger fast enough to prevent multiple firing.

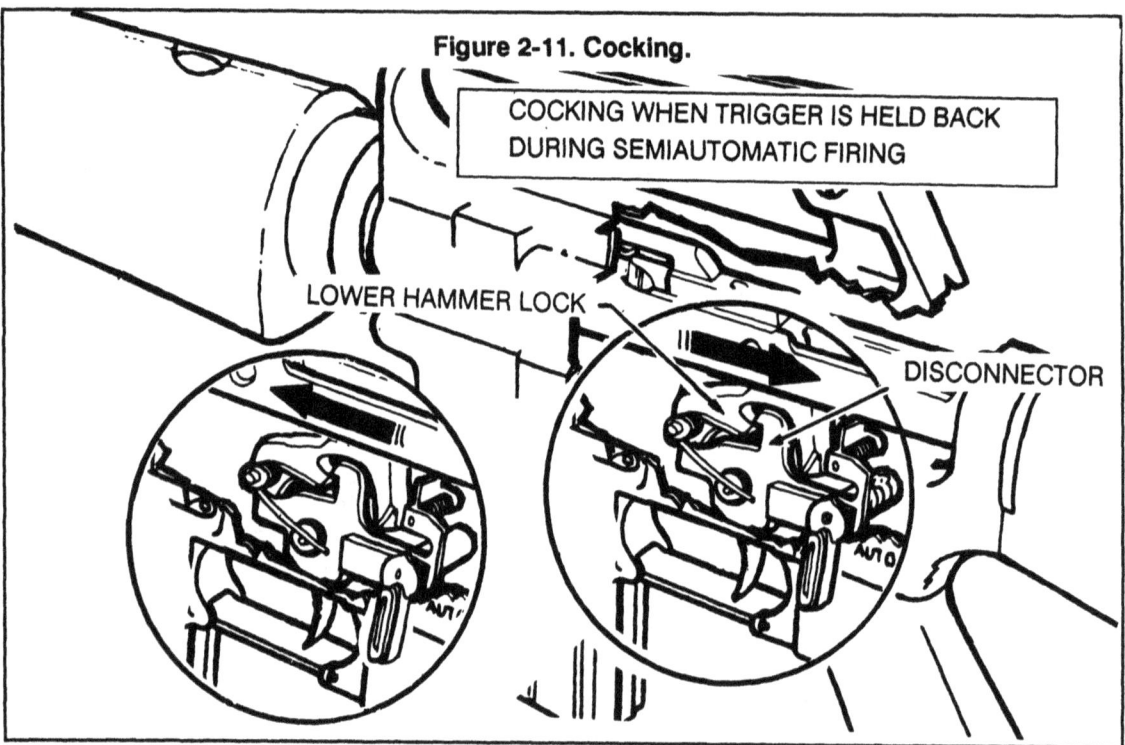

Figure 2-11. Cocking.

SEMIAUTOMATIC MODE (M16A1 AND M16A2)

The disconnector is mechanism installed so that the firer can fire single rounds in the M16A1 and M16A2 rifles. It is attached to the trigger and is rotated forward by action of the disconnector spring. When the hammer is cocked by the recoil of the bolt carrier, the disconnector engages the lower hook of the hammer and holds it until the trigger is released. Then the disconnector rotates to the rear and down, disengaging the hammer and allowing it to rotate forward until caught by the nose of the trigger. This prevents the hammer from following the bolt carrier forward and causing multiple firing. The trigger must be squeezed again before the next round will fire.

AUTOMATIC FIRE MODE (M16A1)

When the selector lever (Figure 2-12) is set on the AUTO position, the rifle continues to fire as long as the trigger is held back and ammunition is in the magazine. The functioning of certain parts of the rifle changes when firing automatically.

Once the trigger is squeezed and the round is fired, the bolt carrier group moves to the rear and the hammer is cocked. The center cam of the selector depresses the rear of the disconnector and prevents the nose of the disconnector from engaging the lower hammer hook. The bottom part of the automatic sear catches the upper hammer hook and holds it until the bolt carrier group moves forward. The bottom part strikes the top of the sear and releases the hammer, causing the rifle to fire automatically.

If the trigger is released, the hammer moves forward and is caught by the nose of the trigger. This ends the automatic cycle of fire until the trigger is squeezed again.

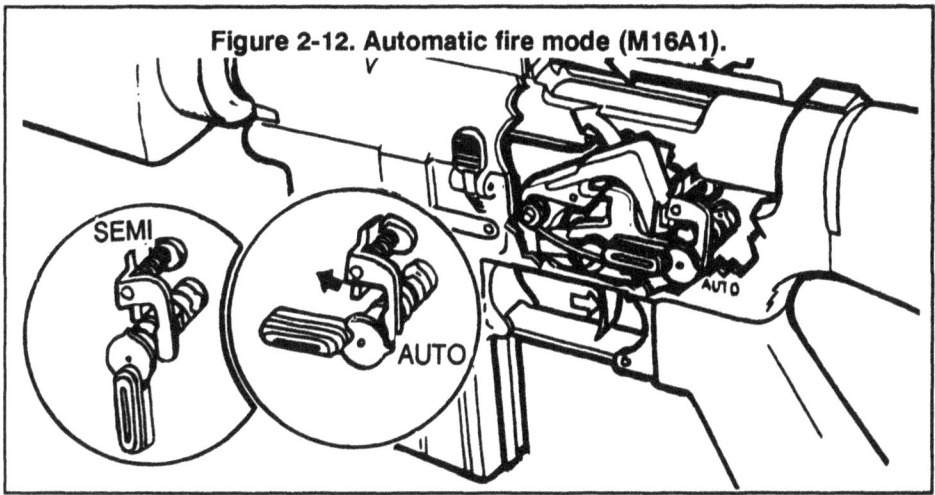

Figure 2-12. Automatic fire mode (M16A1).

BURST FIRE MODE (M16A2)

When the selector lever is set on the BURST position (Figure 2-13), the rifle fires a three-round burst if the trigger is held to the rear during the complete cycle. The weapon continues to fire three-round bursts with each separate trigger pull as long as ammunition is in the magazine. Releasing the trigger or exhausting ammunition at any point in the three-round cycle interrupts fire, producing one or two shots. Reapplying the trigger only completes the interrupted cycle — it does not begin a new one. This is not a malfunction. The M16A2 disconnector has a three-cam mechanism that continuously rotates with each firing cycle. Based on the position of the disconnector cam, the first trigger pull (after initial selection of the BURST position) can produce one, two, or three firing cycles before the trigger must be pulled again. The burst cam rotates until it reaches the stop notch.

NOTE: See the operator's manual for a detailed discussion on the burst position.

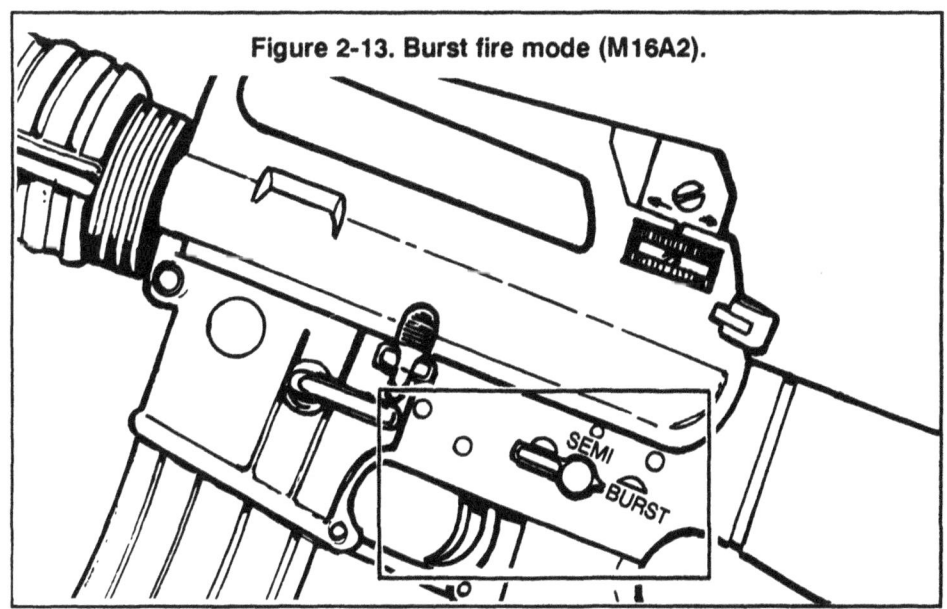

Figure 2-13. Burst fire mode (M16A2).

Section III. MALFUNCTIONS AND CORRECTIONS

Commanders and unit armorers are responsible for the organizational and direct support maintenance of weapons. Soldiers are responsible for keeping their weapons clean and operational at all times—in training and in combat. Therefore, the soldier should be issued an operator's technical manual and cleaning equipment for his assigned weapon.

STOPPAGE

A stoppage is a failure of an automatic or semiautomatic firearm to complete the cycle of operation. The firer can apply immediate or remedial action to clear the stoppage. Some stoppages that cannot be cleared by immediate or remedial action could require weapon repair to correct the problem. A complete understanding of how the weapon functions is an integral part of applying immediate-action procedures.

Immediate Action. This involves quickly applying a possible correction to reduce a stoppage based on initial observation or indicators but without determining the actual cause. To apply immediate action, the soldier would perform these steps: Gently slap upward on the magazine to ensure it is fully seated, and the magazine follower is not jammed. Pull the charging handle fully to the rear and check the chamber (observe for the ejection of a live or expended cartridge). Release the charging handle (do not ride it forward). Strike the forward assist assembly to ensure bolt closure. Try to fire the rifle.

Apply immediate action only one time for a given stoppage. Do not apply immediate action a second time. If the rifle still fails to fire, inspect it to determine the cause of the stoppage or malfunction and take appropriate remedial action.

Remedial Action. Remedial action is the continuing effort to determine the cause for a stoppage or malfunction and to try to clear the stoppage once it has been identified.

> **WARNING**
>
> IF AN AUDIBLE "POP" OR REDUCED RECOIL OCCURS DURING FIRING, IMMEDIATELY CEASE FIRE. THIS POP OR REDUCED RECOIL CAN BE THE RESULT OF A ROUND BEING FIRED WITHOUT ENOUGH FORCE TO SEND THE PROJECTILE OUT OF THE BARREL. DO NOT APPLY IMMEDIATE ACTION. REMOVE THE MAGAZINE, LOCK THE BOLT TO THE REAR, AND PLACE THE SELECTOR LEVER IN THE SAFE POSITION. VISUALLY INSPECT THE BORE TO ENSURE A PROJECTILE IS NOT LODGED IN THE BARREL. IF A PROJECTILE IS LODGED IN THE BARREL, DO NOT TRY TO REMOVE IT. TURN THE RIFLE IN TO THE ARMORER.

MAJOR CATEGORIES OF MALFUNCTIONS

A malfunction is caused by a procedural or mechanical failure of the rifle, magazine, or ammunition. Prefiring checks and serviceability inspections identify potential problems before they become malfunctions. Three primary categories of malfunctions are:

1. Failure to Feed, Chamber, or Lock.

Description. A malfunction can occur when loading the rifle or during the cycle of operation. Once the magazine has been loaded into the rifle, the forward movement of the bolt carrier group could lack enough force (generated by the expansion of the action spring) to feed, chamber, and lock the first round. While firing, the cycle of function is interrupted by a failure to strip a round from the magazine, to chamber the round, and to lock it (Figure 2-14).

Probable causes. The cause could be the result of one or more of the following: excess accumulation of dirt or fouling in and around the bolt and bolt carrier, defective magazine (dented or bulged), magazine improperly loaded. A defective round (projectile forced back into the cartridge case that could result in a "stubbed round") or the base of the previous field cartridge could be separated, leaving the remainder in

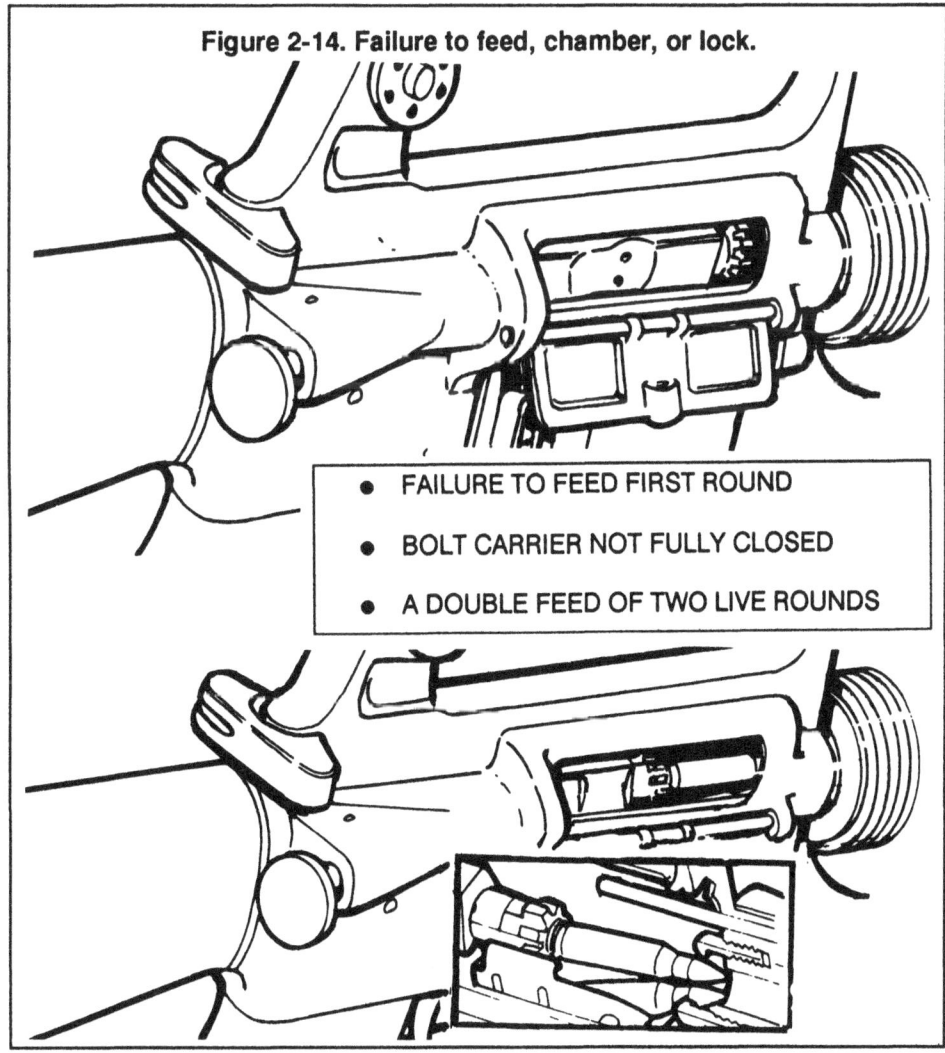

Figure 2-14. Failure to feed, chamber, or lock.

- FAILURE TO FEED FIRST ROUND
- BOLT CARRIER NOT FULLY CLOSED
- A DOUBLE FEED OF TWO LIVE ROUNDS

the chamber. Other causes could be: damaged or broken action spring, exterior accumulation of dirt in the lower receiver extension, or fouled gas tube resulting in short recoil.

Corrective action. Applying immediate action usually corrects the malfunction. However, to avoid the risk of further jamming, the firer should watch for ejection of a cartridge and ensure that the upper receiver is free of any loose rounds. If immediate action fails to clear the malfunction, remedial action must be taken. The carrier should not be forced. If resistance is encountered, which can occur with an unserviceable round, the bolt should be locked to the rear, magazine removed, and malfunction cleared—for example, a bolt override is when a cartridge has wedged itself between the bolt and charging handle. The best way to relieve this problem is by—

- Ensuring that the charging handle is pushed forward and locked in place.

- Holding the rifle securely and pulling the bolt to the rear until the bolt seats completely into the buffer well.

- Turning the rifle upright and allowing the overridden cartridge to fall out.

2. Failure to Fire Cartridge.

Description. Failure of a cartridge to fire despite the fact that a round has been chambered, the trigger is pulled, and the sear has released the hammer. This occurs when the firing pin fails to strike the primer with enough force or when the ammunition is bad.

Probable causes. Excessive carbon buildup on the firing pin (Figure 2-15A) is often the cause, because the full forward travel of the firing pin is restricted. However, a defective or worn firing pin can give the same results. Inspection of the ammunition could reveal a shallow indentation or no mark on the primer, indicating a firing pin problem (Figure 2-15B). Cartridges that show a normal indentation on the primer but did not fire indicate bad ammunition.

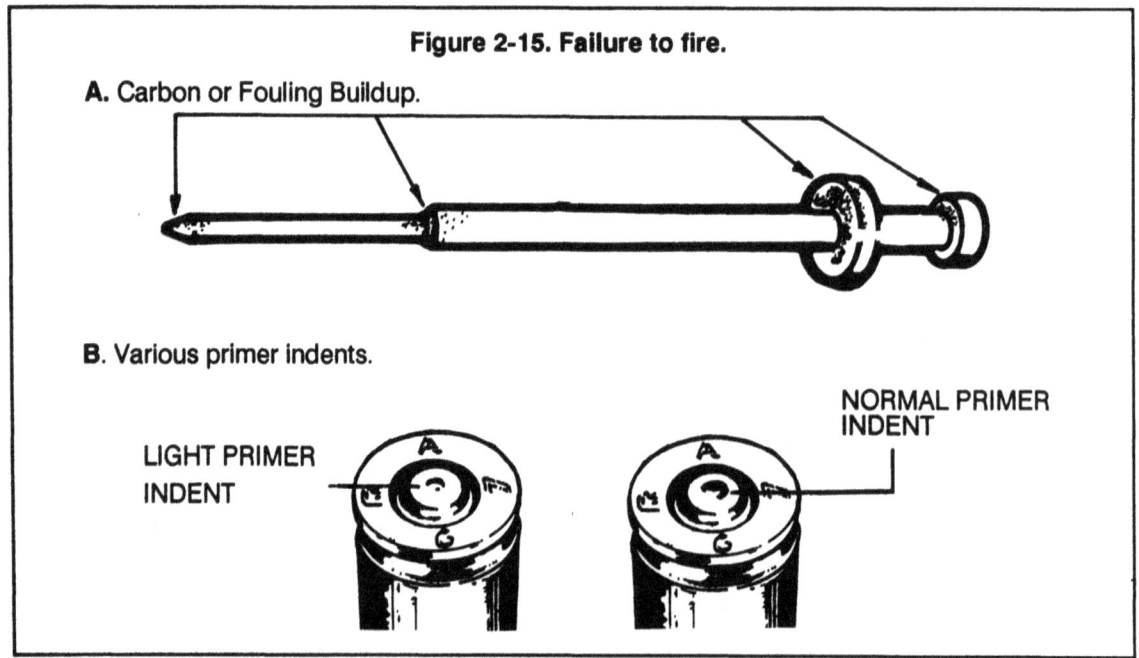

Figure 2-15. Failure to fire.

Corrective action. If the malfunction continues, the firing pin, bolt, carrier, and locking lug recesses of the barrel extension should be inspected, and any accumulation of excessive carbon or fouling should be removed. The firing pin should also be inspected for damage. Cartridges that show a normal indentation on the primer but failed to fire could indicate a bad ammunition lot. Those that show a complete penetration of the primer by the firing pin could also indicate a bad ammunition lot or a failure of the cartridge to fully seat in the chamber.

NOTE: If the round is suspected to be faulty, it is reported and returned to the agency responsible for issuing ammunition

3. Failure to Extract and Eject.

Failure to extract. The cartridge must extract before it can eject.

Description. A failure to extract results when the cartridge case remains in the rifle chamber. While the bolt and bolt carrier could move rearward only a short distance, more commonly the bolt and bolt carrier recoil fully to the rear, leaving the cartridge case in the chamber. A live round is then forced into the base of the cartridge case as the bolt returns in the next feed cycle. This malfunction is one of the hardest to clear.

NOTE: Short recoil can also be caused by a fouled or obstructed gas tube.

> **WARNING**
>
> **A FAILURE TO EXTRACT IS CONSIDERED TO BE AN EXTREMELY SERIOUS MALFUNCTION, REQUIRING THE USE OF TOOLS TO CLEAR. A LIVE ROUND COULD BE LEFT IN THE CHAMBER AND BE ACCIDENTALLY DISCHARGED. IF A SECOND LIVE ROUND IS FED INTO THE PRIMER OF THE CHAMBERED LIVE ROUND, THE RIFLE COULD EXPLODE AND CAUSE PERSONAL INJURY. THIS MALFUNCTION MUST BE PROPERLY IDENTIFIED AND REPORTED. FAILURES TO EJECT SHOULD NOT BE REPORTED AS EXTRACTION FAILURES.**

Probable cause. Short recoil cycles and fouled or corroded rifle chambers are the most common causes of failures to extract. A damaged extractor or weak/broken extractor spring can also cause this malfunction.

Corrective action. The severity of a failure to extract determines the corrective action procedures. If the bolt has moved rearward far enough so that it strips a live round from the magazine in its forward motion, the bolt and carrier must be locked to the rear.

The magazine and all loose rounds must be removed before clearing the stoppage. Usually, tapping the butt of the rifle on a hard surface causes the cartridge to fall out of the chamber. However, if the cartridge case is ruptured, it can be seized. When this occurs, a cleaning rod can be inserted into the bore from the muzzle end. The cartridge case can be forced from the chamber by tapping the cleaning rod against the inside base of the fired cartridge. When cleaning and inspecting the mechanism and chamber reveal no defects but failures to extract persist, the extractor and extractor spring should be replaced. If the chamber surface is damaged, the entire barrel must be replaced.

Failure to Eject. A failure to eject a cartridge is an element in the cycle of functioning of the rifle, regardless of the mode of fire. A malfunction occurs when the cartridge is not ejected through the ejection port and either remains partly in the chamber or becomes jammed in the upper receiver as the bolt closes. When the firer initially clears the rifle, the cartridge could strike an inside surface of the receiver and bounce back into the path of the bolt.

Probable cause. Ejection failures are hard to diagnose but are often related to a weak or damaged extractor spring and/or ejector spring. Failures to eject can also be caused by a buildup of carbon or fouling on the ejector spring or extractor, or from short recoil. Short recoil is usually due to a buildup of fouling in the carrier mechanism or gas tube, which could result in many failures to include a failure to eject. Resistance caused by a carbon-coated or corroded chamber can impede the extraction, and then the ejection of a cartridge.

Corrective action. While retraction of the charging handle usually frees the cartridge and permits removal, the charging handle must not be released until the position of the next live round is determined. If another live round has been sufficiently stripped from the magazine or remains in the chamber, then the magazine and all live rounds could also require removal before the charging handle can be released. If several malfunctions occur and are not corrected by cleaning and lubricating, the ejector spring, extractor spring, and extractor should be replaced.

OTHER MALFUNCTIONS
Some other malfunctions that can occur are as follows.

- Failure of the bolt to remain in a rearward position after the last round in the magazine is fired. Check for a bad magazine or short recoil.

- Failure of the bolt to lock in the rearward position when the bolt catch has been engaged. Check bolt catch; replace as required.

- Firing two or more rounds when the trigger is pulled and the selection lever is in the SEMI position. This indicates a worn sear, cam, or disconnector. Turn in to armorer to repair and replace trigger group parts as required.

- Trigger will not pull or return after release with the selector set in a firing position. This indicates that the trigger pin (Figure 2-16A) has backed out of the receiver or the hammer spring is broken. Turn in to armorer to replace or repair.

- Failure of the magazine to lock into the rifle (Figure 2-16B). Check the magazine and check magazine catch for damage. Turn in to armorer to adjust the catch; replace as required.

- Failure of any part of the bolt carrier group to function (Figure 2-16C). Check for incorrect assembly of components. Correctly clean and assemble the bolt carrier group, or replace damaged parts.
- Failure of the ammunition to feed from the magazine (Figure 2-16D). Check for damaged magazine. A damaged magazine could cause repeated feeding failures and should be turned in to armorer or exchanged.

NOTE: Additional technical information on troubleshooting malfunctions and repairing components is contained in the organizational and DS maintenance publications and manuals.

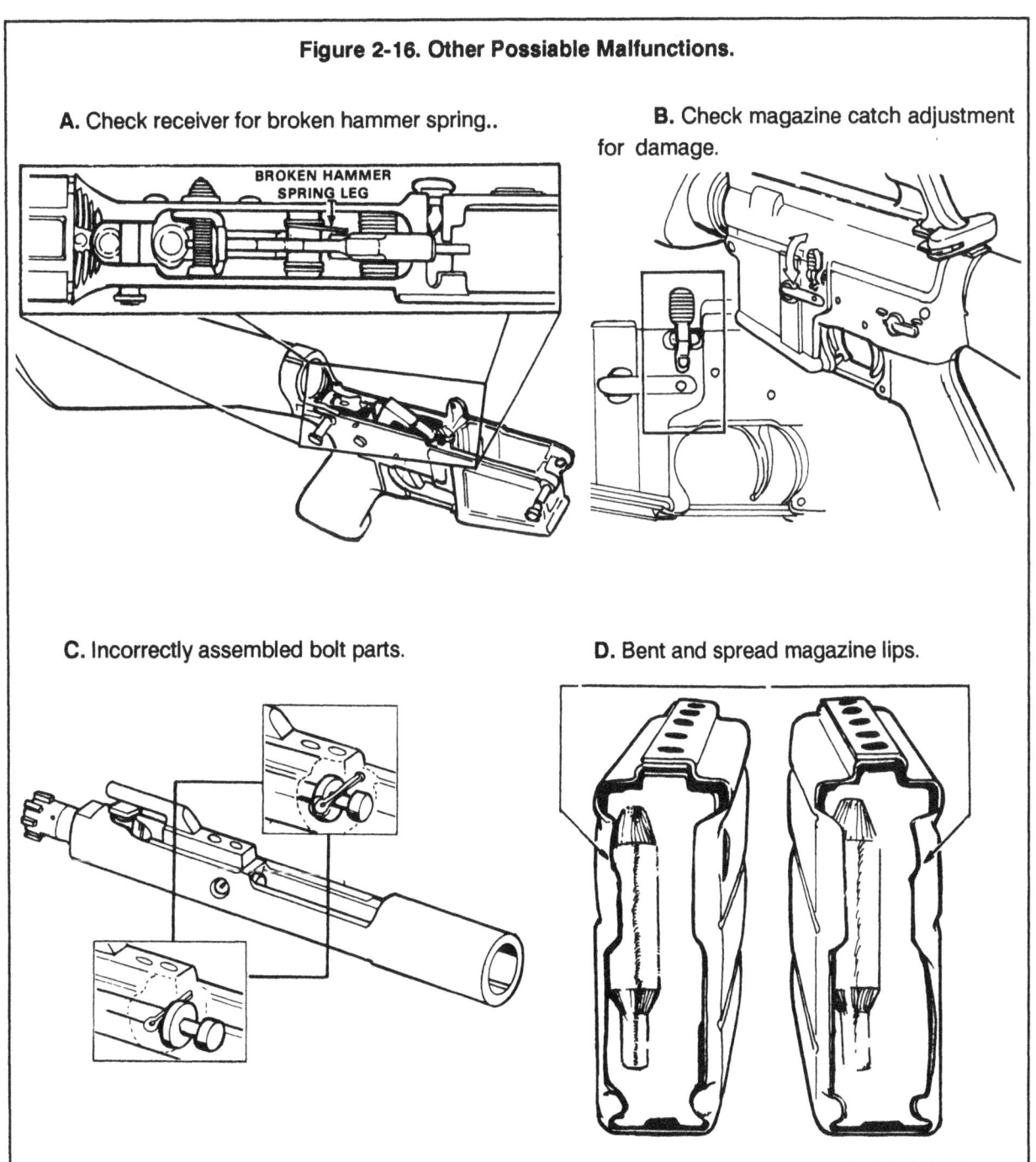

Figure 2-16. Other Possiable Malfunctions.

A. Check receiver for broken hammer spring..

B. Check magazine catch adjustment for damage.

C. Incorrectly assembled bolt parts.

D. Bent and spread magazine lips.

Section IV. AMMUNITION

This section contains information on different types of standard military ammunition used in the M16A1 and M16A2 rifles. Use only authorized ammunition that is manufactured to US and NATO specifications.

TYPES AND CHARACTERISTICS

The characteristics of the M16 family of ammunition are described in this paragraph.

Cartridge, 5.56-mm, Dummy, M199. (Used in both rifles.) The M199 dummy cartridge is used during dry fire and other training (see 3, Figure 2-17). This cartridge can be identified by the six grooves along the side of the case beginning about 1/2 inch from its head. It contains no propellant or primer. The primer well is open to prevent damage to the firing pin.

Cartridge, 5.56-mm, Blank, M200. (Used in the M16A1 or M16A2 rifle.) The M200 blank cartridge has no projectile. The case mouth is closed with a seven-petal rosette crimp and shows a violet tip (see 4, Figure 2-17). (See Appendix C for use of the blank firing attachment.). The original M200 blank cartridge had a white tip. Field use of this cartridge resulted in residue buildup, which caused several malfunctions. Only the violet-tipped M200 cartridge should be used.

Cartridge, 5.56-mm, Plastic Practice Ammunition, M862. (Used in the M16A1 and M16A2 rifles.) The M862 PPA is designed exclusively for training. It can be used in lieu of service ammunition on indoor ranges, and by units that have a limited range fan that does not allow the firing of service ammunition. It is used with the M2 training bolt.

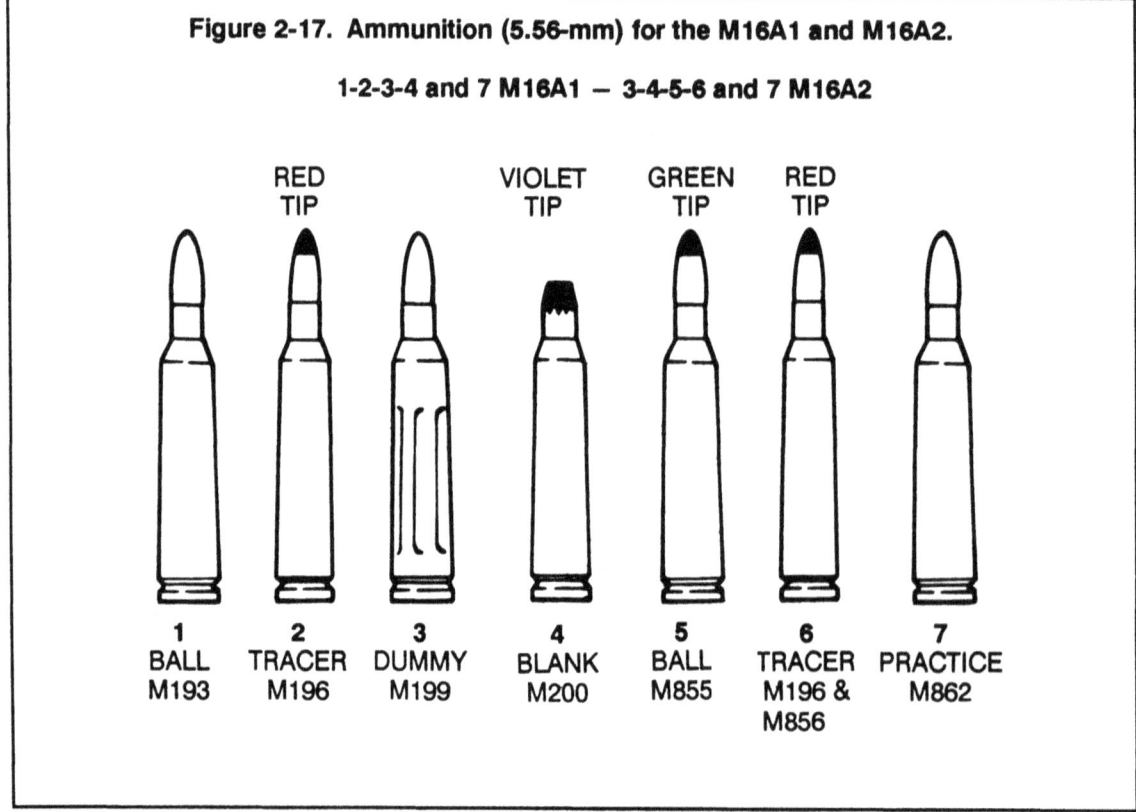

Figure 2-17. Ammunition (5.56-mm) for the M16A1 and M16A2.

Although PPA (see 7, Figure 2-17) closely replicates the trajectory and characteristics of service ammunition out to 25 meters, it should not be used to set the combat battlesight zero of weapons to fire service ammunition. The setting that is placed on the sights for a weapon firing PPA could be different for service ammunition.

If adequate range facilities are not available for sustainment (particularly Reserve Components), PPA can be used for any firing exercises of 25 meters or less. This includes the 25-meter scaled silhouette, 25-meter alternate qualification course, and quick-fire training. Units that have an indoor range with adequate ventilation or MOUT site could use PPA. (See Appendix C for use in training.)

Cartridge, 5.56-mm, Ball, M193. The M193 cartridge is a center-fire cartridge with a 55-grain, gilding-metal, jacketed, lead alloy core bullet. The primer and case are waterproof. The M193 round is the standard cartridge for field use with the M16A1 rifle and has no identifying marks (see 1, Figure 2-17). This cartridge has a projectile weight of 55 grains and is 1.9 cm long, with a solid lead core.

Figure 2-21. Ammunition for 5.56-mm M16A1 (1 through 4 and 7) and M16A2 (3 through 7).

Cartridge, 5.56-mm, Tracer, M196. (Used in the M16A1 rifle.) The M196 cartridge is identified by a red- or orange-painted tip (see 2, Figure 2-17). Its main uses are for observation of fire, incendiary effect, and signaling. Soldiers should avoid long-term use of 100-percent tracer rounds. This could cause deposits of incendiary material/chemical compounds that could cause damage to the barrel. Therefore, when tracer rounds are fired, they are mixed with ball ammunition in a ratio no greater than one-to-one with a preferred ratio of three or four ball rounds to one tracer round.

Cartridge, 5.56-mm, Ball, M855. The M855 cartridge has a 62-grain, gilding-metal, jacketed, lead alloy core bullet with a steel penetrator. The primer and case are waterproof. This is the NATO standard round for the M16A2 rifle (also used in the M249 SAW). It is identified by a green tip (see 5, Figure 2-17). This cartridge has a projectile weight of 62 grains and is 2.3 cm long, with a steel penetrator in the nose.

Cartridge, 5.56-mm, Tracer, M856. (Used in the M16A2 rifle.) The M856 tracer cartridge has similar characteristics as the M196 but slightly longer tracer burnout distance. This cartridge has a 63.7-grain bullet. The M856 does not have a steel penetrator. It is also identified by a red tip (orange when linked 4 and 1) (6, Figure 2-17).

CARE AND HANDLING

When necessary to store ammunition in the open, it must be raised on dunnage at least 6 inches from the ground and protected with a cover, leaving enough space for air circulation. Since ammunition and explosives are adversely affected by moisture and high temperatures, the following must be adhered to:

- Do not open ammunition boxes until ready to use.

- Protect ammunition from high temperatures and the direct rays of the sun.

- Do not attempt to disassemble ammunition or any of its components.

- Never use lubricants or grease on ammunition.

Section V. DESTRUCTION OF MATERIEL

Rifles subject to capture or abandonment in the combat zone are destroyed only by the authority of the unit commander IAW orders of or policy established by the Army commander. The destruction of equipment is reported through regular command channels.

MEANS OF DESTRUCTION

Certain procedures outlined require use of explosives and incendiary grenades. Issue of these and related principles, and specific conditions under which destruction is effected, are command decisions. Of the several means of destruction, the following apply:

- **Mechanical.** Requires axe, pick mattock, sledge, crowbar, or other heavy implement.

- **Burning.** Requires gasoline, oil, incendiary grenades, and other flammables, or welding or cutting torch.

- **Demolition.** Requires suitable explosives or ammunition. Under some circumstances, hand grenades can be used.

- **Disposal.** Requires burying in the ground, dumping in streams or marshes, or scattering so widely as to preclude recovery of essential parts.

It is important that the same parts be destroyed on all like materiel, including spare parts, so that the enemy cannot rebuild one complete unit from several damaged units. If destruction is directed, appropriate safety precautions must be observed.

FIELD-EXPEDIENT METHODS

If destruction of the individual rifle must be performed to prevent enemy use, the rifle must be damaged so it cannot be restored to a usable condition. Expedient destruction requires that key operational parts be separated from the rifle or damaged beyond repair. Priority is given in the following order:

    FIRST: Bolt carrier group; removed and discarded or hidden.

    SECOND: Upper receiver group; separated and discarded or hidden.

    THIRD: Lower receiver group; separated and discarded or hidden.

FM 23-9

CHAPTER 3

Rifle Marksmanship Training

The procedures and techniques for implementing the Army rifle marksmanship training program are based on the concept that all soldiers must understand common firing principles, be proficient marksmen, and be confident in applying their firing skills in combat. This depends on their understanding of the rifle and correct application of marksmanship fundamentals. Proficiency is accomplished through practice that is supervised by qualified instructors/trainers and through objective performance assessments by unit leaders. During preliminary training, instructors/trainers emphasize initial learning, reviewing, reinforcing, and practicing of the basics. Soldiers must master weapon maintenance, functions checks, and firing fundamentals before progressing to advanced skills and firing exercises under tactical conditions. The skills the soldier must learn are developed in the following four phases:

- *PHASE I. Preliminary Rifle Instruction.*
- *PHASE II. Downrange Feedback Range Firing.*
- *PHASE III. Field Firing on Train-Fire Ranges.*
- *PHASE IV. Advanced and Collective Firing Exercises.*

Each soldier progresses through these phases to meet the objective of rifle marksmanship training and sustainment. The accomplishment of these phases are basic and necessary in mastering the correct techniques of marksmanship and when functioning as a soldier in a combat area. (See Chapter 1 and Appendix A.)

Section I. BASIC PROGRAM IMPLEMENTATION

Knowledgeable instructors/cadre are the key to marksmanship performance. All commanders must be aware of maintaining expertise in marksmanship instruction/training. (See Appendix D.)

INSTRUCTOR/TRAINER SELECTION

Institutional and unit instructors/trainers are selected and assigned from the most highly qualified soldiers. These soldiers must have an impressive background in rifle marksmanship; be proficient in applying these fundamentals; know the importance of marksmanship training; and have a competent and professional attitude. The commander must ensure that selected instructors/trainers can effectively train other soldiers. Local instructor/trainer training courses and marksmanship certification programs must be established to ensure that instructor/trainer skills are developed.

Cadre/trainer refers to a marksmanship instructor/trainer that has more experience and expertise than the firer. He trains soldiers in the effective use of the rifle by maintaining strict discipline on the firing line, insisting on compliance with range procedures and program objectives, and enforcing safety regulations. A good instructor/trainer must understand the training phases and techniques for developing marksmanship skills, and he must possess the following qualifications:

Knowledge. The main qualifications for an effective instructor/trainer are thorough knowledge of the rifle, proficiency in firing, and understanding supporting marksmanship manuals.

Patience. The instructor/trainer must relate to the soldier calmly, persistently, and patiently.

Understanding. The instructor/trainer can enhance success and understanding by emphasizing close observance of rules and instructions.

Consideration. Most soldiers enjoy firing regardless of their performance and begin with great enthusiasm. The instructor/trainer can enhance this enthusiasm by being considerate of his soldiers feelings and by encouraging firing abilities throughout training, which can also make teaching a rewarding experience.

Respect. An experienced cadre is assigned the duties of instructor/trainer, which classifies him as a technical expert and authority. The good instructor/trainer is alert for mistakes and patiently makes needed corrections.

Encouragement. The instructor/trainer can encourage his soldiers by convincing them to achieve good firing performance through practice. His job is to impart knowledge and to assist the soldier so he can gain the practical experience needed to become a good firer.

DUTIES OF THE INSTRUCTOR/TRAINER

The instructor/trainer helps the firer master the fundamentals of rifle marksmanship. He ensures that the firer consistently applies what he has learned. Then, it is a matter of practice, and the firer soon acquires good firing skills. When training the beginner, the instructor/trainer could confront problems such as fear, nervousness, forgetfulness, failure to understand, and a lack of coordination or determination. An expert firer is often unaware that some problems are complicated by arrogance and carelessness. With all types of firers, the instructor/trainer must ensure that firers are aware of their firing errors, understand the causes, and apply remedies. Sometimes errors are not evident. The instructor/trainer must isolate errors, explain them, and help the firer concentrate on correcting them.

Observing the Firer. The instructor/trainer observes the firer during drills and in the act of firing to pinpoint errors. If there is no indication of probable error, then the firer's position, breath control, shot anticipation, and trigger squeeze are closely observed.

Questioning the Firer. The firer is asked to detect his errors and to explain his firing procedure to include position, aiming, breath control, and trigger squeeze.

Analyzing the Shot Group. This is an important step in detecting and correcting errors. When analyzing a target, the instructor/trainer critiques and correlates observations of the firer to probable errors in performance, according to the shape and size of shot groups. A poor shot group is usually caused by more than one observable error.

NOTE: To assist instructors/trainers, TVTs 7-1 and 7-2 should be viewed before conducting training.

Section II. CONDUCT OF TRAINING

In the conduct of marksmanship training, the instructor/trainer first discusses an overview of the program to include the progression and step-by-step process in developing firing skills. (This can be accomplished by showing TVT 7-13.) Once the soldier realizes the tasks and skills involved, he is ready to begin. He receives preliminary rifle instruction before firing any course. Also during this initial phase, an understanding of the service rifle developes through review.

MECHANICAL TRAINING

Mechanical training includes characteristics and capabilities, disassembly and assembly, operations and functioning, serviceability checks, and weapons maintenance. It also stresses the performance of immediate action to clear or reduce a stoppage, and the safe handling of rifles and ammunition (see Chapter 2). Examples of mechanical training drills, along with tasks, conditions, and standards, are provided in Appendix A. These examples are also used for initial entry training at the Army training centers. Mechanical training must encompass all related tasks contained in the soldier's manual of common tasks (SMCT) to include the correct procedures for disassembly, cleaning, inspection, and reassembly of the rifle and magazine (Figure 3-1).

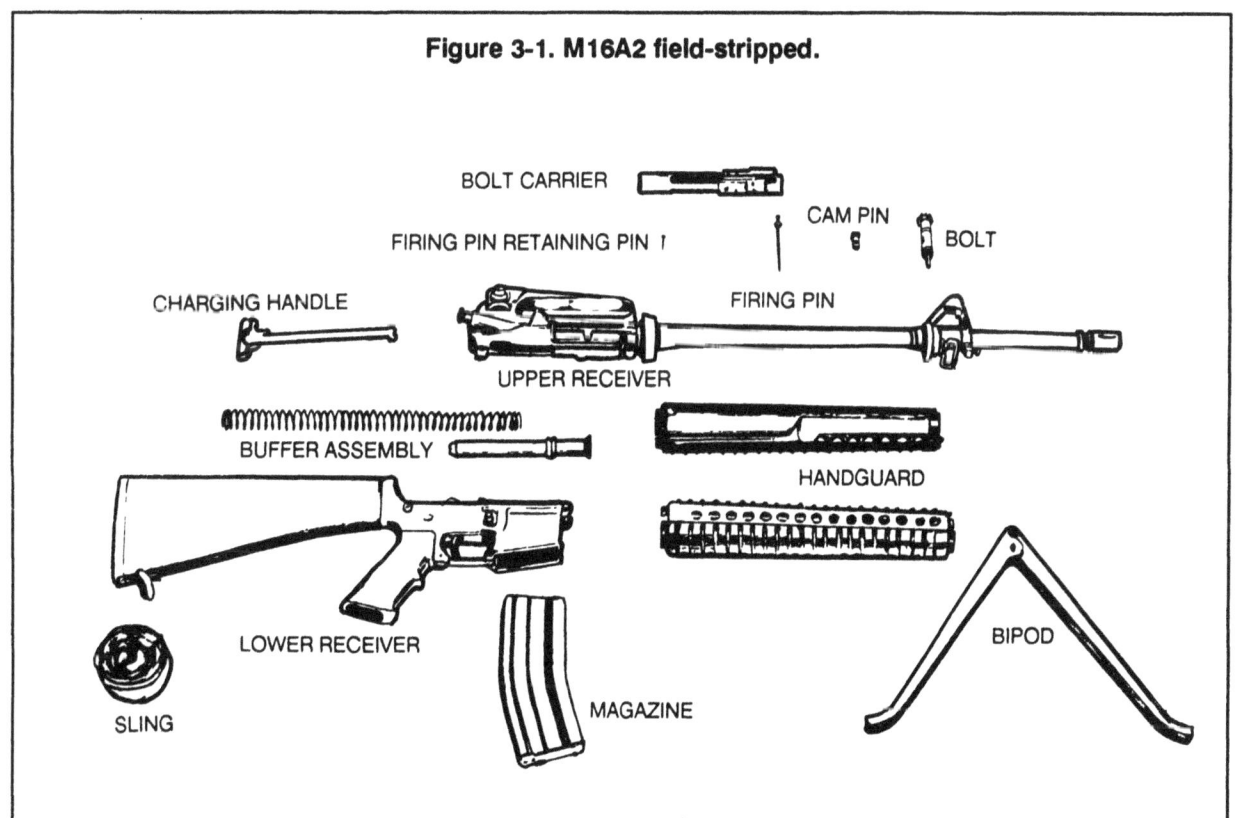

Figure 3-1. M16A2 field-stripped.

Serviceability inspections and preventive maintenance checks must be practiced to ensure soldiers have reliable weapons systems during training and in combat. Technical information necessary to conduct mechanical training is contained in the soldier's operator's manual (M16A1—TM 9-1005-249-10; M16A2—TM 9-1005-319-10). Once the basic procedures have been demonstrated, soldiers should practice the mechanical training skills under varied conditions to include during nighttime, and in MOPP and arctic clothing.

As part of mechanical training, soldiers must be taught and must practice procedures for properly loading ammunition into magazines to include both single loose rounds and speed loading of 10-round clips (Figure 3-2).

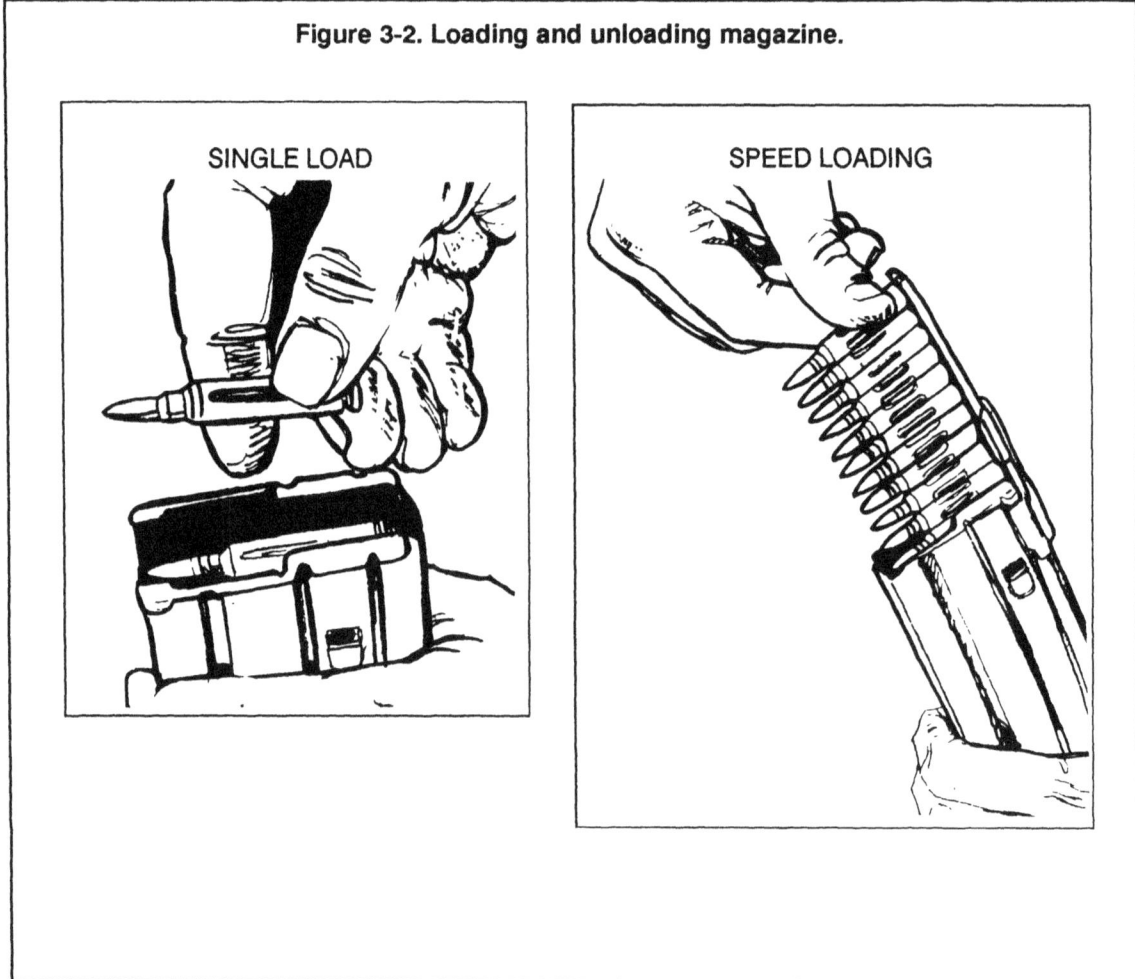

Figure 3-2. Loading and unloading magazine.

Emphasis on maintenance and understanding of the rifle can prevent most problems and malfunctions. However, a soldier could encounter a stoppage or malfunction. The soldier must quickly correct the problem by applying immediate action and continue to place effective fire on the target.

Immediate-action procedures contained in Chapter 2 and the operator's technical manual should be taught and practiced as part of preliminary dry-fire exercises, and should be reinforced during live-fire exercises.

Immediate-action drills should be conducted using dummy ammunition (M199) loaded into the magazine. The soldier chamber the first dummy round and assume a firing position. When he squeezes the trigger and the hammer falls with no recoil, this is the cue to apply the correct immediate-action procedure and to refire. Drill should continue until soldiers can perform the task in three to five seconds.

The word **SPORTS** is a technique for assisting the soldier in learning the proper procedures for applying immediate action to the M16A1 and M16A2 rifles.

First, **THINK,** then:

> **S**lap up on the bottom of the magazine.
>
> **P**ull the charging handle to the rear.
>
> **O**bserve the chamber for an ejection of the round.
>
> **R**elease the charging handle.
>
> **T**ap the forward assist.
>
> **S**queeze the trigger again.

NOTE: When slapping up on the magazine, be careful not to knock a round out of the magazine into the line of the bolt carrier, causing more problems. Slap hard enough only to ensure the magazine is fully seated.

MARKSMANSHIP FUNDAMENTALS

The soldier must understand the four key fundamentals before he approaches the firing line. He must be able to establish a **steady position** that allows observation of the target. He must **aim** the rifle at the target by aligning the sight system, and he must fire the rifle without disturbing this alignment by improper **breathing** or during **trigger squeeze**. The skills needed to accomplish these are known as **rifle marksmanship fundamentals**. These simple procedures aid the firer in achieving target hits under many conditions when expanded with additional techniques and information. Applying these four fundamentals rapidly and consistently is called the **integrated act of firing**.

Steady Position. When the soldier approaches the firing line, he should assume a comfortable, steady firing position in order to hit targets consistently. The time and supervision each soldier has on the firing line are limited (illustrated on the following page in Figure 3-3). Therefore, he must learn how to establish a steady position during dry-fire training. The firer is the best judge as to the quality of his position. If he can hold the front sight post steady through the fall of the hammer, he has a good position. The steady position elements are as follows:

Nonfiring hand grip. The rifle handguard rests on the heel of the hand in the V formed by the thumb and fingers. The grip of the nonfiring hand is light, and slight rearward pressure is exerted.

Rifle butt position. The butt of the stock is placed in the pocket of the firing shoulder. This reduces the effect of recoil and helps ensure a steady position.

Firing hand grip. The firing hand grasps the pistol grip so that it fits the V formed by the thumb and forefinger. The forefinger is placed on the trigger so that the lay of the rifle is not disturbed when the trigger is squeezed. A slight rearward pressure is exerted by the remaining three fingers to ensure that the butt of the stock remains in the pocket of the shoulder, thus minimizing the effect of recoil.

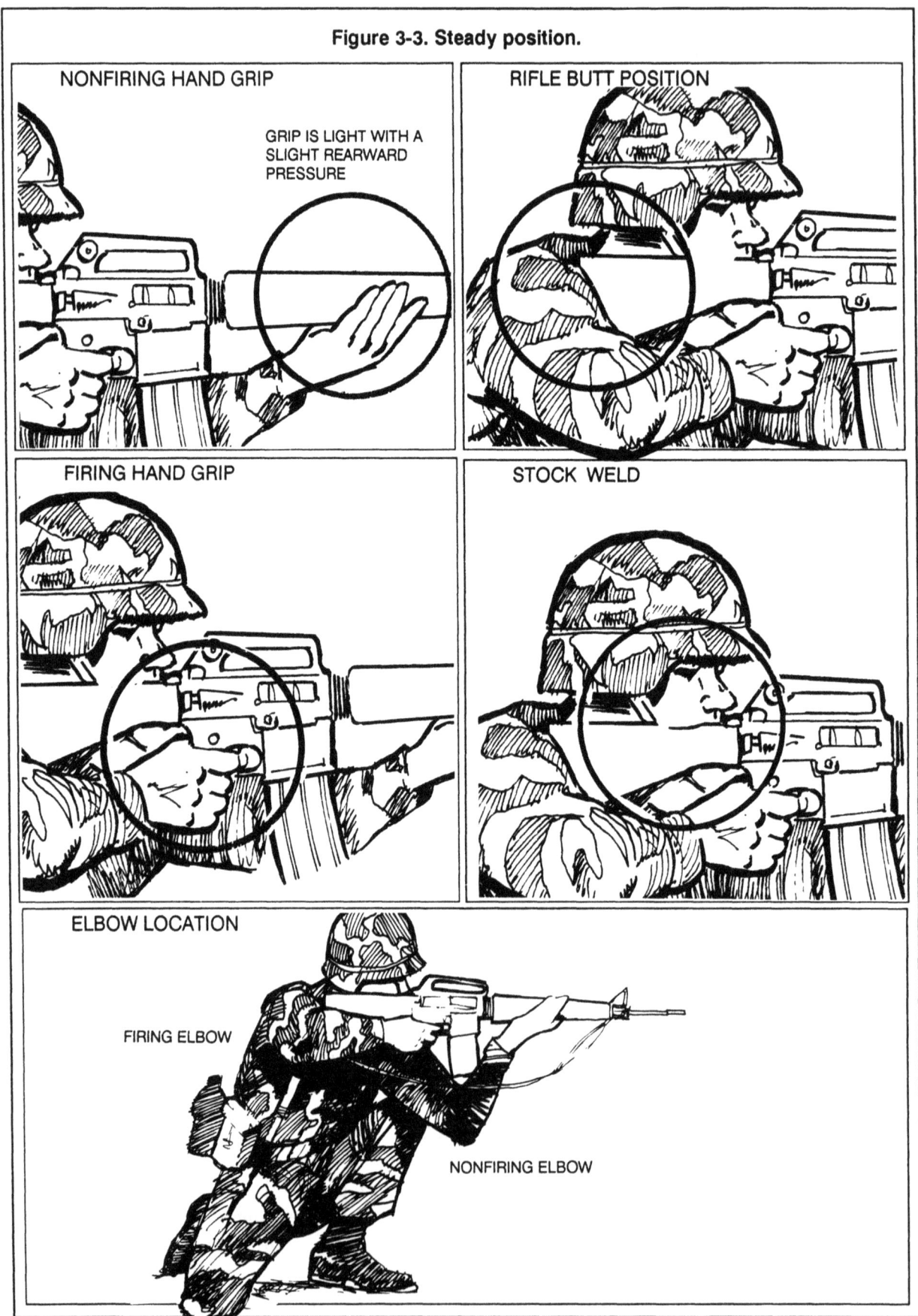

Figure 3-3. Steady position.

Firing elbow placement. The location of the firing elbow is important in providing balance. The exact location, however, depends on the firing/fighting position used — for example, kneeling, prone, or standing. Placement should allow shoulders to remain level.

Nonfiring elbow. The nonfiring elbow is positioned firmly under the rifle to allow for a comfortable and stable position. When the soldier engages a wide sector of fire, moving targets, and targets at various elevations, his nonfiring elbow should remain free from support.

Stock weld. The stock weld is taught as an integral part of various positions. Two key factors emphasized are that the stock weld should provide for a natural line of sight through the center of the rear sight aperture to the front sight post and to the target. The firer's neck should be relaxed, allowing his cheek to fall naturally onto the stock. Through dry-fire training, the soldier is encouraged to practice this position until he assumes the same stock weld each time he assumes a given position. This provides consistency in aiming, which is the purpose of obtaining a correct stock weld. Proper eye relief is obtained when a soldier establishes a good stock weld. There is normally a small change in eye relief each time he assumes a different firing position. Soldiers should begin by trying to touch his nose close to the charging handle when assuming a firing position.

Support. If artificial support (sandbags, logs, stumps) is available, it should be used to steady the position and to support the rifle. If it is not available, then the bones, not the muscles, in the firer's upper body must support the rifle.

Muscle relaxation. If support is properly used, the soldier should be able to relax most of his muscles. Using artificial support or bones in the upper body as support allows him to relax and settle into position. Using muscles to support the rifle can cause it to move.

Natural point of aim. When the soldier first assumes his firing position, he orients his rifle in the general direction of his target. Then he adjusts his body to bring the rifle and sights exactly in line with the desired aiming point. When using proper support and consistent stock weld, the soldier should have his rifle and sights aligned naturally on the target. When this correct body-rifle-target alignment is achieved, the front sight post must be held on target, using muscular support and effort. As the rifle fires, the muscles tend to relax, causing the front sight to move away from the target toward the natural point of aim. Adjusting this point to the desired point of aim eliminates this movement. When multiple target exposures are expected (or a sector of fire must be covered), the soldier should adjust his natural point of aim to the center of the expected target exposure area (or center of sector).

Aiming. Focusing on the front sight post is a vital skill the firer must acquire during practice. Having mastered the task of holding the rifle steady, the soldier must align the rifle with the target in exactly the same way for each firing. The firer is the final judge as to where his eye is focused. The instructor/trainer emphasizes this point by having the firer focus on the target and then focus back on the front sight post. He checks the position of the firing eye to ensure it is in line with the rear sight aperture. He uses the M16 sighting device to see what the firer sees through the sights. (See Appendix C.)

Rifle sight alignment. Alignment of the rifle with the target is critical. It involves placing the tip of the front sight post in the center of the rear sight aperture. (Figure 3-4.) Any alignment error between the front and rear sights repeats itself for every 1/2 meter the bullet travels. For example, at the 25-meter line, any error in rifle alignment is multiplied 50 times. If the rifle is misaligned by 1/10 inch, it causes a target at 300 meters to be missed by 5 feet.

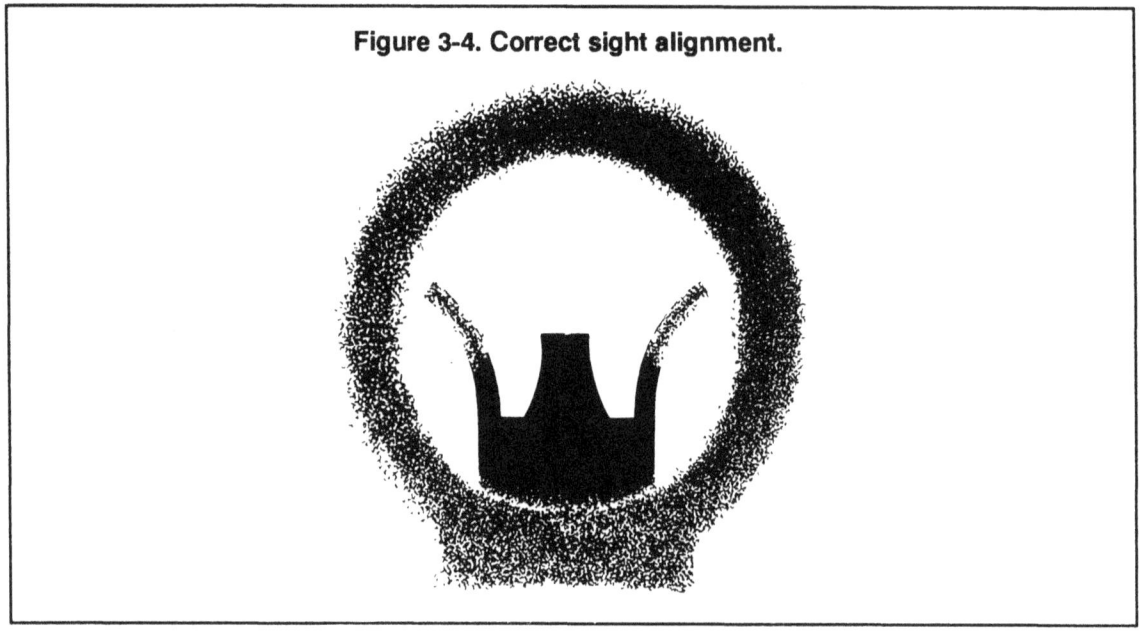

Figure 3-4. Correct sight alignment.

Focus of the eye. A proper firing position places the eye directly on line with the center of the rear sight. When the eye is focused on the front sight post, the natural ability of the eye to center objects in a circle and to seek the point of greatest light (center of the aperture) aid in providing correct sight alignment. For the average soldier firing at combat-type targets, the natural ability of the eye can accurately align the sights. Therefore, the firer can place the tip of the front sight post on the aiming point, but the eye must be focused on the tip of the front sight post. This causes the target to appear blurry, while the front sight post is seen clearly. Two reasons for focusing on the tip of the front sight post are:

- Only a minor aiming error should occur since the error reflects only as much as the soldier fails to determine the target center. A greater aiming error can result if the front sight post is blurry due to focusing on the target or other objects.

- Focusing on the tip of the front sight post aids the firer in maintaining proper sight alignment (Figure 3-4).

Sight picture. Once the soldier can correctly align his sights, he can obtain a sight picture. A correct sight picture has the target, front sight post, and rear sight aligned. The sight picture includes two basic elements: sight alignment and placement of the aiming point.

Placement of the aiming point varies, depending on the engagement range. For example, Figure 3-5 shows a silhouette at 250 meters – the aiming point is the center of mass, and the sights are in perfect alignment; this is a correct sight picture.

Figure 3-5. Correct sight picture.

A technique to obtain a good sight picture is the side aiming technique (Figure 3-6). It involves positioning the front sight post to the side of the target in line with the vertical center of mass, keeping the sights aligned. The front sight post is moved horizontally until the target is directly centered on the front sight post.

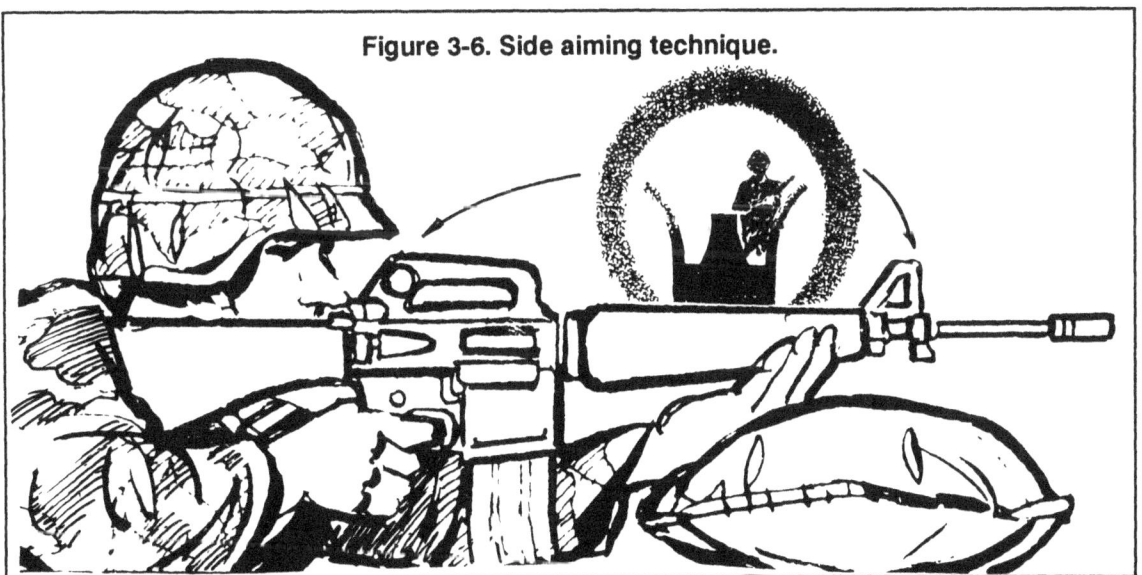

Figure 3-6. Side aiming technique.

Front sight. The front sight post is vital to proper firing and should be replaced when damaged. Two techniques that can be used are the carbide lamp and the burning plastic spoon. The post should be blackened anytime it is shiny since precise focusing on the tip of the front sight post cannot be done otherwise.

Aiming practice. Aiming practice is conducted before firing live rounds. During day firing, the soldier should practice sight alignment and placement of the aiming point. This can be done by using training aids such as the M15A1 aiming card and the Riddle sighting device. (See Appendix C.)

Breath Control. As the firer's skills improve and as timed or multiple targets are presented, he must learn to hold his breath at any part of the breathing cycle. Two types of breath control techniques are practiced during dry fire.

- The first is the technique used during zeroing (and when time is available to fire a shot) (Figure 3-7A. There is a moment of natural respiratory pause while breathing when most of the air has been exhaled from the lungs and before inhaling. Breathing should stop after most of the air has been exhaled during the normal breathing cycle. The shot must be fired before the soldier feels any discomfort.

- The second breath control technique is employed during rapid fire (short-exposure targets) (Figure 3-7B). Using this technique, the soldier holds his breath when he is about to squeeze the trigger.

The coach/trainer ensures that the firer uses two breathing techniques and understands them by instructing him to exaggerate his breathing. Also, the firer must be aware of the rifle's movement (while sighted on a target) as a result of breathing.

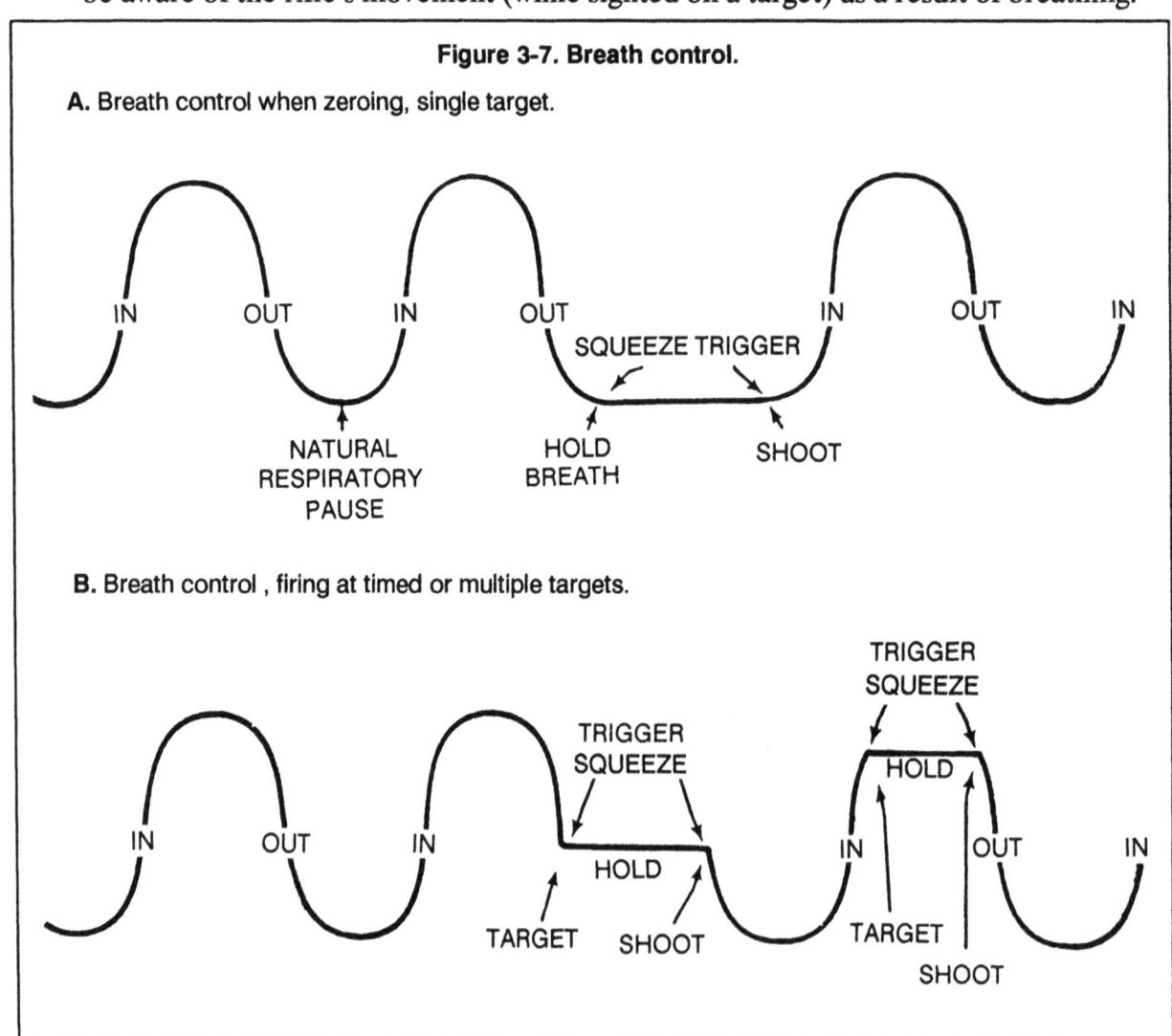

Figure 3-7. Breath control.

Trigger Squeeze. A novice firer can learn to place the rifle in a steady position and to correctly aim at the target if he follows basic principles. If the trigger is not properly squeezed, the rifle is misaligned with the target at the moment of firing.

Rifle movement. Trigger squeeze is important for two reasons:

- First, any sudden movement of the finger on the trigger can disturb the lay of the rifle and cause the shot to miss the target.

- Second, the precise instant of firing should be a surprise to the soldier.

The soldier's natural reflex to compensate for the noise and slight punch in the shoulder can cause him to miss the target if he knows the exact instant the rifle will fire. The soldier usually tenses his shoulders when expecting the rifle to fire, but it is difficult to detect since he does not realize he is flinching. When the hammer drops on a dummy round and does not fire, the soldier's natural reflexes demonstrate that he is improperly squeezing the trigger.

NOTE: See Appendix C for the Weaponeer and ball-and-dummy exercise. They are good training devices in detecting improper trigger squeeze.

Trigger finger. The trigger finger (index finger on the firing hand) is placed on the trigger between the first joint and the tip of the finger (not the extreme end) and is adjusted depending on hand size, grip, and so on. The trigger finger must squeeze the trigger to the rear so that the hammer falls without disturbing the lay of the rifle. When a live round is fired, it is difficult to see what affect trigger pull had on the lay of the rifle. Therefore, it is important to experiment with many finger positions during dry-fire training to ensure the hammer is falling with little disturbance to the aiming process.

As the firer's skills increase with practice, he needs less time spend on trigger squeeze. Novice firers can take five seconds to perform an adequate trigger squeeze, but, as skills improve, he can squeeze the trigger in a second or less. The proper trigger squeeze should start with slight pressure on the trigger during the initial aiming process. The firer applies more pressure after the front sight post is steady on the target and his is holding his breath.

The coach/trainer observes the trigger squeeze, emphasizes the correct procedure, and checks the firer's applied pressure. He places his finger on the trigger and has the firer squeeze the trigger by applying pressure to the coach/trainer's finger. The coach/trainer ensures that the firer squeezes straight to the rear on the trigger avoiding a left or right twisting movement. A steady position reduces disturbance of the rifle during trigger squeeze.

From an unsupported position, the firer experiences a greater wobble area than from a supported position. Wobble area is the movement of the front sight around the aiming point when the rifle is in the steadiest position. If the front sight strays from the target during the firing process, pressure on the trigger should be held constant and resumed as soon as sighting is corrected. The position must provide for the smallest possible wobble area. From a supported position, there should be minimal wobble area and little reason to detect movement. If movement of the rifle causes the front sight to leave the target, more practice is needed. The firer should never try to quickly squeeze the trigger while the sight is on the target. The best firing performance results when the trigger is squeezed continuously, and the rifle is fired without disturbing its lay.

FIRING POSITIONS

All firing positions are taught during basic rifle marksmanship training. During initial fundamental training, the basic firing positions are used. The other positions are added later in training to support tactical conditions.

Basic Firing Positions. Two firing positions are used during initial fundamental training: the individual supported fighting position and prone unsupported position. Both offer a stable platform for firing the rifle. They are also the positions used during basic record fire.

Supported fighting position. This position provides the most stable platform for engaging targets (Figure 3-8). Upon entering the position, the soldier adds or removes dirt, sandbags, or other supports to adjust for his height. He then faces the target, executes a half-face to his firing side, and leans forward until his chest is against the firing-hand corner of the position. He places the rifle handguard in a V formed by the thumb and fingers of his nonfiring hand, and rests the nonfiring hand on the material (sandbags or berm) to the front of the position. The soldier places the stock butt in the pocket of his firing shoulder and rests his firing elbow on the ground outside the position. (When prepared positions are not available, the prone supported position can be substituted.)

Figure 3-8. Supported fighting position.

Once the supported fighting position has been mastered, the firer should practice various unsupported positions to obtain the smallest possible wobble area during final aiming and hammer fall. The coach/trainer can check the steadiness of the position by observing movement at the forward part of the rifle, by looking through the M16 sighting device, or by checking to see that support is being used.

NOTE: The objective is to establish a steady position under various conditions. The ultimate performance of this task is in a combat environment. Although the firer must be positioned high enough to observe all targets, he must remain as low as possible to provide added protection from enemy fire.

Prone unsupported position. This firing position (Figure 3-9) offers another stable firing platform for engaging targets. To assume this position, the soldier faces his target, spreads his feet a comfortable distance apart, and drops to his knees. Using the butt of the rifle as a pivot, the firer rolls onto his nonfiring side, placing the nonfiring elbow close to the side of the magazine. He places the rifle butt in the pocket formed by the firing shoulder, grasps the pistol grip with his firing hand, and lowers the firing elbow to the ground. The rifle rests in the V formed by the thumb and fingers of the nonfiring hand. The soldier adjusts the position of his firing elbow until his shoulders are about level, and pulls back firmly on the rifle with both hands. To complete the position, he obtains a stock weld and relaxes, keeping his heels close to the ground.

Advanced Positions. After mastering the four marksmanship fundamentals in the

Figure 3-9. Prone unsupported position.

two basic firing positions, the soldier is taught the advanced positions. He is trained to assume different positions to adapt to the combat situation.

Alternate prone position (Figure 3-10). This position is an alternative to both prone supported and unsupported fighting positions, allowing the firer to cock his firing leg. The firer can assume a comfortable position while maintaining the same relationship between his body and the axis of the rifle. This position relaxes the stomach muscles and allows the firer to breathe naturally.

Kneeling supported position (Figure 3-11). This position allows the soldier to obtain the height necessary to better observe many target areas, taking advantage of available cover. Solid cover that can support any part of the body or rifle assists in firing accuracy.

Kneeling unsupported position (Figure 3-12). This position is assumed quickly, places the soldier high enough to see over small brush, and provides for a stable firing position. The nonfiring elbow should be pushed forward of the knee so that the upper arm is resting on a flat portion of the knee to provide stability. The trailing foot can be placed in a comfortable position.

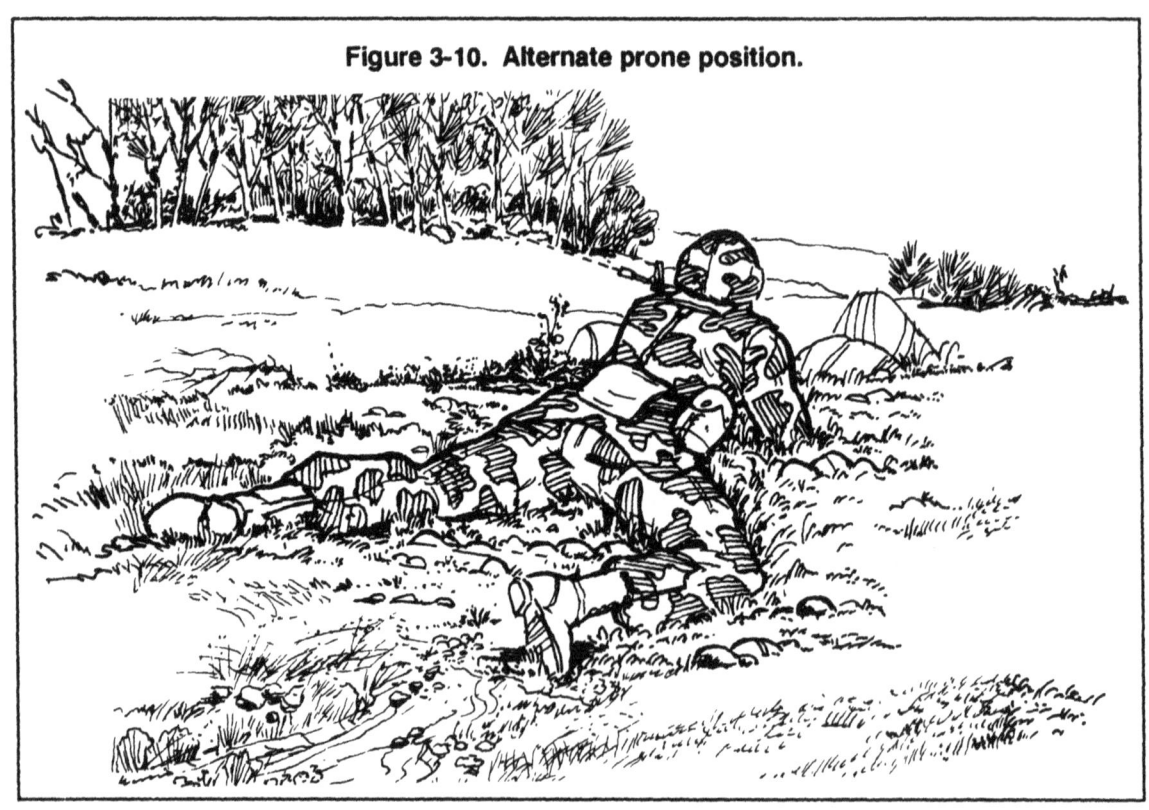

Figure 3-10. Alternate prone position.

Figure 3-11. Kneeling supported position.

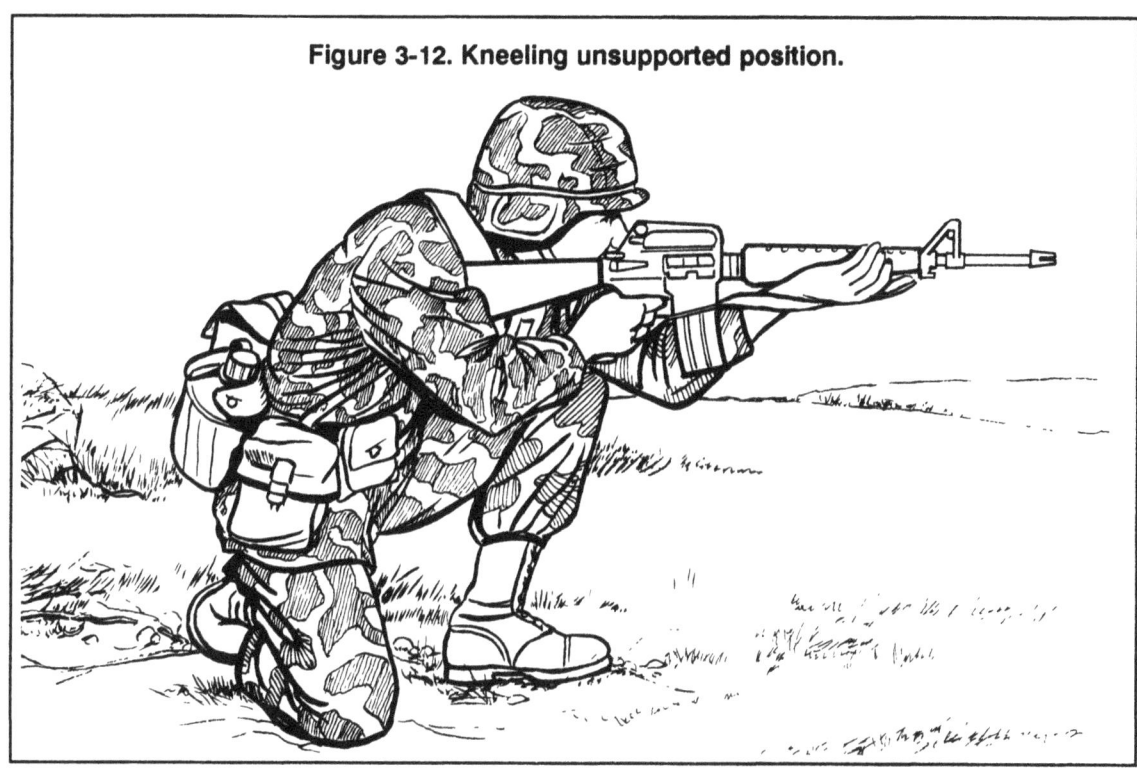

Figure 3-12. Kneeling unsupported position.

Standing position (Figure 3-13). To assume the standing position, the soldier faces his target, executes a facing movement to his firing side, and spreads his feet a comfortable distance apart. With his firing hand on the pistol grip and his nonfiring hand on either the upper handguard or the bottom of the magazine, the soldier places

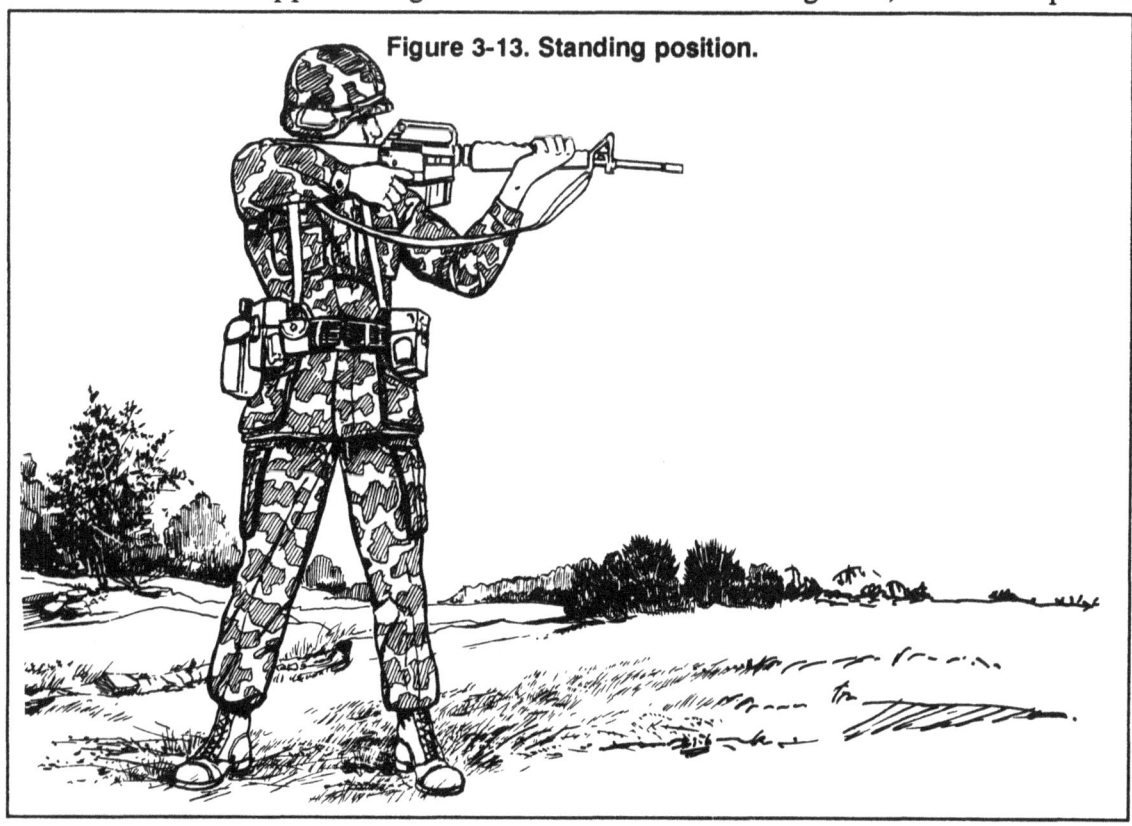

Figure 3-13. Standing position.

the butt of the rifle in the pocket formed by his firing shoulder so that the sights are level with his eyes. The weight of the rifle is supported by the firing shoulder pocket and nonfiring hand. The soldier shifts his feet until he is aiming naturally at the target and his weight is evenly distributed on both feet. The standing position provides the least stability but could be needed for observing the target area since it can be assumed quickly while moving. Support for any portion of the body or rifle improves stability. More stability can be obtained by adjusting the ammunition pouch to support the nonfiring elbow, allowing the rifle magazine to rest in the nonfiring hand.

Modified Firing Positions. Once the basic firing skills have been mastered during initial training, the soldier should be encouraged to modify positions, to take advantage of available cover, to use anything that helps to steady the rifle, or to make any change that allows him to hit more combat targets. The position shown in Figure 3-14 uses sandbags to support the handguard and frees the nonfiring hand to be used on any part of the rifle to hold it steady.

NOTE: Modified positions can result in small zero changes due to shifting pressure and grip on the rifle.

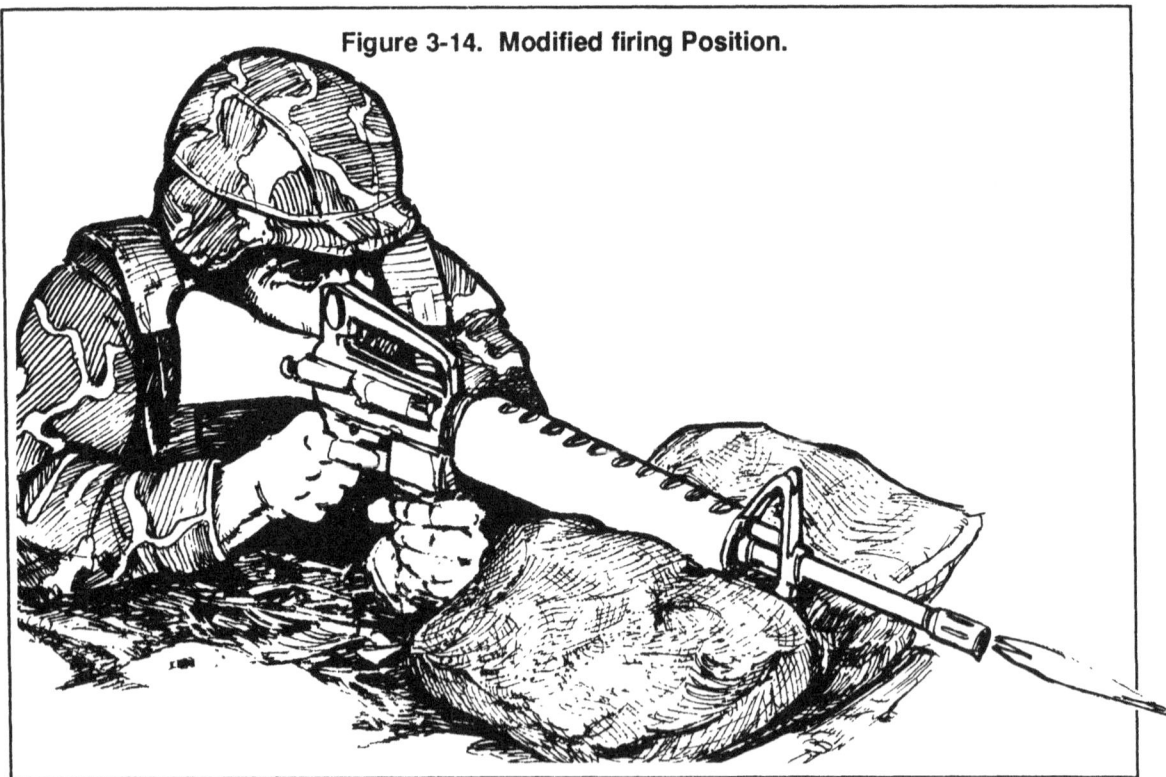

Figure 3-14. Modified firing Position.

MOUT Firing Positions. Although the same principles of rifle marksmanship apply, the selection and use of firing positions during MOUT requires some special considerations. Firing from around corners could require the soldier to fire from the opposite shoulder to avoid exposing himself to enemy fire.

The requirement for long-range observation can dictate that positions be occupied that are high above ground. Figure 3-15 shows a soldier firing over rooftops, exposing only the parts of his body necessary to engage a target. Figure 3-16 shows a soldier firing around obstacles. Figure 3-17 highlights the need to stay in the shadows while firing from windows, and the requirements for cover and rifle support.

Figure 3-15. Firing over rooftops.

Figure 3-16. Firing around obstacles.

Figure 3-17. Firing from windows.

FM 23-9

Section III. DRY FIRE

Dry-fire exercises are conducted as they relate to each of the fundamentals of rifle marksmanship. The standard 25-meter zero targets (Figures 3-18 and 3-19) are mounted as illustrated, because they provide the consistent aiming point the soldier must use throughout preparatory marksmanship training.

CONDUCT OF DRY-FIRE TRAINING

A skilled instructor/trainer should supervise soldiers on dry-fire training. Once an explanation and demonstration are provided, soldiers should be allowed to work at their own pace, receiving assistance as needed. The peer coach-and-pupil technique can be effectively used during dry-fire training with the coach observing performance and offering suggestions. Several training aids are available to correctly conduct initial dry-fire training of the four fundamentals (Appendix C).

A supported firing position should be used to begin dry-fire training. Sandbags and chest-high support are used to effectively teach this position. While any targets at any range can be used, the primary aim point should be a standard silhouette zeroing target placed at a distance of 25 meters from the firing position. The other scaled-silhouette targets—slow fire and timed fire—are also excellent for advanced dry-fire training.

After the soldier understands and has practiced the four fundamentals, he proceeds to integrated dry-fire exercises. The objective of integrated dry fire is to master the four fundamentals of marksmanship in a complete firing environment. With proper dry-fire training, a soldier can assume a good, comfortable, steady firing position when he moves to the firing line. He must understand the aiming process, breath control is second nature, and correct trigger squeeze has been practiced many times. Also, by adding dummy ammunition to the soldier's magazine, other skills can be integrated into the dry-fire exercise to include practicing loading and unloading, reinforcing immediate-action drills, and using the dime (washer) exercise.

When correctly integrated, dry fire is an effective procedure to use before firing live bullets for grouping and zeroing, scaled silhouettes, field firing, or practice record fire. It can be used for remedial training or opportunity training, or as a primary training technique to maintain marksmanship proficiency.

PEER COACHING

Peer coaching is using two soldiers of equal firing proficiency and experience to assist (coach) each other during marksmanship training. Some problems exist with peer coaching. If the new soldier does not have adequate guidance, a "blind-leading-the-blind" situation results, which can lead to negative training and safety violations. However, when adequate instruction is provided, peer coaching can be helpful even in the IET environment. Since all soldiers in units have completed BRM, peer coaching should yield better results.

Benefits. The pairing of soldiers can enhance learning for both of them. The coach learns what to look for and what to check as he provides guidance to the firer. Communication between peers is different than communication between a firer and

Figure 3-18. The M16A1 and M16A2 zero targets.

(NSN 6920-01-167-1392)

(NSN 6920-01-253-4005)

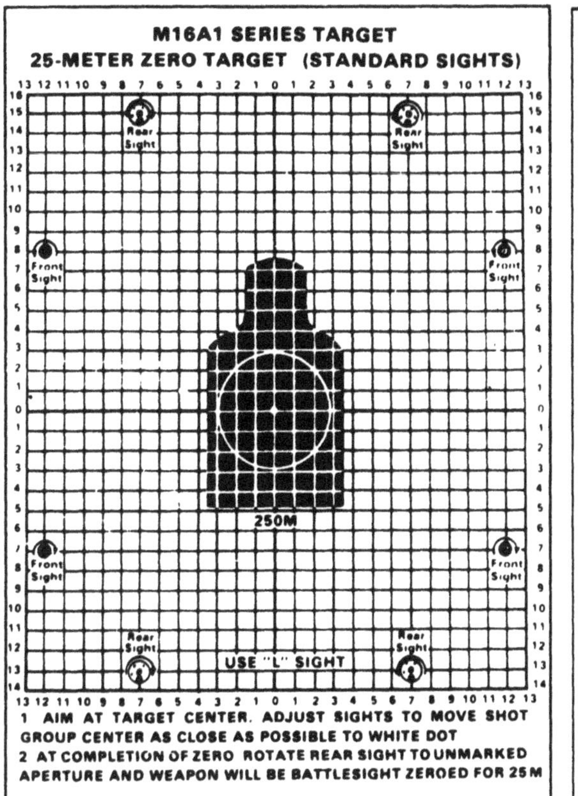

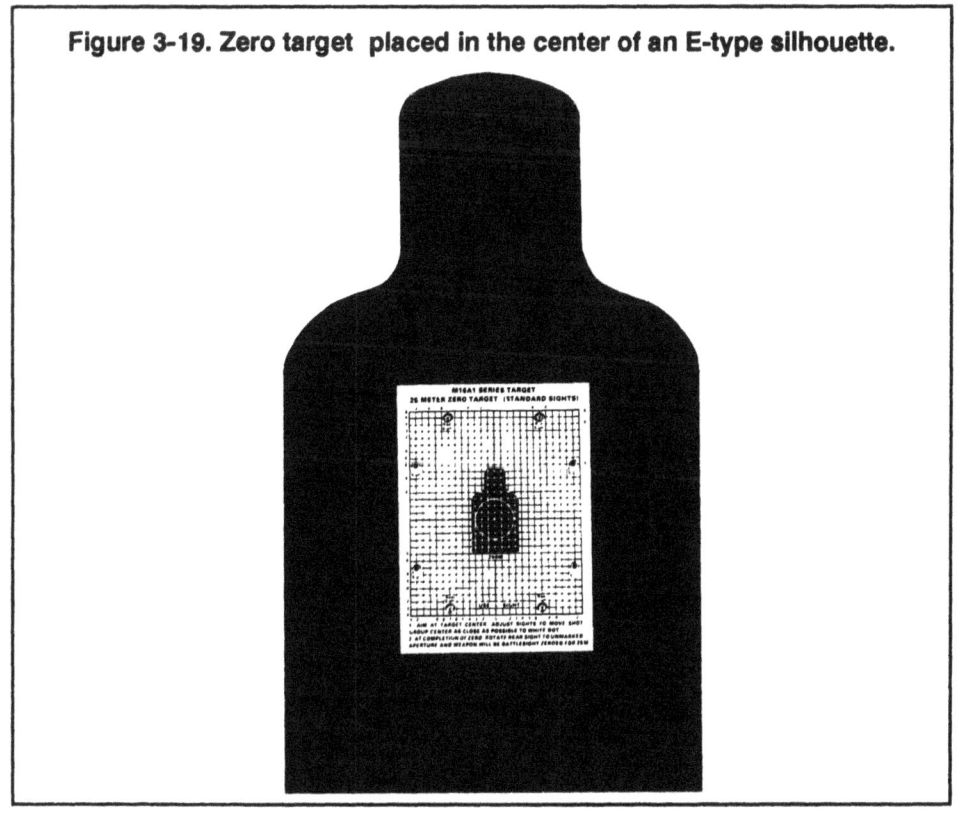

Figure 3-19. Zero target placed in the center of an E-type silhouette.

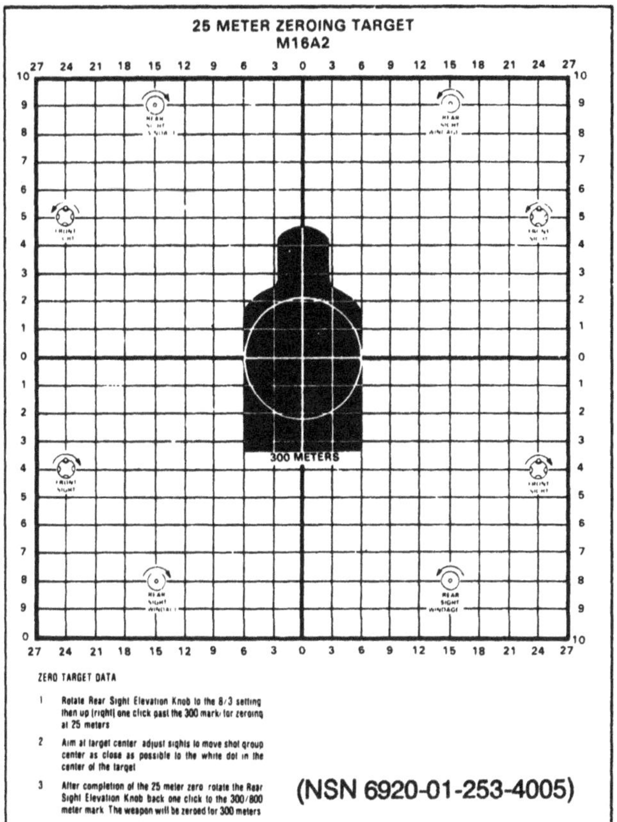

drill sergeant or senior NCO. Peers have the chance to ask simple questions and to discuss areas that are not understood. Pairing soldiers who have demonstrated good firing proficiency with those who have firing problems can improve the performance of problem firers

Duties. The peer coach assists the firer in obtaining a good position and in adjusting sandbags. He watches the firer—**not the target**—to see that the firer maintains a proper, relaxed, steady position; that he holds his breath before the final trigger squeeze; that he applies initial pressure to the trigger; and that no noticeable trigger jerk, flinch, eye blink, or other reaction can be observed in anticipating the rifle firing. The peer coach can use an M16 sighting device, allowing him to see what the firer sees through the sights. (Appendix C.)

The peer coach can load magazines, providing a chance to use ball and dummy. At other times, he could be required to observe the target area—for example, when field-fire targets are being engaged and the firer cannot see where he is missing targets. The peer coach can add to range safety procedures by helping safety personnel with preliminary rifle checks.

NOTE: When a peer coach is used during M16A1 live-fire exercises, a brass deflector should be attached to the rifle and eye protection should be worn.

CHECKLIST FOR THE COACH

The procedures to determine and eliminate rifle and firer deficiencies follows.

The coach checks to see that the—

- Rifle is cleared and defective parts have been replaced.

- Ammunition is clean, and the magazine is properly placed in the pouch.

- Sights are blackened and set correctly for long/short range.

The coach observes the firer to see that he—

- Uses the correct position and properly applies the steady-position elements.

- Properly loads the rifle.

- Obtains the correct sight alignment (with the aid of an M16 sighting device).

- Holds his breath correctly (by watching his back at times).

- Applies proper trigger squeeze; determines whether he flinches or jerks by watching his head, shoulders, trigger finger, and firing hand and arm.

- Is tense and nervous. If the firer is nervous, the coach has the firer breathe deeply several times to relax.

Supervisory personnel and peer coaches correct errors as they are detected. If many common errors are observed, it is appropriate to call the group together for more discussion and demonstration of proper procedures and to provide feedback.

POSITION OF THE COACH

The coach constantly checks and assists the firer in applying marksmanship fundamentals during firing. He observes the firer's position and his application of the steady position elements. The coach is valuable in checking factors the firer is unable to observe for himself and in preventing the firer from repeating errors.

During an exercise, the coach should be positioned where he can best observe the firer when he assumes position. He then moves to various points around the firer (sides and rear) to check the correctness of the firer's position. The coach requires the firer to make adjustments until the firer obtains a correct position.

When the coach is satisfied with the firing position, he assumes a coaching position alongside the firer. The coach usually assumes a position like that of the firer (Figure 3-20), which is on the firing side of the soldier.

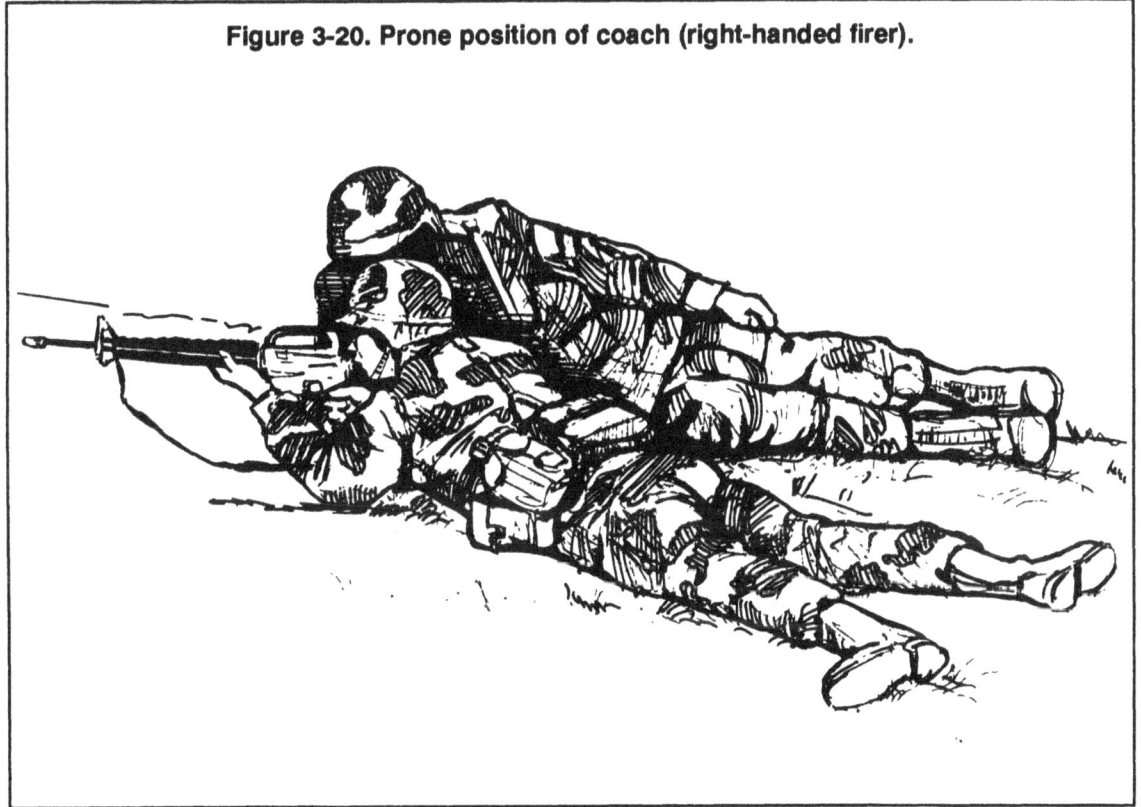

Figure 3-20. Prone position of coach (right-handed firer).

GROUPING

Shot grouping is a form of practice firing with two primary objectives: firing tight shot groups and consistently placing those groups in the same location. Shot grouping should be conducted between dry-fire training and zeroing. The initial live-fire training should be a grouping exercise with the purpose of practicing and refining marksmanship fundamentals. Since this is not a zeroing exercise, few sight changes are made. Grouping exercises can be conducted on a live-fire range that provides precise location of bullet hits and misses such as a 25-meter zeroing range or KD range.

CONCEPT OF ZEROING

The purpose of battlesight zeroing is to align the **fire control system** (sights) with the rifle barrel, considering the given ammunition ballistics. When this is accomplished

correctly, the fire control and point of aim are point of impact at a **standard battlesight zero range** such as 250 (300) meters.

When a rifle is zeroed, the sights are adjusted so that bullet strike is the same as point of aim at some given range. A battlesight zero (250 meters, M16A1; 300 meters, M16A2) is the sight setting that provides the highest hit probability for most combat targets with minimum adjustment to the aiming point.

When standard zeroing procedures are followed, a rifle that is properly zeroed for one soldier is close to the zero for another soldier. When a straight line is drawn from target center to the tip of the front sight post and through the center of the rear aperture, it makes little difference whose eye is looking along this line. There are many subtle factors that result in differences among individual zeros; however, the similarity of individual zeros should be emphasized instead of the differences.

Most firers can fire with the same zeroed rifle if they are properly applying marksmanship fundamentals. If a soldier is having difficulty zeroing and the problem cannot be diagnosed, having a good firer zero the rifle could find the problem. When a soldier must fire another soldier's rifle without opportunity to verify the zero by firing—for example, picking up another man's rifle on the battlefield—it is closer to actual zero if the rifle sights are left unchanged. This information is useful in deciding initial sight settings and recording of zeros. All rifles in the arms room, even those not assigned, should have their sights aligned (zeroed) for battlesight zero.

There is no relationship between the specific sight setting a soldier uses on one rifle (his zero) to the sight setting he needs on another rifle. For example, a soldier could be required to move the rear sight of his assigned rifle 10 clicks left of center for zero, and the next rifle he is assigned could be adjusted 10 clicks right of center for zero. This is due to the inherent variability from rifle to rifle, which makes it essential that each soldier is assigned a permanent rifle on which all marksmanship training is conducted. Therefore, all newly assigned personnel should be required to fire their rifle for zero as soon as possible after assignment to the unit. The same rule must apply anytime a soldier is assigned a new rifle, a rifle is returned from DS or GS maintenance, or the zero is in question.

M16A1 STANDARD SIGHTS AND ZEROING

To battlesight zero the rifle, the soldier must understand sight adjustment procedures. The best possible zero is obtained by zeroing at actual range. Because facilities normally do not exist for zeroing at 250 meters, most zeroing is conducted at 25 meters. By pushing the rear sight forward so the L is exposed, the bullet crosses line of sight at 25 meters, reaches a maximum height above line of sight of about 11 inches at 225 meters, and crosses line of sight again at 375 meters (Figure 3-21).

To gain the many benefits associated with having bullets hit exactly where the rifle is aimed during 25-meter firing, the long-range sight is used on the zero range. Therefore, when bullets are adjusted to hit the same place the rifle is aimed at 25 meters, the bullet also hits where the rifle is aimed at 375 meters. After making this adjustment and flipping back to the short-range sight and aiming center of mass at a 42-meter target, the bullet crosses the line of sight at 42 meters and again at 250 meters as shown in Figure 3-22.

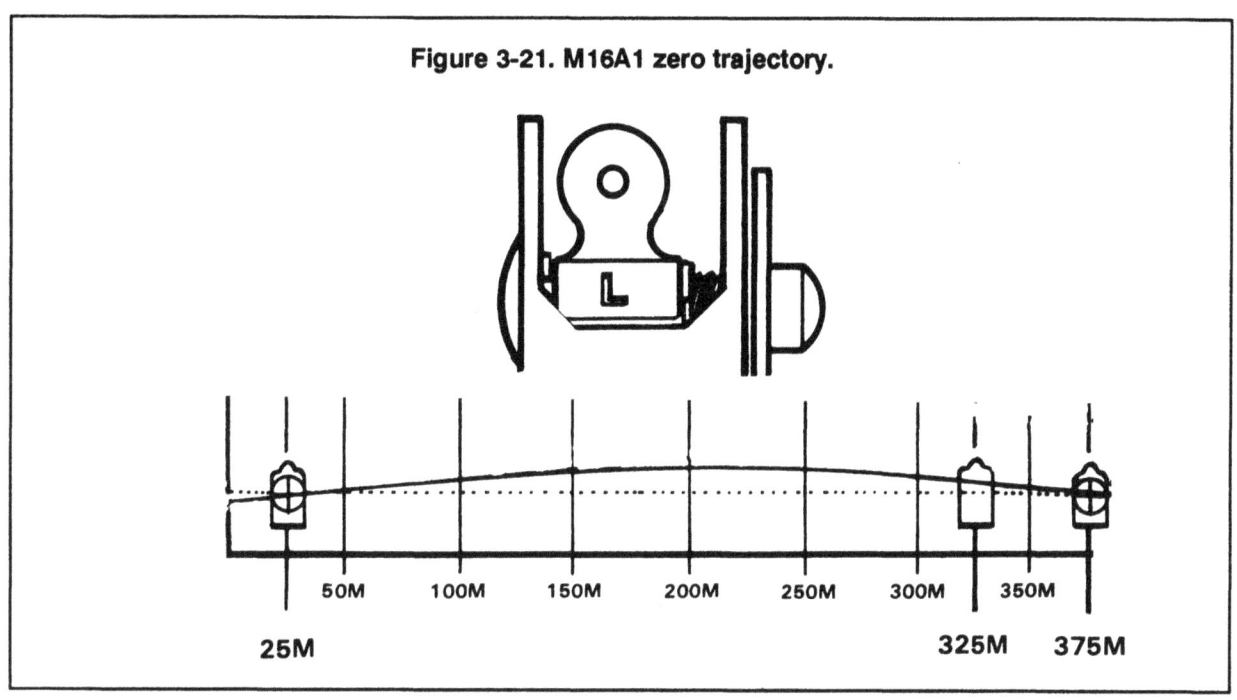

Figure 3-21. M16A1 zero trajectory.

Figure 3-22. M16A1 250-meter trajectory.

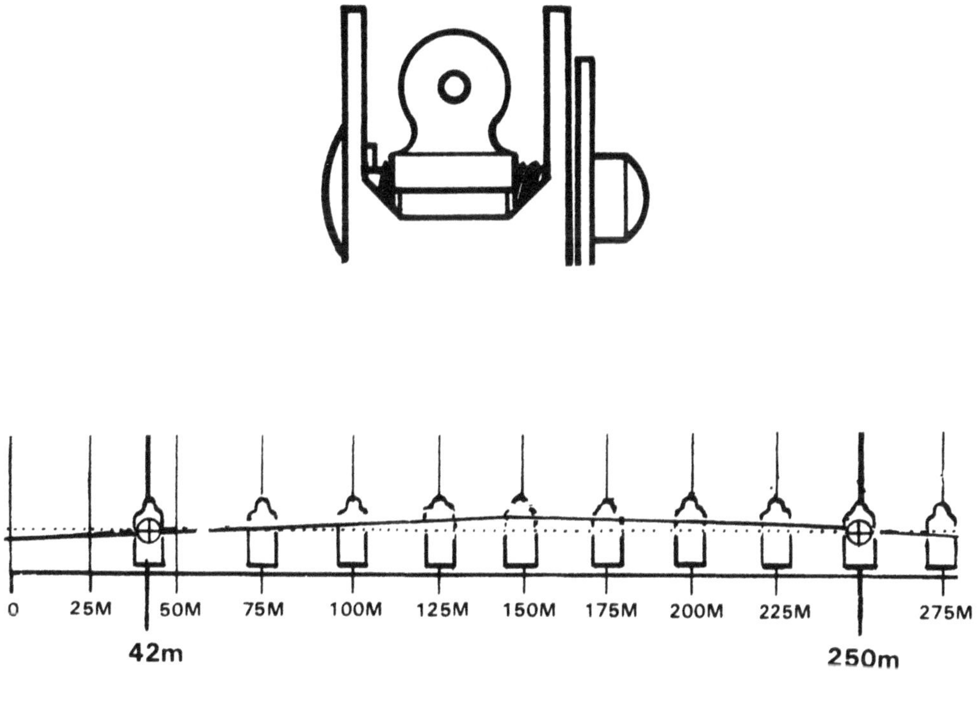

Most combat targets are expected to be engaged in the ranges from 0 to 300 meters; therefore, the 250-meter battlesight zero is the setting that remains on the rifle. At 25 meters, the bullet is about 1 inch below line of sight, crossing line of sight at 42 meters. It reaches its highest point above the line of sight (about 5 inches) at a range of about 175 meters, crosses line of sight again at 250 meters, and is about 7 inches below line of sight at 300 meters. Targets can be hit out to a range of 300 meters with no adjustments to point of aim. (A somewhat higher hit probability results with minor adjustments to the aiming point.)

Sights. The sights are adjustable for both elevation and windage. Windage adjustments are made on the rear sight; elevation adjustments on the front sight.

Rear sight. The rear sight consists of two apertures and a windage drum with a spring-loaded detent (Figure 3-23). The aperture marked L is used for ranges beyond 300 meters, and the unmarked or short-range aperture is used for ranges up to 300 meters. Adjustments for windage are made by pressing in on the spring-loaded detent with a sharp instrument (or the tip of a cartridge) and rotating the windage drum in the desired direction of change (right or left) in the strike of the bullet.

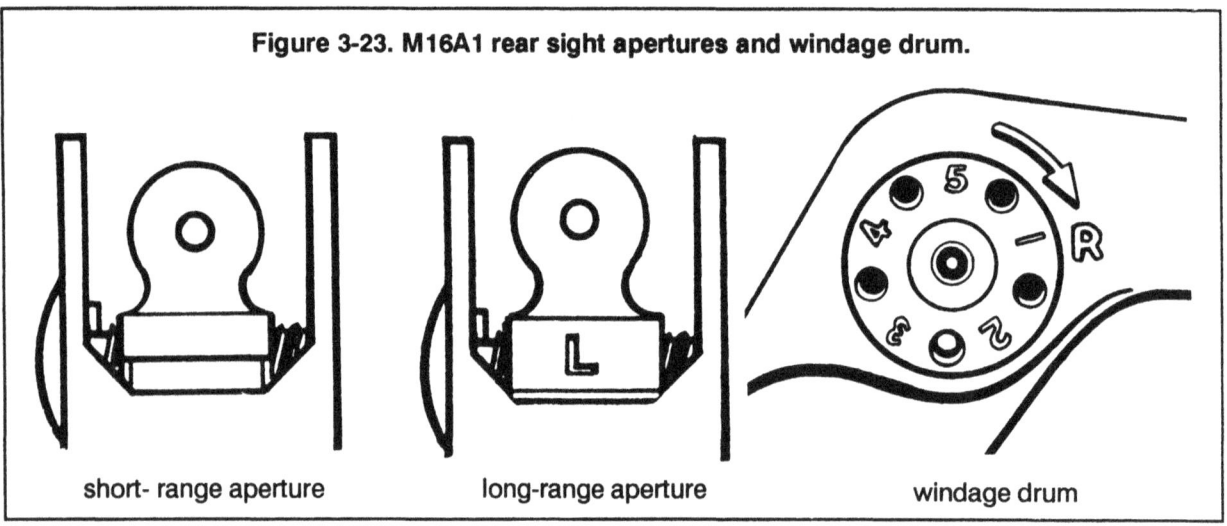

Figure 3-23. M16A1 rear sight apertures and windage drum.

Front sight. The front sight consists of a round rotating sight post with a five-position, spring-loaded detent (Figure 3-24). Adjustments are made by using a sharp instrument (or the tip of a cartridge). To move the front sight post, the spring-loaded detent is depressed, and the post is rotated in the desired direction of change (up or down) in the strike of the bullet.

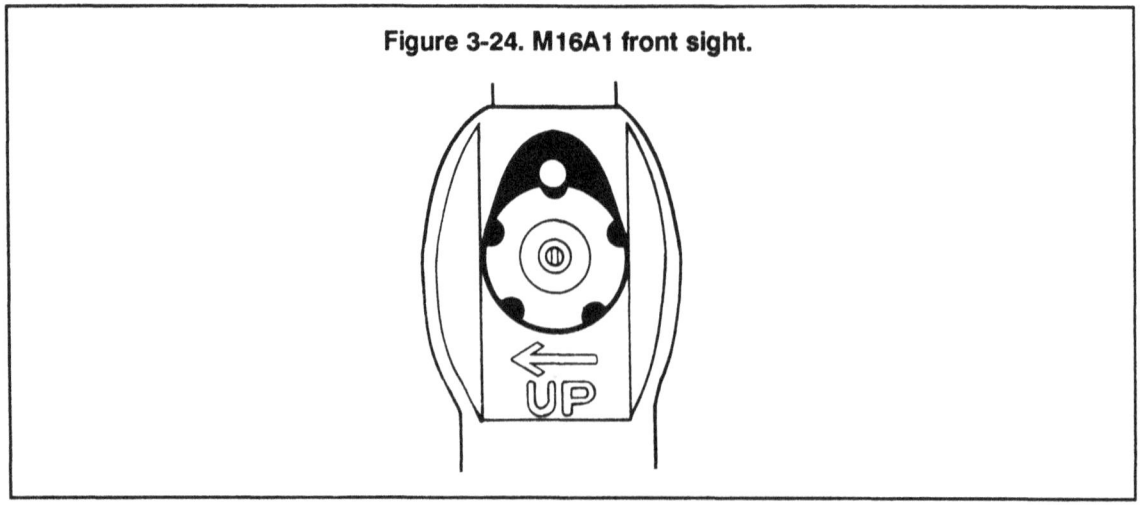

Figure 3-24. M16A1 front sight.

Sight Changes. To make sight changes, the firer first locates the center of his three-round shot group and then determines the distance between it and the desired location. An error in elevation is measured vertically, while a windage error is measured horizontally. When using standard zero targets or downrange feedback targets, sight adjustment guidance on the target is provided. (See Appendix F for the elevation and windage rule.)

To raise the strike of the bullet, the firer rotates the front sight post the desired number of clicks clockwise (in the direction of the arrow marked UP in Figure 3-24). Thus, the strike of the bullet is raised but the post is lowered. He reverses the direction of rotation to move the strike of the bullet down.

To move the strike of the bullet to the right, the windage drum is rotated the desired number of clicks clockwise (in the direction of the arrow marked R, Figure 3-23). The firer reverses the direction of rotation to move the strike of the bullet to the left.

NOTE: Before making any sight changes, the firer should make a serviceability check of the sights, looking for any bent, broken, or loose parts. The firer must also be able to consistently fire 4-cm shot groups.

M16A2 STANDARD SIGHTS AND ZEROING

When the soldier can consistently place three rounds within a 4-cm circle at 25 meters, regardless of group location, he is ready to zero his rifle.

The front and rear sights are set as follows:

Rear sight. The rear sight consists of two sight apertures, a windage knob, and an elevation knob (Figure 3-25).

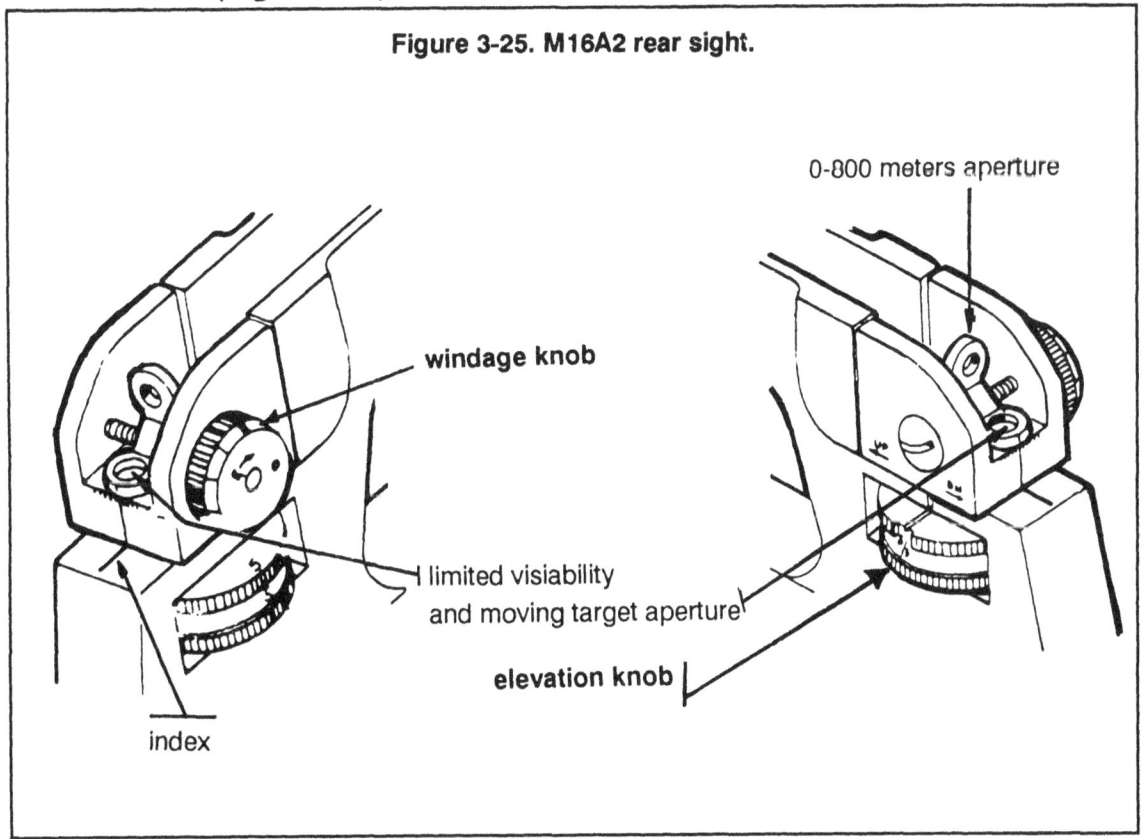

Figure 3-25. M16A2 rear sight.

FM 23-9

The larger aperture, marked 0-2, is used for moving target engagement and during limited visibility. The unmarked aperture is used for normal firing situations, zeroing, and with the elevation knob for target distances up to 800 meters. The unmarked aperture is used to establish the battlesight zero.

After the elevation knob is set, adjustments for elevation are made by moving the front sight post up or down to complete zeroing the rifle. Adjustments for windage are made by turning the windage knob.

The rear windage knob start point is when the index mark on the 0-2 sight is aligned with the rear sight base index (Figure 3-26).

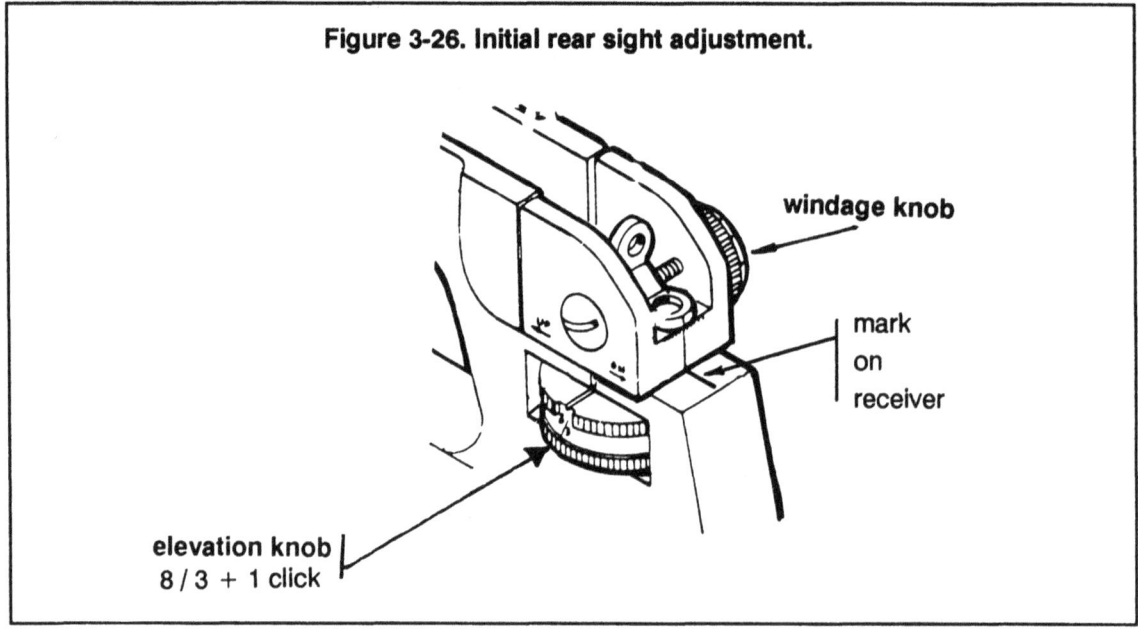

Figure 3-26. Initial rear sight adjustment.

Front sight. The front sight is adjusted the same as the front sight of the M16A1. It consists of a square, rotating sight post with a four-position, spring-loaded detent (Figure 3-27). Adjustments are made by using a sharp instrument or the tip of a cartridge. To raise or lower the front sight post, the spring-loaded detent is depressed, and the post is rotated in the desired direction of change. (Figure 3-28).

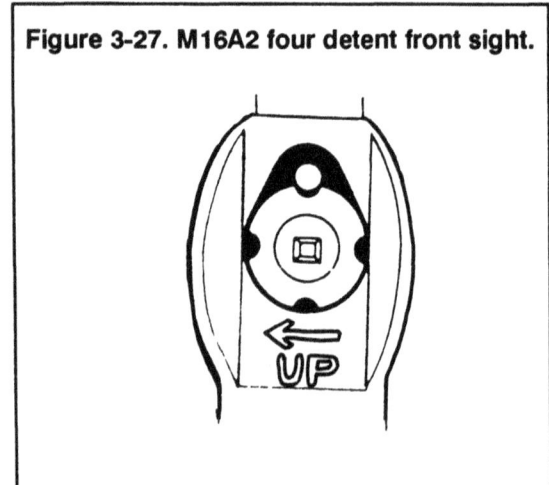

Figure 3-27. M16A2 four detent front sight.

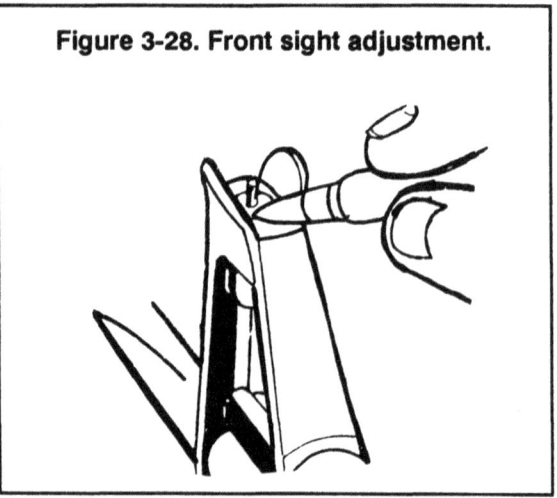

Figure 3-28. Front sight adjustment.

DOWNRANGE FEEDBACK TRAINING

The term downrange feedback describes any training method that provides precise knowledge of bullet strike (exactly where bullets hit or miss the intended target) at ranges beyond 25 meters. The soldier gains confidence in his firing abilities by knowing what happens to bullets at range. The inclusion of downrange feedback during the initial learning process and during refresher training improves the soldier's firing proficiency and record fire scores. Downrange feedback can be incorporated into any part of a unit's marksmanship program. However, an ideal sequence is to conduct downrange feedback following 25-meter firing and before firing on the field fire range. (See Appendix G.)

The use of a KD firing range is an excellent way of providing downrange feedback. Also a good way to obtain downrange feedback is to modify existing field fire ranges by constructing target-holding frames, which requires the soldier to walk from the firing line to the target to locate bullet strike.

Units can design their own downrange feedback training to accommodate available facilities. Any silhouette target with a backing large enough to catch all bullet misses can be set up at any range. For example, it would be ideal if the confirmation of weapon zero could be conducted at the actual zero range of 250 meters/300 meters.

FIELD FIRE TRAINING

Field fire training provides the transition from unstressed slow firing at known-distance/feedback targets to engaging fleeting combat-type pop-up silhouettes. Two basic types of field firing exercises are single-target and multiple-target engagements, which use 75-, 175-, and 300-meter targets. Once the soldier has developed the unstressed firing skills necessary to hit single KD targets, he must learn to detect and quickly engage combat-type targets at various ranges. Time standards are provided during this instruction to add stress and to simulate the short exposure times of combat targets. The soldier must, therefore, detect, acquire, and engage the target before the exposure ends. During field fire training, the firer learns to quickly detect and apply the fundamentals at the same time. (See Appendix G.)

PRACTICE RECORD FIRE

Practice record fire is a training exercise designed to progressively develop and refine the soldiers combat firing skills. During this exercise, the soldier is exposed to a more difficult course of fire with increased time stress to include single and multiple target engagements at six distances ranging from 50 to 300 meters. This exercise also provides the opportunity to practice and demonstrate skills learned during target detection. To perform well, a soldier must integrate all the tasks learned from previous training. When firing exercises are properly organized, conducted, and critiqued, the soldier gains knowledge and confidence in his firing performance. Through close observation, coaching, and critiquing, instructors/trainers can base remedial training on specific needs. (See Appendix G.)

RECORD FIRE

Qualification ratings and first-time GO rates are important during record fire, if properly used. They provide goals for the soldier and aid the commander in identifying

FM 23-9

the quality of his training. This should be considered in the assignment of priorities, instructor personnel, and obtaining valuable training resources. The objective of record firing is to access and confirm the individual proficiency of firers and the effectiveness of the training program. (See Appendix A for information on unit training and Appendix G for detailed information on record fire.)

FM 23-9

CHAPTER 4

Combat Fire Techniques

The test of a soldier's training is applying the fundamentals of marksmanship and firing skills in combat. The marksmanship skills mastered during training, practice, and record fire exercises must be applied to many combat situations (attack, assault, ambush, MOUT). Although these situations present problems, only two modifications of the basic techniques and fundamentals are necessary (see Chapter 3): changes to the rate of fire and alterations in weapon/ target alignment. The necessary changes are significant and must be thoroughly taught and practiced before discussing live-fire exercises.

NOTE: For tactical applications of fire see FM 7-8.

Section I. SUPPRESSIVE FIRE

In many tactical situations, combat rifle fire will be directed to suppress enemy personnel or weapons positions. Rifle fire, which is precisely aimed at a definite point or area target, is suppressive fire. Some situations may require a soldier to place suppressive fire into a wide area such as a wood line, hedgerow, or small building. While at other times, the target may be a bunker or window. Suppressive fire is used to control the enemy and the area he occupies. Suppressive fire is employed to kill the enemy or to prevent him from observing the battlefield or effectively using his weapons. When a sustained volume of accurate suppressive fire is placed on enemy locations to contain him, it can be effective even though he cannot be seen. When the enemy is effectively pinned down behind cover, this reduces his ability to deliver fire and allows friendly forces to move.

NATURE OF THE TARGET
Many soldiers have difficulty delivering effective suppressive fire when they cannot see a definite target. They must fire at likely locations or in a general area where the enemy is known to exist. Even though definite targets cannot be seen, most suppressive fire should be well aimed. Figure 4-1, page 4-2, shows a landscape target suitable for suppressive fire training. When this type target is used, trainers must develop a firing program to include areas of engagement and designated target areas that will be credited as sustained effective suppressive fire. At 25 meters, this target provides the firer with an area to suppress without definite targets to engage.

POINT OF AIM
Suppressive fire should be well-aimed, sustained, semiautomatic fire. Although lacking a definite target, the soldier must be taught to control and accurately deliver fire within the limits of the suppressed area. The sights are used as when engaging a point-type target — with the front sight post placed so that each shot impacts within the desired area (window, firing portal, tree line).

4-1

Figure 4-1. Landscape target.

RATE OF FIRE

During most phases of live fire (grouping, zeroing, qualifying), shots are delivered using the slow semiautomatic rate of fire (one round every 3 to 10 seconds). During training, this allows for a slow and precise application of the fundamentals. Successful suppressive fire requires that a faster but sustained rate of fire be used. Sometimes firing full automatic bursts (13 rounds per second) for a few seconds may be necessary to gain initial fire superiority. Rapid semiautomatic fire (one round every one or two seconds) allows the firer to sustain a large volume of accurate fire while conserving ammunition. The tactical situation dictates the most useful rate of fire, but the following must be considered:

Applying Fundamentals. As the stress of combat increases, some soldiers may fail to apply the fundamentals of marksmanship. This factor contributes to soldiers firing less accurately and without obtaining the intended results. While some modifications are appropriate, the basic fundamentals should be applied and emphasized regardless of the rate of fire or combat stress.

Making Rapid Magazine Changes. One of the keys to sustained suppressive fire is rapidly reloading the rifle. Rapid magazine changes must be correctly taught and practiced during dry-fire and live-fire exercises until the soldier becomes proficient. Small-unit training exercises must be conducted so that soldiers who are providing suppressive fire practice magazine changes that are staggered. Firing is, therefore, controlled and coordinated so that a continuous volume of accurate suppressive fire is delivered to the target area.

Conserving Ammunition. Soldiers must be taught to make each round count. Automatic fire should be used sparingly and only to gain initial fire superiority. Depending on the tactical situation, the rate of fire should be adjusted so that a minimum number of rounds are expended. Accurate fire conserves ammunition, while preventing the enemy from placing effective fire on friendly positions.

Section II. RAPID SEMIAUTOMATIC FIRE

Rapid semiautomatic fire delivers a large volume of accurate fire into a target or target area. Increases in speed and volume should be sought only after the soldier has demonstrated expertise and accuracy during slow semiautomatic fire. The rapid application of the four fundamentals will result in a well-aimed shot every one or two seconds. This technique of fire allows a unit to place the most effective volume of fire in a target area while conserving ammunition. It is the most accurate means of delivering suppressive fire.

EFFECTIVENESS OF RAPID FIRE

When a soldier uses rapid semiautomatic fire, he is sacrificing accuracy to deliver a greater volume of fire. The difference in accuracy between slow and rapid semiautomatic fire diminishes with proper training and repeated practice. Training and practice improve the soldier's marksmanship skills to the point that accuracy differences become minimal. There is little difference in the volume of effective fire that would be delivered by units using much less accurate automatic fire.

NOTE: Learning rapid fire techniques also improves the soldier's response time to short-exposure, multiple, and moving targets.

MODIFICATIONS FOR RAPID FIRE

Trainers must consider the impact of the increased rate of fire on the soldier's ability to properly apply the fundamentals of marksmanship and other combat firing skills. These fundamentals/skills include:

Immediate Action. To maintain an increased rate of suppressive fire, immediate action must be applied quickly. The firer must identify the problem and correct the stoppage immediately. Repeated dry-fire practice, using blanks or dummy rounds, followed by live-fire training and evaluation ensures that soldiers can rapidly apply immediate action while other soldiers initiate fire.

Marksmanship Fundamentals. The four fundamentals are used when firing in the rapid semiautomatic mode. The following differences apply:

Steady position. Good support improves accuracy and reduces recovery time between shots. somewhat tighter grip on the handguards assists in recovery time and in rapidly shifting or distributing fire to subsequent targets. When possible, the rifle should pivot at the point where the nonfiring hand meets the support. The soldier should avoid changing the position of the nonfiring hand on the support, because it is awkward and time-consuming when rapidly firing a series of shots.

Aiming. The aiming process does not change during rapid semiautomatic fire. The firer's head remains on the stock, his firing eye is aligned with the rear aperture, and his focus is on the front sight post.

Breath control. Breath control must be modified because the soldier does not have time to take a complete breath between shots. He must hold his breath at some point in the firing process and take shallow breaths between shots.

Trigger squeeze. To maintain the desired rate of fire, the soldier has only a short period to squeeze the trigger (one well-aimed shot every one or two seconds).

The firer must cause the rifle to fire in a period of about one-half of a second or less and still not anticipate the precise instant of firing. Rapid semiautomatic trigger squeeze is difficult to master. It is important that initial trigger pressure be applied as soon as a target is identified and while the front sight post is being brought to the desired point of aim. When the post reaches the point of aim, final pressure must be applied to cause the rifle to fire almost at once. This added pressure, or final trigger squeeze, must be applied without disturbing the lay of the rifle.

Repeated dry-fire training, using the Weaponeer device, and live-fire practice ensure the soldier can squeeze the trigger and maintain a rapid rate of fire consistently and accurately.

NOTE: When presented with multiple targets, the soldier may fire the first round, release pressure on the trigger to reset the sear, then reapply more pressure to fire the next shot. This technique eliminates the time used in releasing all the trigger pressure. It allows the firer to rapidly deliver subsequent rounds. Training and practice sessions are required for soldiers to become proficient in the technique of rapid trigger squeeze.

Magazine Changes. Rapid magazine changes are an integral part of sustaining rapid semiautomatic suppressive fire. Soldiers must quickly reload their rifles and resume accurate firing.

Magazine handling. Most units establish the soldier's basic load of ammunition and loaded magazines. The number of magazines vary based on the mission and tactical situation. During combat, some magazines are lost, but it is the soldier's responsiblility to keep this loss to a minimum. While training a soldier to reload his magazines, the trainer must emphasize the need to account for these magazines.

The sequence for magazine handling during rapid changes is illustrated for right- and left-handed firers in Figure 4-2.

Rifle loading. Removing a magazine from the firing side ammunition pouch is the same for both right- and left-handed firers. Empty magazines must be removed from the rifle before performing the following.

To remove a magazine from the pouch, the magazine is grasped on the long edge with the thumb, and the first and second fingers are placed on the short edge.

The magazine is withdrawn from the ammunition pouch, and the arm is extended forward, rotating the hand and wrist so that the magazine is in position (open end up and long edge to the rear) to load into the rifle. It is loaded into the rifle by inserting the magazine straight up into the magazine well until it is seated. The base of the magazine is tapped with the heel of the hand to ensure the magazine is fully seated.

Removing a magazine from the nonfiring side of the ammunition pouch requires the firer to support the rifle with his firing hand. His nonfiring hand grasps the magazine and loads it into the rifle.

Rapid magazine changing. Training and repeated practice in this procedure improves soldier proficiency. The firer does not move the selector lever to SAFE during a rapid magazine change, but he must maintain a safe posture during the change.

FM 23-9

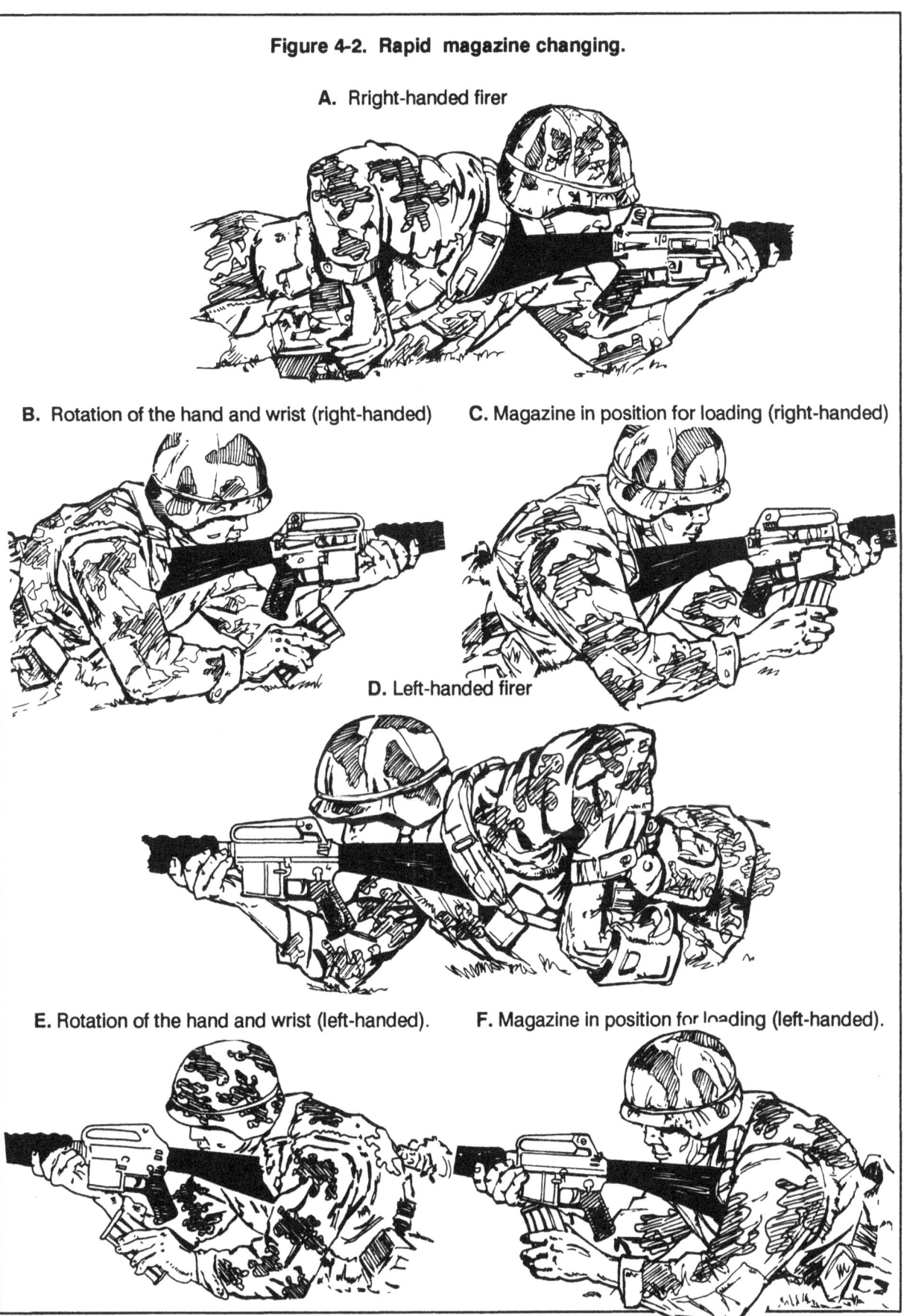

Figure 4-2. Rapid magazine changing.

A. Rright-handed firer

B. Rotation of the hand and wrist (right-handed)

C. Magazine in position for loading (right-handed)

D. Left-handed firer

E. Rotation of the hand and wrist (left-handed).

F. Magazine in position for loading (left-handed).

The following is a step-by-step sequence for rapid magazine changing.

- *Right-handed firer.* Remove the index finger from the trigger and depress the magazine catch button while keeping a secure grip on the rifle with the nonfiring hand (Figure 4-3). Release the pistol grip, grasp and remove the empty magazine with the right (firing) hand, and secure it. Grasp the loaded magazine with the right hand (rounds up and forward). Insert the loaded magazine into the magazine well and tap upward with the palm of the right hand. This ensures that the magazine is fully seated and locked into the rifle. Depress the upper half of the bolt catch with the fingers of the right hand. This allows the bolt to go forward, chambering the first round. If necessary, use the right hand to tap the forward assist to fully chamber the first round. Return the right hand to its original firing position on the pistol grip. Return the index finger to the trigger.

Figure 4-3. Magazine release catch button being depressed (right-handed firer).

- *Left-handed firer.* Remove the index finger from the trigger and release the pistol grip. Depress the magazine catch button with the index finger of the left (firing) hand. Remove the empty magazine with the left hand and secure it. Grasp the loaded magazine with the left hand (rounds up, bullets forward). Insert the loaded magazine into the magazine well and tap upward with the palm of the left hand. This ensures that the magazine is fully seated and locked into the rifle. Depress the upper half of the bolt catch with a finger of the left hand. This allows the bolt to go forward, chamberng the first round. If necessary, use the right hand to tap the forward assist to fully chamber the first round. Return the left hand to its original firing position on the pistol grip. Return the index finger to the trigger. The firer must maintain a safe posture during the change.

When loading from the nonfiring side, the previous steps are followed with with this exception: the loaded magazine is secured and inserted into the magazine well with the nonfiring hand. The firing hand supports the rifle at the pistol grip. After the magazine is inserted, the firer should shift the rifle's weight to his nonfiring hand and continue with the recommended sequence.

RAPID-FIRE TRAINING

Soldiers should be well trained in all aspects of slow semiautomatic firing before attempting any rapid-fire training. Those who display a lack of knowledge of the fundamental skills should not advance to rapid semiautomatic training until these skills are learned. Initial training should focus on the modifications to the fundamentals and other basic combat skills necessary during rapid semiautomatic firing.

Dry-Fire Exercises. Repeated dry-fire exercises are the most efficient means available to ensure soldiers can apply modifications to the fundamentals. Multiple dry-fire exercises are needed, emphasizing a rapid shift in position and point of aim, followed by breath control and fast trigger squeeze. Blanks or dummy rounds may be used to train rapid magazine changes and the application of immediate action. The soldier should display knowledge and skill during these dry-fire exercises before attempting live fire.

Live-Fire Exercises. There are two types of live-fire exercises.

Individual. Emphasis is on each soldier maintaining a heavy volume of accurate fire. Weapon down time (during immediate action and rapid magazine changes) is kept to a minimum. Firing should begin at shorter ranges, progressing to longer ranges as soldiers display increased proficiency. Exposure or engagement times are shortened and the number of rounds increased to simulate the need for a heavy volume of fire. Downrange feedback is necessary to determine accuracy of fire.

Unit. Rapid semiautomatic fire should be the primary means of delivering fire during a unit LFX. It is the most accurate technique of placing a large volume of fire on poorly defined targets or target areas. Emphasis should be on staggered rapid magazine changes, maintaining a continuous volume of fire and conserving ammunition.

Section III. AUTOMATIC FIRE

Automatic fire delivers the maximum amount of rounds into a target area. It should be trained only after the soldier has demonstrated expertise during slow and rapid semiautomatic fire. Automatic fire involves the rapid application of the four fundamentals while delivering from 3 to 13 rounds per second into a designated area. This technique of fire allows a unit to place the most fire in a target area (when conserving ammunition is not a consideration). It is a specialized technique of delivering suppressive fire and may not apply to most combat engagements. The M16A1 rifle has a full automatic setting. (The M16A2 uses a three-round burst capability.) Soldiers must be taught the advantages and disadvantages of automatic firing so they know when it should be used. Without this knowledge, in a life-threatening situation the soldier will tend to switch to the automatic/burst mode. This fire can be effective in some situations. It is vital for the unit to train and practice the appropriate use of automatic fire.

EFFECTIVENESS OF AUTOMATIC FIRE

Automatic fire is inherently less accurate than semiautomatic fire. The first automatic shot fired may be on target, but recoil and high-cyclic rate of fire often combine to place subsequent rounds far from the desired point of impact. Even controlled (three-round burst) automatic fire may place only one round on the target. Because of these inaccuracies, it is difficult to evaluate the effectiveness of automatic fire, and even more difficult to establish absolute guidelines for its use.

Closely spaced multiple targets, appearing at the same time at 50 meters or closer, may be engaged effectively with automatic/burst fire. More widely spaced targets appearing at greater distances should be engaged with semiautomatic fire.

The M16A1 and M16A2 rifles should normally be employed in the semiautomatic mode. Depending on the tactical situation, the following conditions would be factors against the use of automatic fire:

- Ammunition is in short supply or resupply may be difficult.

- Single targets are being engaged.

- Widely spaced multiple targets are being engaged.

- The distance to the target is beyond 50 meters.

- The effect of bullets on the target cannot be observed.

- Artificial support is not available.

- Targets may be effectively engaged using semiautomatic fire.

In some combat situations, the use of automatic fire can improve survivability and enhance mission accomplishment. Clearing buildings, final assaults, FPF, and ambushes may require the limited use of automatic fire. Depending on the tactical situation, the following conditions may favor the use of automatic fire:

- Enough available ammunition. Problems are not anticipated with resupply.

- Closely spaced multiple targets appear at 50 meters or less.

- Maximum fire is immediately required at an area target.

- Tracers or some other means can be used to observe the effect of bullets on the target.

- Leaders can maintain adequate control over rifles firing on automatic.

- Good artificial support is available.

- The initial sound of gunfire disperses closely spaced targets.

Trainers must ensure soldiers understand the capabilities and limitations of automatic fire. They must know when it should and should not be used.

MODIFICATIONS FOR AUTOMATIC FIRE POSITIONS

Trainers must consider the impact of the greatly increased rate of fire on the soldier's ability to properly apply the fundamentals of marksmanship and other combat firing skills. These fundamentals/skills include:

Immediate Action. To maintain automatic fire, immediate action must be applied quickly. The firer must identify the problem and correct it immediately. Repeated dry-fire practice, using blanks or dummy rounds, followed by live-fire training and evaluation ensures that soldiers can rapidly apply immediate action.

Marksmanship Fundamentals. The four fundamentals are used when firing in the automatic mode. The following differences apply:

Steady position (Figure 4-4). Maximum use of available artificial support is necessary during automatic fire. The rifle should be gripped more firmly and pulled into the shoulder more securely than when firing in the semiautomatic mode. This support and increased grip help to offset the progressive displacement of weapon/target alignment caused by recoil. To provide maximum stability, prone and supported positions are best. One possible modification involves forming a 5-inch loop with the

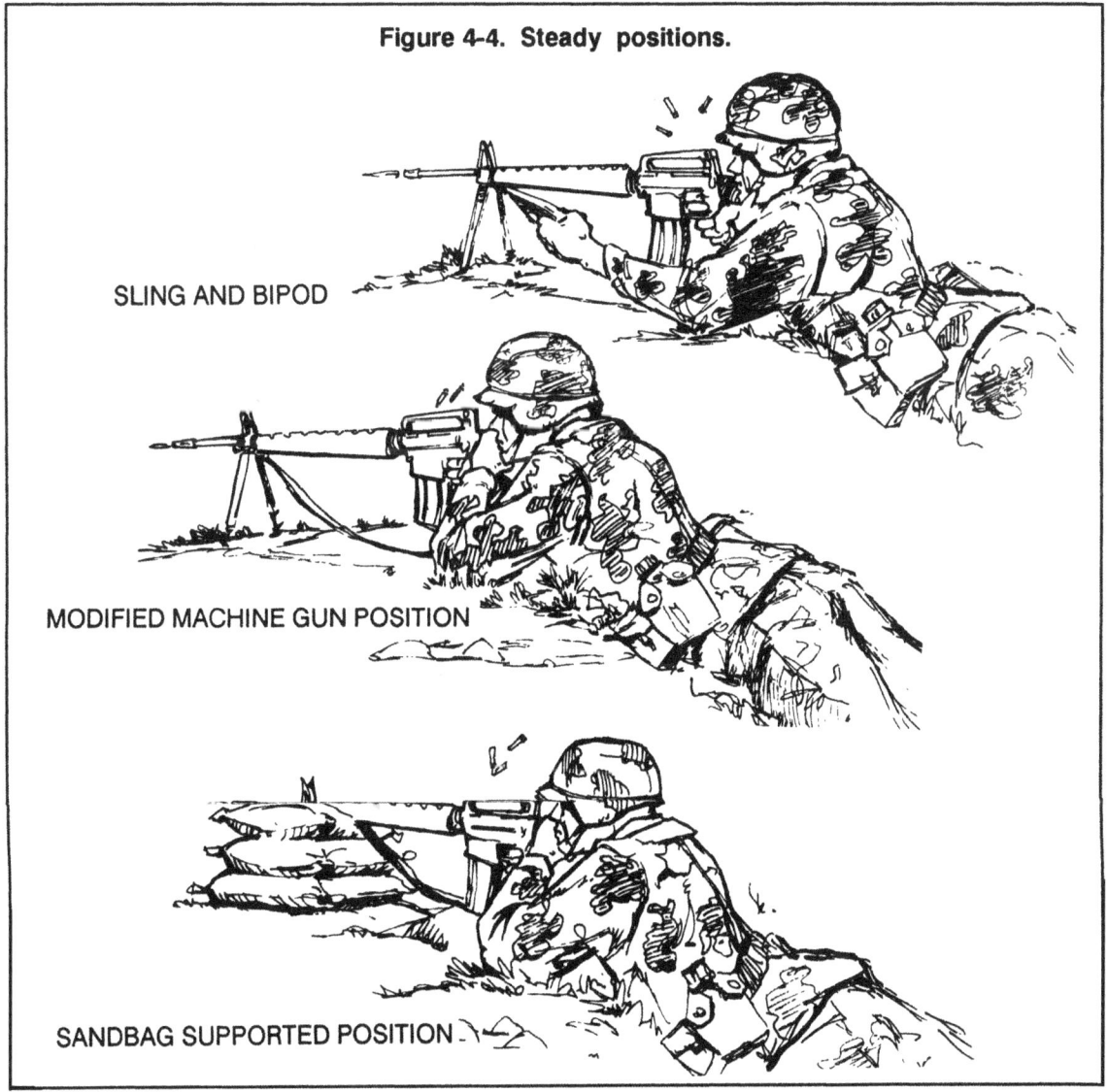

Figure 4-4. Steady positions.

sling at the upper sling swivel, grasping this loop with the nonfiring hand, and pulling down and to the rear while firing. Another modification involves grasping the small of the stock with the nonfiring hand, and applying pressure down and to the rear while firing. If a bipod is not available, sandbags may be used to support the rifle. The nonfiring hand may be positioned on the rifle wherever it provides the most stability and flexibility. The goal is to maintain weapon stability and minimize recoil.

Aiming. The aiming process does not change during automatic fire. The firer's head remains on the stock, his firing eye stays aligned with the rear sight aperture, and his focus is on the front sight post. Although recoil may disrupt this process, the firer must try to apply the aiming techniques throughout recoil.

Breath control. Breath control must be modified because the firer will not have the time to breathe between shots. He must hold his breath for each burst and adapt his breathing cycle, taking breaths between bursts.

Trigger squeeze. Training and repeated dry-fire practice will aid the soldier in applying proper trigger squeeze during automatic firing. Live-fire exercises will enable him to improve this skill.

NOTE: The trigger is not slapped or jerked. It is squeezed and pressure is quickly released.

- **M16A1.** Trigger squeeze is applied in the normal manner up to the instant the rifle fires. Because three-round bursts are the most effective rate of fire, pressure on the trigger should be released as soon as possible. The index finger should remain on the trigger, but a quick release of pressure is necessary to prevent an excessive amount of rounds from being fired in one burst. With much dry-fire practice, the soldier can become proficient at delivering three-round bursts with the squeeze/release technique.

- **M16A2.** Trigger squeeze is applied in the normal manner up to the instant the rifle fires. Using the burst-mode, the firer holds the trigger to the rear until three rounds are fired. He then releases pressure on the trigger until it resets, then reapplies pressure for the next three-round burst.

NOTE: Depending on the position of the burst cam when the selector is moved to the burst mode, the rifle may fire one, two, or three rounds when the trigger is held to the rear the first time. If the rifle fires only one or two rounds, the firer must quickly release pressure on the trigger and squeeze again, holding it to the rear until a three-round burst is completed.

Magazine Changes. Rapid magazine changes are vital in maintaining automatic fire. (See SECTION II. RAPID SEMIAUTOMATIC FIRE, Magazine Handling, for detailed information on rapid magazine changes.)

TRAINING OF AUTOMATIC FIRE TECHNIQUES

Soldiers should be well trained in all aspects of slow semiautomatic firing before attempting any automatic training. Those who display a lack of knowledge of the fundamental skills should not advance to automatic fire training until these skills are learned. Initial training should focus on the modifications to the fundamentals and other basic combat skills necessary during automatic firing.

Dry-Fire Exercises. Repeated dry-fire exercises are the most efficient means available to ensure soldiers can apply these modifications. Multiple dry-fire exercises are needed, emphasizing a stable position and point of aim, followed by breath control and the appropriate trigger squeeze. Blanks or dummy rounds may be used to train trigger squeeze, rapid magazine changes, and application of immediate action. The soldier should display knowledge and skill during these exercises before attempting live fire.

Live-Fire Exercises. There are two types of live-fire exercises.

Individual. Emphasis is on each individual maintaining a heavy volume of fire. Weapon down time (during immediate action and rapid magazine changes) is held to a minimum. Firing can begin at 25 meters, progressing to 50 meters as soldiers display increased proficiency. Exposure or engagement times, as well as ranges, are varied to best simulate the need for a heavy volume of fire. Downrange feedback is necessary to determine effectiveness of fire. The course of fire should allow the soldier to decide whether he should engage a given target or area with automatic or semiautomatic fire.

A soldier's zero during automatic fire may be different than his semiautomatic (battlesight) zero. This is due to the tendency of the lightweight M16 barrel to respond to external pressure such as the bipod or pulling on the sling. However, it is recommended that the battlesight zero be retained on the rifle and holdoff used to place automatic fire on the target. This holdoff training requires downrange feedback and should be conducted before other live-fire exercises.

The soldier can begin by loading and firing one round from an automatic fire position. Three of these rounds, treated as a single group, can establish where the first shot of a three-round burst will probably strike. Loading and firing two rounds simulates the dispersion of the second shot of a three-round burst. Finally, several three-round bursts should be fired to refine any necessary holdoff to center these larger groups on the desired point of impact.

Unit. Unit LFXs should include the careful use of automatic fire. Emphasis should be on staggered rapid magazine changes, maintaining a continuous volume of heavy fire, and conserving ammunition.

Section IV. QUICK FIRE

The two main techniques of directing fire with a rifle are to aim using the sights; and to use weapon alignment, instinct, bullet strike, or tracers to direct the fire. The preferred technique is to use the sights, but sometimes quick reflex action is needed to survive. Quick fire is a technique used to deliver fast, effective fire on surprise personnel targets at close ranges (25 meters or less). Quick-fire procedures have also been referred to as "instinct firing" or "quick kill."

EFFECTIVENESS OF QUICK FIRE

Quick-fire techniques are appropriate for soldiers who are presented with close, suddenly appearing, surprise enemy targets; or when close engagement is imminent. Fire may be delivered in the SEMIAUTO or BURST/AUTO mode. For example, a point man in a patrol may carry the weapon on BURST/AUTO. This may also be required when clearing a room or bunker. Initial training should be in the SEMI mode. Two techniques of delivering quick fire are—

Aimed. When presented with a target, the soldier brings the rifle up to his shoulder and quickly fires a single shot. His firing eye looks through or just over the rear sight aperture, and he uses the front sight post to aim at the target (Figure 4-5). Using this technique, a target at 25 meters or less may be accurately engaged in one second or less.

Pointed. When presented with a target, the soldier keeps the rifle at his side and quickly fires a single shot or burst. He keeps both eyes open and uses his instinct and peripheral vision to line up the rifle with the target (Figure 4-6). Using this technique, a target at 15 meters or less may be engaged in less than one second.

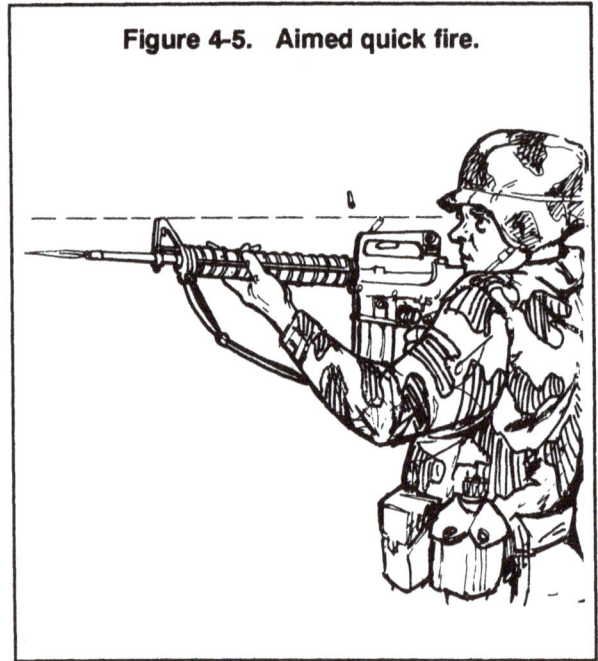

Figure 4-5. Aimed quick fire.

Figure 4-6. Pointed quick fire.

The difference in speed of delivery between these two techniques is small. Pointed quick fire can be used to fire a shot about one-tenth of a second faster than aimed quick fire. The difference in accuracy, however, is more pronounced. A soldier well trained in pointed quick fire can hit an E-type silhouette target at 15 meters, although the shot may strike anywhere on the target. A soldier well trained in aimed quick fire can hit an E-type silhouette target at 25 meters, with the shot or burst striking 5 inches from the center of mass.

The key to the successful employment of either technique is practice. Both pointed and aimed quick fire must be repeatedly practiced during dry-fire training. Live-fire exercises provide further skill enhancement and illustrate the difference in accuracy between the two techniques. Tactical considerations dictate which technique is most effective in a given situation, and when single shot versus burst fire is used.

Pointed and aimed quick fire should be used only when a target cannot be engaged fast enough using the sights in a normal manner. These techniques should be limited to targets appearing at 25 meters or less.

MODIFICATIONS FOR QUICK- FIRE TECHNIQUES

Quick-fire techniques require major modifications to the four fundamentals of marksmanship. These modifications represent a significant departure from the normal applications of the four fundamentals. Initial training in these differences, followed by repeated dry-fire exercises, will be necessary to prepare the soldier for live fire.

Steady Position. The quickness of shot delivery prevents the soldier from assuming a stable firing position. He must fire from his present position when the target appears. If the soldier is moving, he must stop. Adjustments for stability and support cannot be made before the round being fired.

Aimed. The butt of the rifle is pulled into the pocket of the shoulder as the cheek comes in contact with the stock. Both hands firmly grip the rifle, applying rearward pressure. The firing eye looks through or just over the rear sight aperture (Figure 4-5, page 4-12). The firer's sight is in focus and placed on the target.

Pointed. The rifle is pulled into the soldier's side and both hands firmly grip the rifle, applying rearward pressure (Figure 4-6, page 4-12).

Aiming. This fundamental must be highly modified because the soldier may not have time to look through the rear sight, find the front sight, and align it with the target.

Aimed. The soldier's initial focus is on the target. As the rifle is brought up, the firing eye looks through or just over the rear sight aperture at the target. Using his peripheral vision, the soldier locates the front sight post and brings it to the center of the target. When the front sight post is in focus, the shot is fired. Focus remains on the front sight post throughout the aiming process.

Pointed. The soldier's focus is placed on the center or slightly below the center of the target as the rifle is aligned with it and is fired. The soldier's instinctive pointing ability and peripheral vision are used to aid in proper alignment.

NOTE: When using either aiming technique, bullets may tend to impact above the desired location. Repeated live-fire practice is necessary to determine the best aim point on the target or the best focus. Such practice should begin with the soldier using a center mass arms/focus.

Breath Control. This fundamental has little application to the first shot of quick fire. The round must be fired before a conscious decision can be made about breathing. If subsequent shots are necessary, breathing must not interfere with the necessity to fire quickly. When possible, use short, shallow breaths.

Trigger Squeeze. Initial pressure is applied as weapon alignment is moved toward the target. Trigger squeeze is exerted so that when weapon/target alignment is achieved, the round is fired at once. The soldier requires much training and practice to perfect this rapid squeezing of the trigger.

TRAINING OF QUICK- FIRE TECHNIQUES

Initial training should focus on the major modifications to the fundamentals during quick fire.

Dry-Fire Exercises. This dry-fire exercise requires no elaborate preparations or range facilities, yet it provides the soldier with an opportunity to learn and practice quick-fire techniques. Repeated dry-fire exercises ensure soldiers can apply the modifications to the fundamentals. Multiple dry-fire exercises are needed, emphasizing a consistent firing position and weapon alignment with the target, followed by rapid trigger squeeze. No more than one second should elapse between the appearance of the target and a bullet striking it. One example of a dry-fire exercise is:

The trainer/coach places an E-type silhouette target 15 meters in front of the soldier. The soldier stands facing the general direction of the target (vary direction to simulate targets appearing at different locations), holding his rifle at or above waist level. His firing hand should be on the pistol grip; the nonfiring hand cradling the rifle under the handguards.

The trainer/coach should stand slightly behind the soldier, out of his field of view. The trainer/coach claps his hands, signaling target appearance. Immediately after clapping his hands, the trainer/coach counts out loud "one thousand one."

The soldier must either point or aim, squeeze the trigger, and hear the hammer fall before the trainer/coach finishes speaking (about one second or less).

NOTE: When using the aiming technique, the soldier holds his aim and confirms alignment of the rifle with the target. He keeps the rifle pointed toward the target after the hammer falls and looks through the sights to check his actual point of aim for that shot.

Live-Fire Exercises. There are two types of live-fire exercises.

Individual. Emphasis is on engaging each target in one second or less. The previously described timing technique may be used, or pop-up targets set to lock in the full upright position may be used. Pop-up targets require about one second to move from the down to the full up position. Targets set to lock in the upright position must be engaged as they are being raised to "kill" them. This gives the soldier a one-second time limit. At 15 meters (the maximum recommended range), an E-type silhouette engaged using pointed quick fire may be hit anywhere. Using aimed quick fire at the same target, hits should fall within a 10-inch circle located center of target.

NOTE: Repeated live-fire exercises are necessary to train the soldier. If 5.56-mm service ammunition is in short supply, the 5.56-mm practice ammunition and M2 bolt or the .22-caliber rim fire adapter device may be used.

Unit. Unit MOUT LFXs should include the use of quick fire. Targets should be presented at 25 meters or less and soldiers must engage them within one second.

Section V. MOPP FIRING

All soldiers must effectively fire their weapons to accomplish combat missions in an NBC environment. With proper training and practice, soldiers can gain confidence in their ability to effectively hit targets in full MOPP equipment. MOPP firing proficiency must be a part of every unit's training program.

EFFECTS OF MOPP EQUIPMENT ON FIRING

Firing weapons is only part of overall NBC training. Soldiers must first be familiar with NBC equipment, its use, and proper wear before they progress to learning the techniques of MOPP firing. Trainers must consider the impact of MOPP equipment (hood/ mask, gloves, overgarments) on the soldier's ability to properly apply the fundamentals of marksmanship and combat firing skills.

Immediate Action. Under normal conditions a soldier should be able to clear a stoppage in three to five seconds. Under full MOPP, however, this may take as long as ten seconds to successfully complete. Dry-fire practice under these conditions is necessary to reduce time and streamline actions. Hood/mask and gloves must be worn. Care must be taken not to snag or damage the gloves or dislodge the hood/mask during movements. Applying immediate action to a variety of stoppages during dry fire must be practiced using dummy or blank ammunition until such actions can be performed by instinct.

Target Detection. Techniques and principles outlined in Chapter 3 remain valid for target detection while in MOPP, but considerations must be made for limiting factors imposed by MOPP equipment.

Vision is limited to what can be seen through the mask lenses/faceplate. Peripheral vision is severely restricted. The lenses/faceplate may be scratched or partly fogged, thus further restricting vision. Soldiers requiring corrective lenses must be issued insert lenses before training.

Scanning movement may be restricted by the hood/mask. Any of these factors could adversely affect the soldier's ability to quickly and accurately detect targets. Additional skill practice should be conducted.

Marksmanship Fundamentals. Although the four marksmanship fundamentals remain valid during MOPP firing, some modifications may be needed to accommodate the equipment.

Steady position. Due to the added bulk of the overgarments, firing positions may need adjustment for stability and comfort. Dry and live firing while standing, crouching, or squatting may be necessary to reduce body contact with contaminated ground or foliage. A consistent spot/stock weld is difficult to maintain due to the shape of the protective masks. This requires the firer to hold his head in an awkward position to place the eye behind the sight.

Aiming. The wearing of a protective mask may force firers to rotate (cant) the rifle a certain amount to see through the rear aperture. The weapon should be rotated the least amount to properly see through and line up the sights, as previously discussed in Chapter 3. The center tip of the front sight post should be placed on the ideal aiming point. This ideal aiming procedure (Figure 4-7, page 4-16) should be the initial procedure taught and practiced. If this cannot be achieved, a canted sight picture may be practiced.

Breath control. Breathing is restricted and more difficult while wearing the protective mask. Physical exertion can produce labored breathing and make settling down into a normal breath control routine much more difficult. More physical effort is needed to move around when encumbered by MOPP equipment, which can increase the breath rate. All of these factors make holding and controlling the breath to produce a well-aimed shot more energy- and time-consuming. Emphasis must be placed on rapid target engagement during the limited amount of time a firer can control his breath.

Trigger squeeze. Grasping the pistol grip and squeezing the trigger with the index finger are altered when the firer is wearing MOPP gloves. The action of the trigger finger is restricted, and the fit of the glove may require the release of the swing-down trigger guard. Because the trigger feels different, control differs from that used in bare-handed firing. This difference cannot be accurately predicted. Dry-fire training using dime (washer) exercises is necessary to ensure the firer knows the changes he will encounter during live fire.

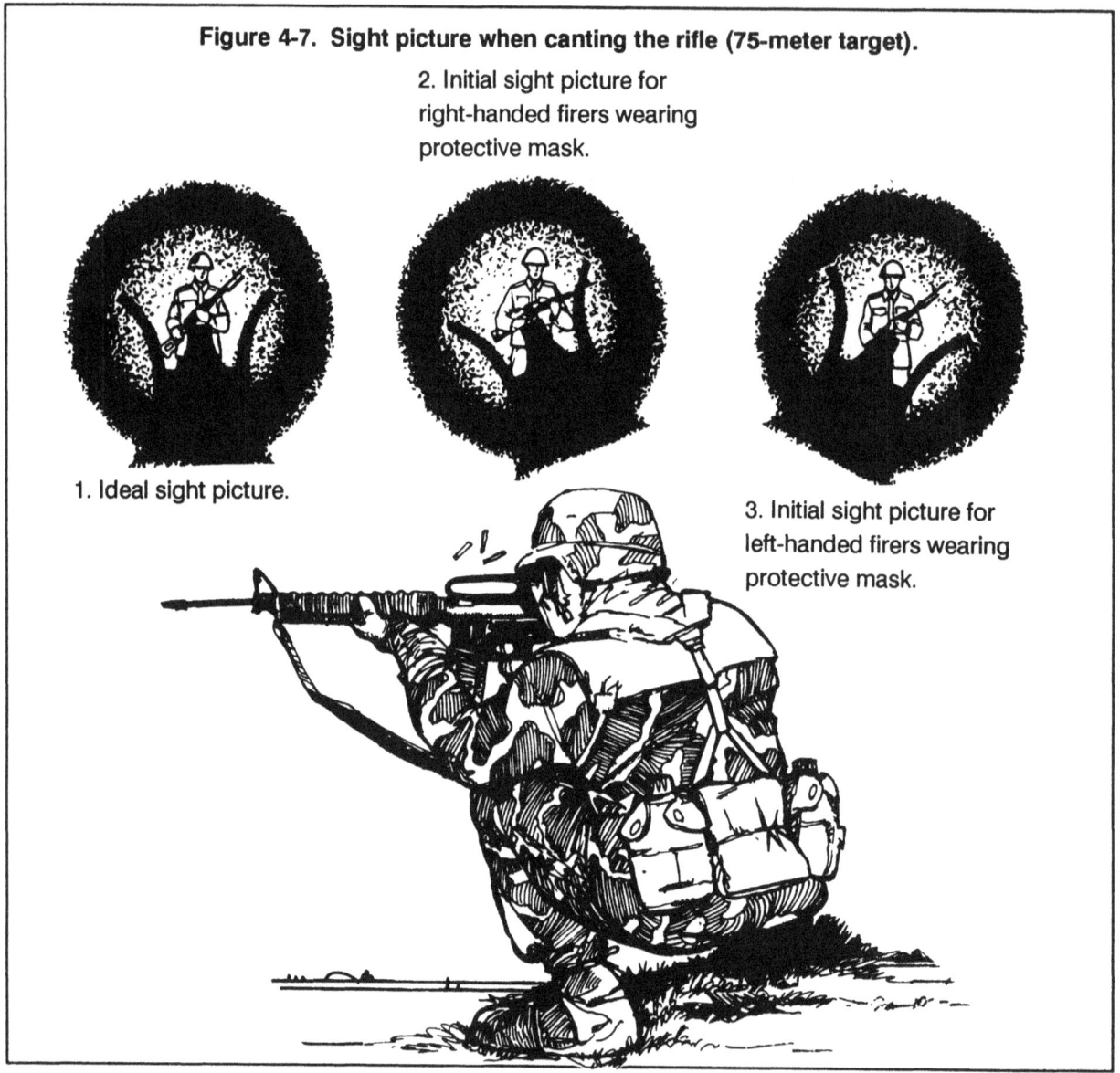

Figure 4-7. Sight picture when canting the rifle (75-meter target).

EFFECTS OF AIMING MODIFICATIONS

The normal amount of cant needed by most firers to properly see through the sights has a limited influence on rounds fired at ranges of 75 meters or less. At longer ranges, however, the change in bullet strike becomes more pronounced.

Rifle ballistics (Appendix F) causes the strike of the bullet to impact low in the direction of the cant (when a cant is used) at longer ranges. Due to this shift in bullet strike and the many individual differences in sight alignment when wearing a protective mask, it is important to conduct downrange feedback training (Appendix G) at ranges beyond 75 meters. This allows soldiers to determine what aiming adjustments are needed to achieve center target hits. Figure 4-8 shows what might be expected for a right-handed firer engaging a target at 175 meters with no cant, a certain amount of cant, and the adjustment in point of aim needed to move the bullet strike to the center of the target. Figure 4-9 shows what might be expected for a right-handed firer engaging a 300-meter target. (The adjustments in point of aim for left-handed firers are the opposite of those shown in Figures 4-8 and 4-9.)

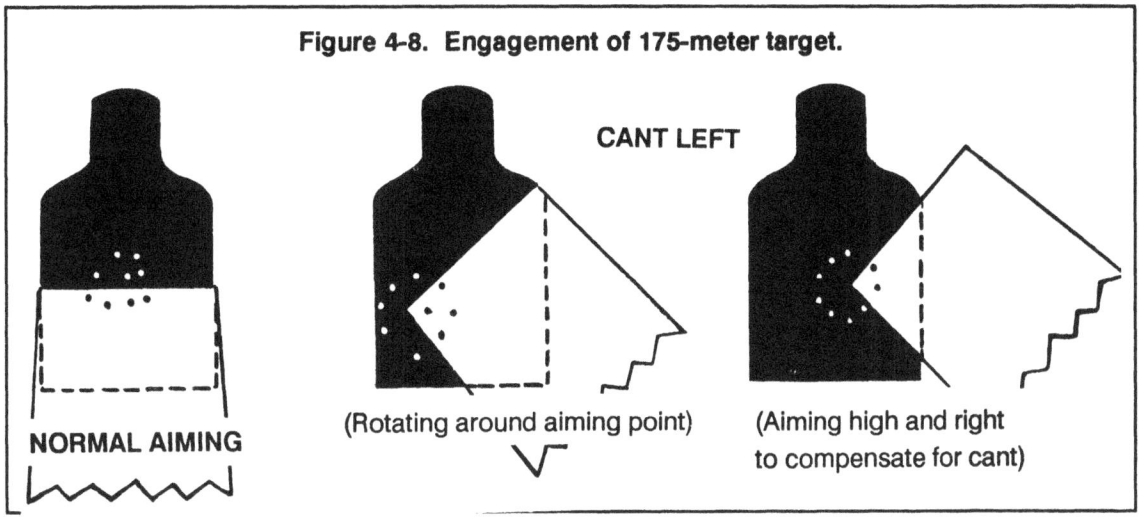

Figure 4-8. Engagement of 175-meter target.

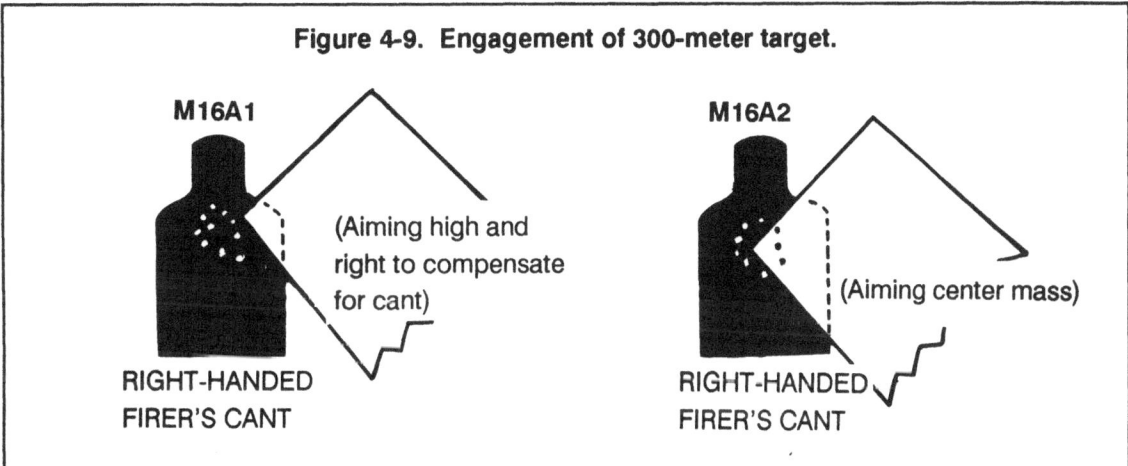

Figure 4-9. Engagement of 300-meter target.

Although bullet strike is displaced when using a cant, individual differences are such that center-of-mass aiming should be used until the individual knows what aiming adjustment is needed. When distant targets are missed, a right-handed firer should usually adjust his point of aim to the right and high; a left-handed firer should adjust to the left and high. Then, the aiming rules are clear. All targets should initially be engaged by aiming center mass, regardless of cant. When targets are missed while using a cant, firers should adjust the point of aim higher and opposite the direction of the cant. Actual displacement of the aiming point must be determined by using downrange feedback targets at ranges beyond 75 meters.

OPERATION AND FUNCTION MODIFICATIONS

Handling the rifle, performing operation and function checks, loading and unloading, and cleaning are affected by MOPP equipment. Movements are slowed, tasks take longer to complete and often require more effort, vision is impaired, and care is needed to avoid damaging MOPP equipment and possible exposure to lethal agents. Because of the great differences between no MOPP and MOPP4, soldiers must be trained in all aspects of operation and maintenance of the weapon while practicing at the highest MOPP level. Only through repeated training and practice can the soldier be expected to perform all tasks efficiently.

MOPP FIRE EXERCISES

The many difficulties the soldier encounters while firing with MOPP gear must be experienced and overcome during training.

Dry-Fire MOPP Exercises. Repeated dry-fire exercises covering all aspects of MOPP firing are the most effective means available to ensure all soldiers can function during a live-fire MOPP situation. Multiple dry-fire exercises must be conducted before the first live round is fired. Otherwise, valuable ammunition and training time are wasted in trying to teach soldiers the basics. The soldier is trained in the fundamentals; repeated dry-fire or Weaponeer exercises are conducted; grouping, zeroing, qualifying, and evaluating are performed using standard non-MOPP firing; the differences and modifications are trained for MOPP firing; and repeated MOPP dry-fire exercises are conducted. The soldier is now ready to move on to MOPP live fire.

Live-Fire MOPP Exercises. These exercises further develop the learned firing skills and allow the soldier to experience the effects of wearing MOPP equipment on downrange performance.

Individual. Application of immediate action, rapid magazine changes, grouping, and adjusted point of aim at 25 meters should all be tested and evaluated for further training. After soldiers exhibit proficiency at these tasks, further training and evaluation at extended ranges are indicated.

Unit. Parts of unit LFXs should be conducted in the highest MOPP level with a planned system of target hit evaluation. As in all aspects of marksmanship training, the emphasis is on soldier knowledge and skills displayed.

Basic 25-meter proficiency course. Initial live-fire exercises are conducted at 25 meters. This training provides all soldiers the basic techniques and introduces firing the rifle in MOPP equipment. This basic proficiency exercise must be fired while wearing gloves and protective mask with hood. The basic 25-meter proficiency exercise is fired to standard and is an annual/semiannual GO/NO-GO requirement for most soldiers. It is entered on the record fire scorecard when completed.

The course of fire can be conducted on any range equipped with mechanical target lifters. Soldiers are given initial instruction and a demonstration of the techniques of firing in MOPP equipment.

Each soldier is issued 20 rounds of 5.56-mm ball ammunition to engage 20 three-to-five-second exposures of F-type silhouette targets at 25 meters. Initial firing is performed with 10 rounds from the individual fighting position (supported), and 10 rounds from a prone unsupported position. Each soldier must obtain a minimum of 11 target hits out of 20 exposures to meet the basic requirement. This initial basic 25-meter exercise prepares soldiers for future individual and unit training in full MOPP gear.

Downrange feedback. Once the soldier has mastered basic marksmanship proficiency, he should be introduced to firing at range. This phase of firing should provide the maximum hit-and-miss performance feedback; it can be conducted on a KD or modified field fire range at 75, 175, and 300 meters.

Practice firing under full MOPP can also be conducted on the standard RETS ranges—for example, the standard record fire tables may be fired in MOPP. MOPP fire must also be part of unit tactical exercises, which are fired on MPRC as part of STXs.

NOTE: The .22-caliber rimfire adapter or plastic practice ammunition may be used during live-fire practice at scaled 25-meter targets when 5.56-mm ammunition is not available.

When the rimfire adapter, plastic ammunition, or live-fire range is not available, the Weaponeer device may be used. Scaled silhouette targets may also be used at this distance to introduce the many target sizes common at longer ranges. The slow-fire target and course outlined in Appendix E are appropriate.

Having mastered the 25-meter firing phase, the soldier is then introduced to firing at range, using the standard 75-, 175-, and 300-meter downrange feedback targets (Chapter 3). Adjusted point of aim, for individual differences of cant, is first used during this training. Live-fire training is conducted on a KD or modified field fire range, giving the soldier feedback on targets engaged at many ranges.

Section VI. MOVING TARGET ENGAGEMENT

The enemy normally moves by rushes from one covered or concealed position to another. While making the rush, the enemy soldier presents a rapidly moving target. However, for a brief time as he begins, movement is slow since many steps are needed to gain speed. Many steps are needed to slow down at the new position. A moving target is open to aimed fire both times.

MOVING TARGET TECHNIQUES

There are two primary techniques of engaging moving targets.

Tracking. Tracking is a more accurate technique of engaging targets by experienced firers. It involves the establishment and maintaining of the aiming point in relationship to the target and maintaining that sight picture (moving with the target) while squeezing the trigger. As the target moves, this technique puts the firer in position for a second shot if the first one misses.

Trapping. Trapping is the setting up of an aiming point forward of the target and along the target path. The trigger is squeezed as the target comes into the sights. This is a technique that works on targets with slow lateral movement. It does not require tracking skills. It does require that the firer know precisely when the rifle is going to fire. Some soldiers can squeeze the trigger without reacting to the rifle firing, and they may fire better using this technique.

Another technique is to use a modified 25-meter scaled timed-fire silhouette (see Figure 4-10). Trainers evaluate performance based on where shot groups are placed when the lead rule is applied. This target can be used for both the M16A1 and M16A2 rifles.

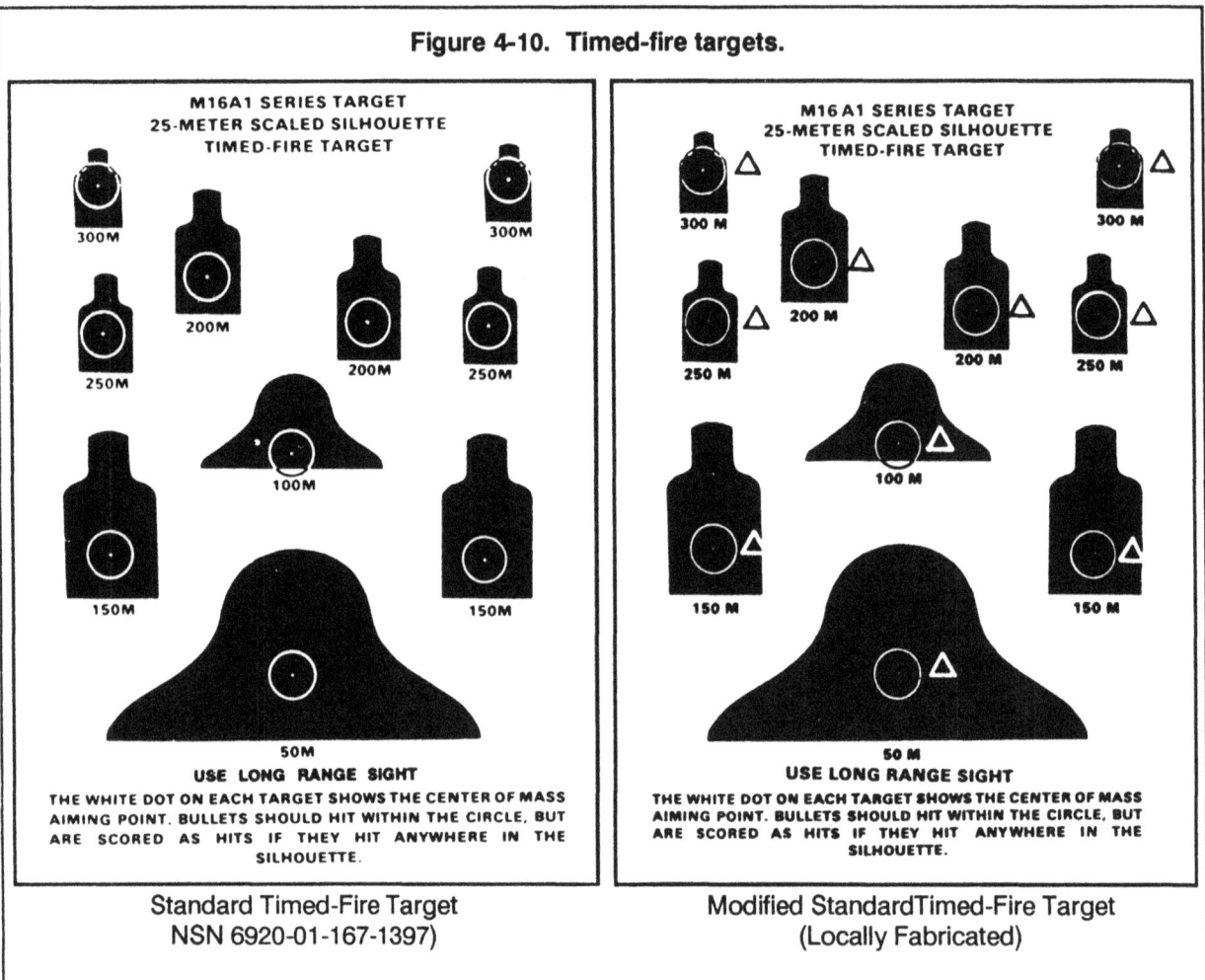

Figure 4-10. Timed-fire targets.

MOVING TARGET FUNDAMENTALS

The fundamentals needed to hit moving targets are similar to those needed to hit stationary targets. The main skill is to engage moving targets with the least changes to procedures. Another consideration is that soldiers in a combat defensive position do not know if their next target will be stationary or moving — they must fire immediately at whatever targets occur.

The fundamentals for engaging stationary targets are steady position, aiming, breath control, and trigger squeeze. They are also used to engage moving targets. Considering the environment and the variables of the rifle and ammunition, the well-trained soldier should be able to hit 300-meter stationary silhouette targets with a .5 PH. When the target has lateral movement, hits at 150 meters may be seven out of ten times, which is a good performance. Therefore, twice as much variability, twice as much dispersion, and a few more erratic shots are expected when soldiers are trained to hit moving targets.

The procedures used to engage moving targets vary as the angle and speed of the target vary. For example, when a moving target is moving directly at the firer, the same procedures are used as would be used if the target were stationary. However, if it is a close, fast-moving target at a 90-degree angle, the rifle and entire upper body of the firer must be free from support so that the target can be tracked. To hit moving targets, the firer must move the rifle smoothly and steadily as the target moves. The front sight post is placed with the trailing edge at target center, breath is held, and the trigger is squeezed. Several factors complicate this process.

Steady position. When firing from a firing position, the firer is in the standard supported position and is flexible enough to track any target in his sector. When a moving target is moving directly at the firer, directly away, or at a slight angle, the target is engaged without changing the firing position. When targets have much lateral movement, only minor changes are needed to allow for effective target engagement. Most moving targets are missed in the horizontal plane (firing in front of or behind the target) and not in the vertical plane (firing too low or too high). Therefore, a smooth track is needed on the target, even if the support arm must be lifted. Other adjustments include the following:

- *Nonfiring hand.* The grip of the nonfiring hand may need to be increased and more pressure applied to the rear. This helps to maintain positive control of the rifle and steady it for rapid trigger action.

- *Nonfiring elbow.* The elbow is lifted from the support position only to maintain a smooth track.

- *Grip of the right hand.* Rearward pressure may be applied to the pistol grip to steady the rifle during trigger squeeze.

- *Firing elbow.* The firing elbow is lifted from support only to help maintain a smooth track.

NOTE: The rifle pocket on the shoulder and the stock weld are the same for stationary targets.

Aiming. The trailing edge of the front sight post is at target center.

Breath control. Breathing is locked at the moment of trigger squeeze.

Trigger squeeze. Rearward pressure on the handguard and pistol grip is applied to hold the rifle steady while pressure is applied to the trigger. The trigger is squeezed fast (almost a controlled jerk). Heavy pressure is applied on the trigger (at least half the pressure it takes to make the rifle fire) before squeezing the trigger.

SINGLE-LEAD RULE FOR MOVING TARGETS

A target moving directly toward the firer can be engaged the same way as a stationary target. However, to hit a target moving laterally, the firer places the trailing edge of the front sight sight post at target center. The sight-target relationship is shown in Figure 4-11(page 4-22). The single-lead rule automatically increases the lead as the range to the target increases.

Figure 4-12 (page 4-22) shows how this works, with the front sight post covering about 1.6 inches at 15 meters and about 16 inches at 150 meters. Since the center of the front sight post is the actual aiming point, this technique of placing the trailing edge of the front sight post at target center provides for an .8-inch lead on a 15-meter target, and an 8-inch lead on a target at 150 meters.

This rules provides for a dead-center hit on a 15-meter target that is moving at 7 mph at a 25-degree angle because the target moves .8 inch between the time the rifle is fired and the bullet arrives at the target. A 150-meter target moving at 7 mph at a 25-degree angle moves 8 inches between the time the weapon is fired and the bullet arrives. This rule provides for hits on the majority of high-priority combat targets.

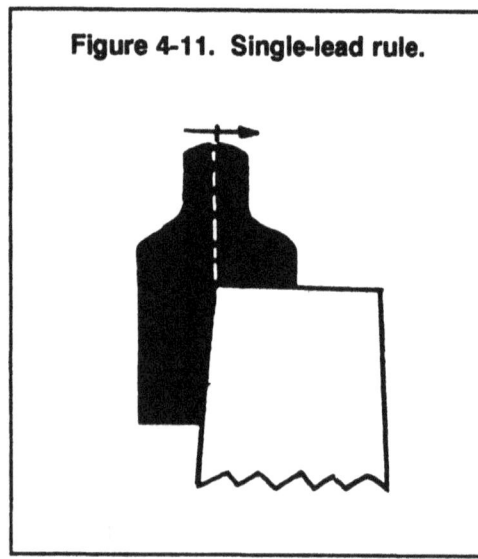

Figure 4-11. Single-lead rule.

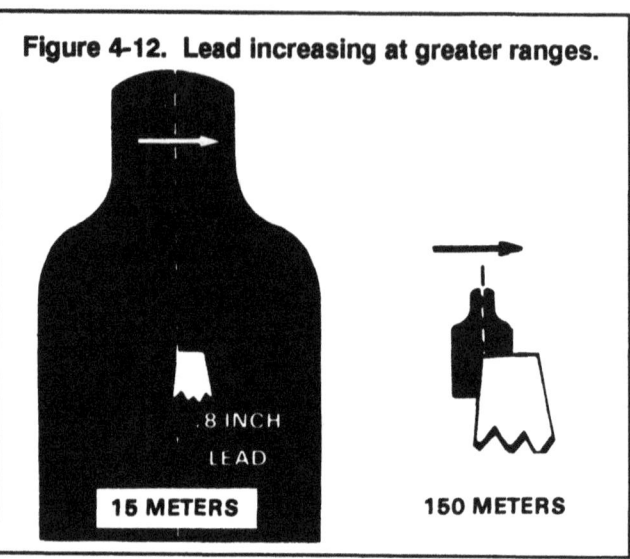

Figure 4-12. Lead increasing at greater ranges.

LEAD REQUIREMENTS

To effectively engage moving targets on the battlefield, soldiers must understand lead requirements. Figure 4-13 shows the amount of lead required to hit a 300-meter target when it is moving 8 mph at an angle of 90 degrees. Aiming directly at the target would result in missing it. When an enemy soldier is running 8 mph, 90 degrees to the firer, and at a range of 300 meters, he covers 4 1/2 feet while the bullet is traveling toward him. To get a hit, the firer must aim and fire at position D when the enemy is at position A. This indicates the need for target lead and for marksmanship trainers to know bullet speed and how it relates to the range, angle, and speed of the target. Soldiers must understand that targets moving fast and laterally are led by some distance if they are to be hit.

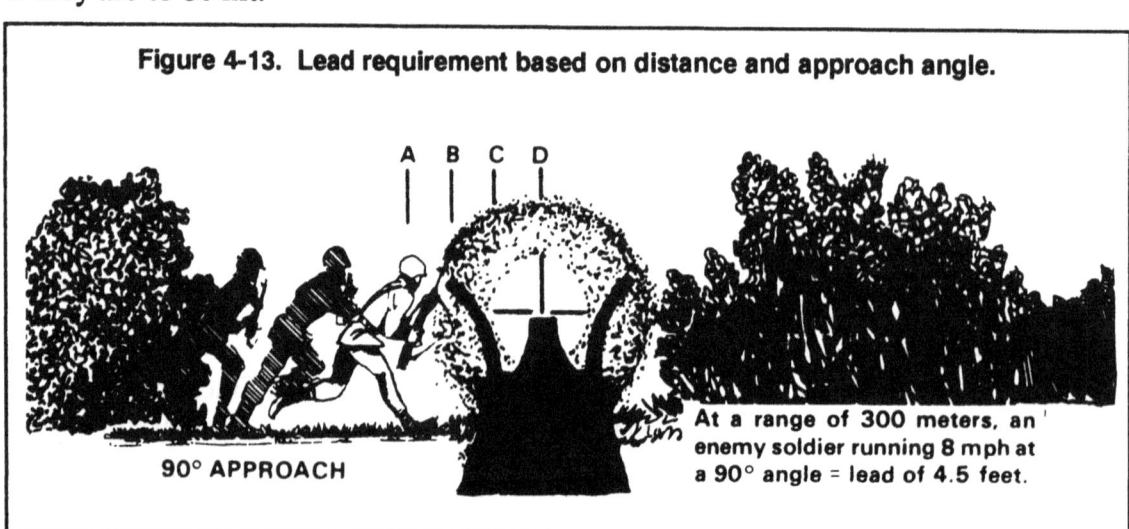

Figure 4-13. Lead requirement based on distance and approach angle.

Target Speed. Figure 4-14 reflects the differences in lateral speed for various angles of target movement for a target that is traveling at 8 mph at a distance of 150 meters from the firer. The angle of target movement is the angle between the target-firer line and the target's direction of movement. An 8-mph target moves 24 inches during the bullet's flight time. If the target is moving on a 15-degree angle, it moves 6 inches (the equivalent of 2 mph). For the firer to apply precise lead rules, he must accurately estimate speed, angle, and range to the target during the enemy soldier's brief exposure. The single-lead rule (place the trailing edge of the front sight post at target center) places effective fire on most high-priority combat targets. At 100 meters, the rule begins to break down for targets moving at slight and large angles.

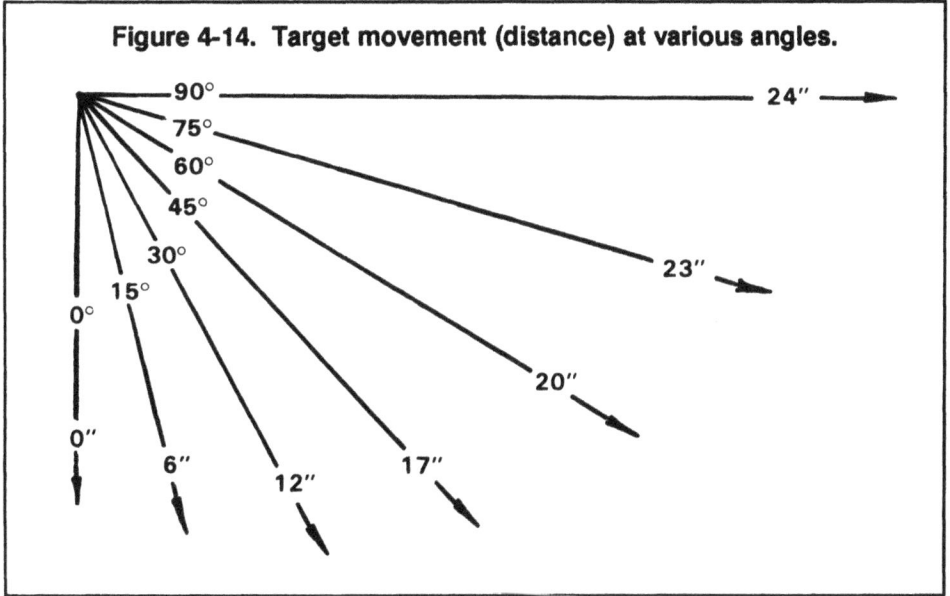

Figure 4-14. Target movement (distance) at various angles.

Since the target lead is half the perceived width of the front sight post, at 100 meters the standard sight provides for 5.4 inches of lead for the M16A1 and M16A2 front sights (Figure 4-15, page 4-24).

Target Distance. The front sight post covers only a small part of close-in targets, providing for target hits on close targets moving at any angle and any speed. However, if the lead rule is applied on more distant targets moving at a slight angle — for example, 5 degrees at 100 meters — the bullet strikes forward of target center, about 4 inches with standard sights and about 7 inches with LLLSS sights. Therefore, soldiers are taught to fire at targets as though they are stationary until lateral movement is observed (15 degrees).

The rule provides for many speed-angle combinations that place the bullet within 2 inches of target center (Figure 4-16, page 4-25). Since the soldier is expected to fire a 12-inch group on moving targets at 100 meters, the rule provides for hits on the majority of targets. Even the worst case (a 90-degree target moving at 8 mph) would result in the shot-group center being located 9.8 inches behind target center. If bullets were evenly distributed within a 12-inch group, this would result in hitting the target 40 percent of the time.

Soldiers should be taught to increase their lead when targets are missed. This increases their probability of hitting all targets. The amount of additional lead required

FM 23-9

Figure 4-15. Angle of target movement.

ANGLE OF TARGET MOVEMENT	RANGE: 100 METERS		
	(STANDARD SIGHT) TARGET SPEED		
	4 MPH	6 MPH	8 MPH
5°	+4.9"	+4.5"	+4.3"
10°	+4.1"	+3.5"	+2.7"
15°	+3.5"	+2.5"	+1.5"
20°	+2.8"	+1.5"	+.2"
25°	+2.2"	+.7"	-1.0"
30°	+1.7"	-.2"	-2.0"
35°	+1.1"	-1.1"	-3.2"
40°	+.6"	-1.9"	-4.3"
45°		-2.7"	-5.4"
50°	-.4"	-3.3"	-6.2"
55°	-.8"	-4.0"	-7.0"
60°	-1.2"	-4.5"	-7.7"
65°	-1.5"	-4.9"	-8.4"
70°	-1.7"	-5.3"	-8.8"
75°	-1.9"	-5.6"	-9.2"
80°	-2.0"	-5.9"	-9.6"
85°	-2.1"	-5.9"	-9.7"
90°	-2.1"	-6.0"	-9.8"

NOTE: Plus (+) indicates bullet strike in the direction of movement; minus (−) indicates bullet strike behind the target center.

should be developed through experience with only general guidance provided. For example, if there is much lateral movement of the target and the soldier feels by applying the lead rule and firing fundamentals he has missed the target, then he should increase his lead.

The training program must be simple and provide soldiers with only relevant information to improve their performance in combat. First, all soldiers should understand and apply the single-lead rule in the absence of more information. Second, soldiers should understand that moving targets coming toward them or on a slight angle (0 to 15 degrees) should be engaged as stationary targets. Third, information should be presented and practice allowed on applying additional lead to targets for soldiers who demonstrate an aptitude for this skill.

Target Angle. The rule does not apply to targets moving at small and large angles (Figure 4-16). For example, a walking enemy soldier at 250 meters is hit dead center when he is moving at 40 degrees. Hits can be obtained if he is moving on any angle between 15 and 75 degrees. When he is running (a center hit is obtained when the target is on an angle of 18 degrees), misses occur when he exceeds an angle of 30 to 35 degrees. The information provided in Figures 4-13, page 4-23, and 4-14, page 4-24, is designed to enhance instructor understanding so proper concepts are presented during instruction. For example, a target at 100 meters moving at 6 mph receives a center hit when moving at 29 degrees. When moving at an angle less than 29 degrees, the bullet strikes somewhat in front of target center. When moving at an angle of more than 29 degrees, the bullet strikes somewhat behind target center.

Figure 4-16. Target angle when dead center: hits accure using single lead rule.

(STANDARD SIGHT)

RANGE	4 MPH	6 MPH	8 MPH
25M	48°	30°	22°
50M	47°	30°	22°
100M	45°	29°	21°
150M	44°	28°	20°
200M	41°	27°	19°
250M	40°	26°	18°
300M	33°	21°	16°
350M	38°	24°	18°
400M	35°	22°	17°
450M	33°	21°	16°

MULTIPURPOSE RANGE COMPLEX TRAIN-UP

MPRCs require soldiers to hit moving targets. Ranges are used for collective training. Commanders should try to use the MPRCs for individual training and to teach the individual to engage moving targets. If no MPRCs are available for individual training, any range can be used that will support any type of moving target. Building a moving target range is limited only by the imagination of the trainer, but always within safety constraints. The following are examples that can be incorporated on many ranges.

Popsicle Sticks. This requires placing an E-type silhouette on a long stick and having an individual walk back and forth behind a high berm (high enough to protect the individual from fire) the length of the berm. Feedback should be made available for the firer such as for lowering the target when a hit is scored or reversing direction upon a hit.

Sled Targets. This requires constructing a simple sled that has one or more targets attached. The sled is pulled by a rope or cable across and off the range safely by a vehicle.

FM 23-9

CHAPTER 5

Night Firing

All units must be able to fight during limited visibility. All soldiers should know the procedures for weapons employment during such time. Soldiers must experience the various conditions of night combat—from total darkness, to the many types of artificial illumination, to the use of surveillance aids. All units must include basic, unassisted night fire training annually in their unit marksmanship programs. Combat units should conduct tactical night fire training at least quarterly. This tactical training should include MILES during force-on-force training as well as live fire. Night-fire training must include the use of applicable night vision devices when this equipment is part of a TOE. The many effects darkness has on night firing are discussed herein.

NOTE: Although this chapter addresses night firing, the appropriate modifications to the fundamentals of firing may be applied whenever visibility is limited.

CONSIDERATIONS

Trainers must consider the impact of limited visibility on the soldier's ability to properly apply the fundamentals of marksmanship and combat firing skills. These fundamentals/skills include:

Operation and Maintenance of the Weapon. Handling the weapon, performing operation and function checks, loading and unloading, and maintenance are affected by nighttime conditions. Movements are slowed, tasks take longer to complete, vision is impaired, and equipment is more easily misplaced or lost. Because combat conditions and enforcement of noise and light discipline restrict the use of illumination, soldiers must be trained to operate (load, unload, and clear), service, and clean their weapons using the lowest lighting conditions. Although initial practice of these tasks should occur during daylight (using simulated darkness) to facilitate control and error correction, repeated practice during actual nighttime conditions should be integrated with other training. Only through repeated practice and training can the soldier be expected to perform all tasks efficiently.

Immediate Action. Under normal conditions, a soldier should clear a stoppage in three to five seconds. After dark, this task usually takes longer. Identifying the problem may be frustrating and difficult for the soldier. A tactile (hands only) technique of identifying a stoppage must be taught and practiced. Clearing the stoppage using few or no visual indicators must also be included. The firer must apply immediate action with his eyes closed. Dry-fire practice using dummy or blank rounds under these conditions is necessary to reduce time and build confidence. Training should be practiced first during daylight for better control and error correction by the trainer. Practice during darkness can be simulated by closing the eyes or using a blindfold. Once the soldier is confident in applying immediate action in daylight or darkness, he can perform such actions rapidly on the firing line.

Target Detection. Many of the skills discussed in Appendix B apply to target detection after dark. Light from a cigarette or flashlight, discharge of a rifle (muzzle flash), or reflected moonlight/starlight are the main means of target location. Sounds may also be indicators of target areas. Because the other techniques of detection (movement, contrast) are less apparent at night, light and sound detection must be taught, trained, and reviewed repeatedly in practice exercises. Exercises should also emphasize shortened scanning ranges, night vision adaptation, and use of off-center vision. Target detection exercises should be integrated into all collective training tasks.

NOTE: Binoculars are often overlooked as night vision aids. Because they amplify the available light, binoculars or spotting/rifle scopes can provide the firer with another means to locate targets during limited visibility. Also, the use of MILES equipment is effective for use in engaging detected targets.

Marksmanship Fundamentals. The four marksmanship fundamentals apply to night firing. Some modifications are needed depending on the conditions. The firer must still place effective fire on the targets or target areas that have been detected.

Steady position. When the firer is firing unassisted, changes in his head position/stock weld will be necessary, especial when using weapon-target alignment techniques. When using rifle-mounted night vision devices, head position/stock weld must be changed to bring the firing eye in line with the device. Also, such mounted devices alter the rifle's weight and center of gravity, forcing a shift in placement of the support (nonfiring arm or sandbags). Repeated dry-fire practice, followed by live-fire training, is necessary to learn and refine these modifications and still achieve the most steady position.

Aiming. Modifications to the aiming process vary from very little (when using LLLSSs) to extensive (when using modified quick-fire techniques). When firing unassisted, the firer's off-center vision is used instead of pinpoint focus. When using a mounted night vision device, the firer's conventional iron sights are not used. The soldier uses the necessary aiming process to properly use the device.

Breathing. Weapon movement caused by breathing becomes more apparent when using night vision devices that magnify the field of view. This fundamental is not greatly affected by night fire conditions.

Trigger squeeze. This important fundamental does not change during night fire. The objective is to not disrupt alignment of the weapon with the target.

PRINCIPLES OF NIGHT VISION

For a soldier to effectively engage targets at night, he must apply the three principles of night vision:

Dark Adaptation. Moving from lighted to darkened areas (as in leaving a tent) can be temporarily blinding. After several minutes have passed, the soldier can slowly see his surroundings. If he remains in this completely darkened environment, he adapts to the dark in about 30 minutes. This does not mean he can see in the dark at the end of this time. After about 30 minutes, his visibility reaches its maximum level. If light is encountered, the eyes must adapt again. The fire on the end of a cigarette or a red-lensed flashlight can degrade night vision; larger light sources cause more severe losses.

Off-Center Vision. During the day, the soldier focuses his vision on the object he wants to see. Shifting this pinpoint focus slightly to one side causes the object to become blurry or lose detail. At night, the opposite is true. Focusing directly on an object after dark results in that object being visible for only a few seconds. After that, the object becomes almost invisible. To view an object at night, the soldier must shift his gaze slightly to one side. This allows the light-sensitive parts of the eye (parts not used during daylight) to be used. These can detect faint light sources or reflections and relay their image to the brain. (Figures 5-1 and 5-2.)

NOTE: Vision is shifted slightly to one side, but attention is still on the object. Because of the blind spot at the center of vision, directing attention to an off-centered objective is possible (with practice).

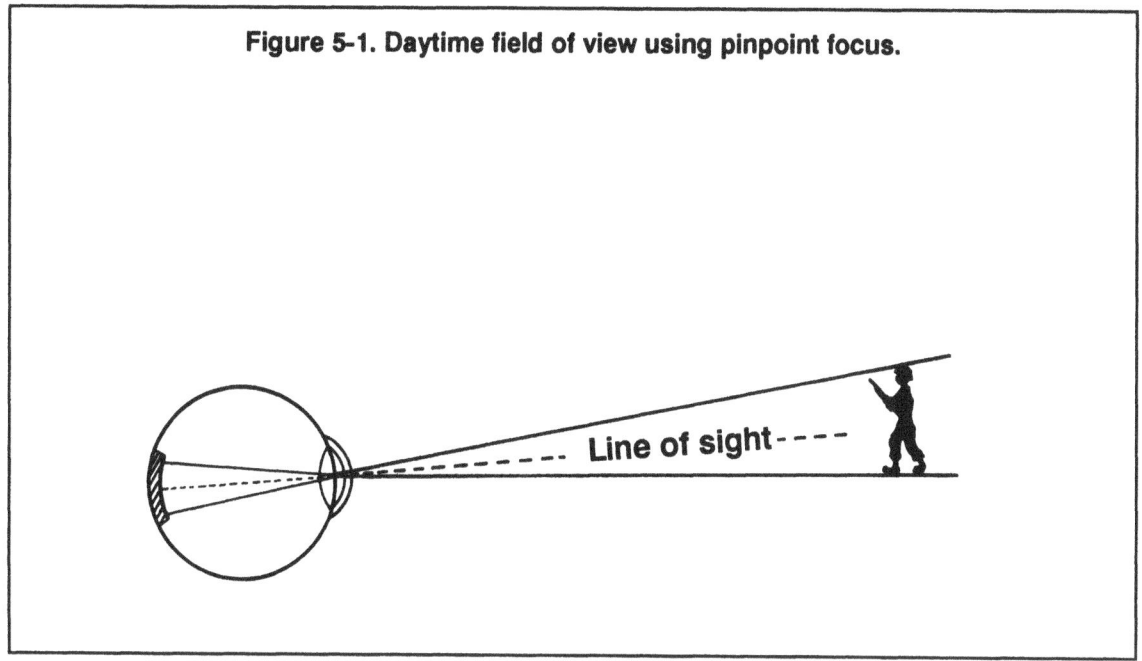

Figure 5-1. Daytime field of view using pinpoint focus.

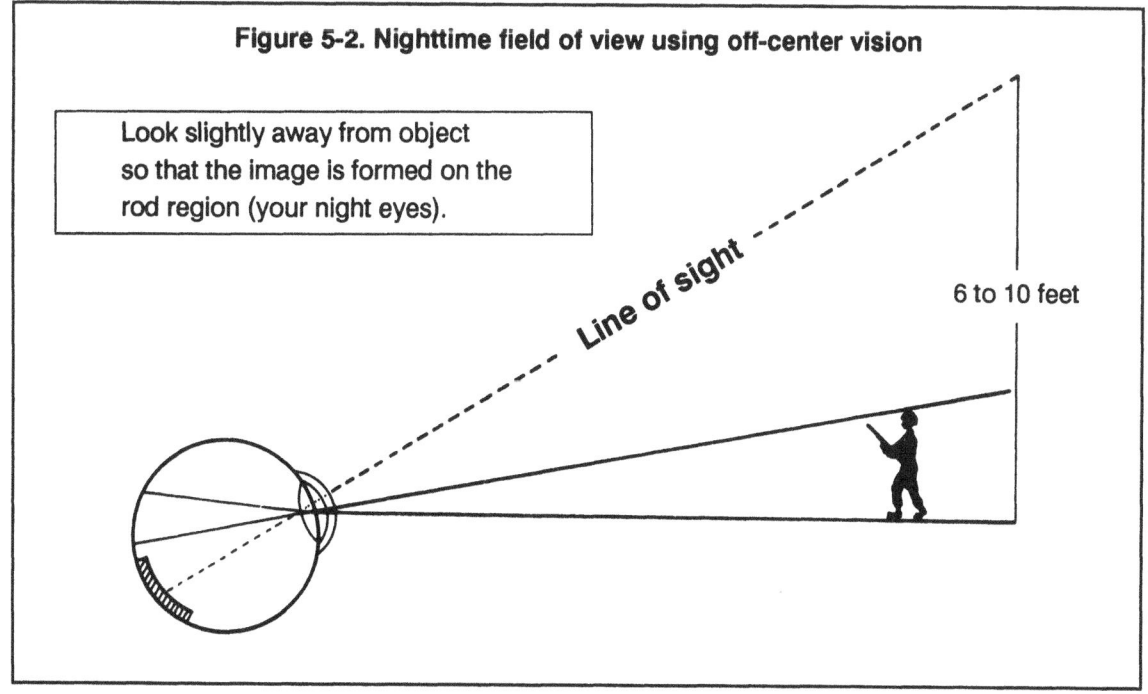

Figure 5-2. Nighttime field of view using off-center vision

Look slightly away from object so that the image is formed on the rod region (your night eyes).

6 to 10 feet

Scanning. Scanning is the short, abrupt, irregular movement of the soldier's eyes around an object or area every 4 to 10 seconds. Off-center vision is used. Scanning ranges vary according to visibility.

NOTE: For detailed information on the three principles, see FM 21-75.

TARGET ENGAGEMENT TECHNIQUES

Night fire usually occurs under three general conditions.

Unassisted Firing Exercise. The firer must detect and engage targets without artificial illumination or night vision devices. Potential target areas are scanned. When a target is detected, the firer should engage it using a modified quick-fire position. His head is positioned high so that he is aligning the weapon on the target and looking just over the iron sights. His cheek should remain in contact with the stock.

The firer should take a few seconds to improve weapon/target alignment by pointing slightly low to compensate for the usual tendency to fire high. Both eyes are open to the maximum advantage of any available light, and the focus is downrange. Off-center vision is used to keep the target in sight. Tracer ammunition may provide feedback on the line of trajectory and facilitate any adjustments in weapon/target alignment.

Repeated dry-fire training, target detection, and proper aiming practice are the most efficient means to ensure the soldier can successfully engage short-range targets (50 meters or closer) unassisted during MILES exercises, and then live-fire training.

Artificial Illumination. Targets as distant as 175 meters can be engaged successfully with some type of artificial illumination. Illumination may be from hand flares, mortar or artillery fire, or bright incandescent lights such as searchlights.

When artificial illumination is used, the eyes lose most of their night adaption, and off-center vision is no longer useful. Aiming is accomplished as it is during the day. Artificial illumination allows the firer to use the iron sights as he does during the day. (M16A2 users should keep the large rear sight aperture flipped up during darkness.)

Engaging targets under artificial illumination allows for better target detection and long-range accuracy than the unassisted technique. When the light is gone, time must be spent in regaining night vision and adaptation. Only when the light level drops enough so that the target cannot be seen through the iron sights should the firer resume short-range scanning, looking just over the sights.

Soldiers have sometimes been taught to close their eyes during artificial illumination to preserve their night vision. This technique is effective but also renders the soldier (or entire unit) blind for the duration of the illumination. Keeping one eye closed to preserve its night vision results in a drastically altered sense of perception when both eyes are opened, following the illumination burnout. Tactical considerations should be the deciding factor as to which technique to use. Repeated dry-fire training and target detection practice are the keys to successful engagement of targets out to 150 meters or more during live fire under artificial illumination.

Night Vision Devices. Rifle-mounted night vision devices are the most effective night fire aids. By using these devices, the firer can observe the area, detect and engage any suitable targets, and direct the fire of soldiers who are firing unassisted.

NVDs can be used to engage targets out to 300 meters. Repeated training, dry-fire practice, and correct zeroing are vital to the proper employment of NVDs during live-fire training.

TRAINING

Dry-fire training and live-fire training are necessary to mastering basic rifle marksmanship. The soldier must adhere to the following procedures and applications to be effective in combat.

Dry-Fire Exercises. Repeated training and dry-fire practice are the most effective means available to ensure all soldiers can function efficiently after dark.

Target detection and dry-fire exercises must be conducted before the first live round is fired. They can take place almost anywhere—elaborate live-fire range facilities are not needed. Modified fundamentals can be taught in a classroom/practical exercise situation. Further training in the proper zeroing and engagement techniques can take place anywhere that targets can be set up and darkness can be expected.

Without extensive dry-fire training, soldiers do not perform to standards during live fire. Valuable range time and ammunition are wasted in a final attempt to teach the basics.

The soldier must demonstrate skill during daylight live fire. Next, he is trained in the differences and modifications needed for successful night firing. Many dry-fire exercises are conducted until skill at night firing is displayed. Only then is the soldier ready to move on to the night live-fire exercises.

Live-Fire Exercises. These exercises continue to develop the firing skills acquired during dry-fire exercises, and they allow the soldier to experience the effects of darkness on downrange performance.

The basic unassisted live-fire exercise allows all soldiers to apply night-fire principles, and to gain confidence in their abilities to effectively engage targets at 25 and 50 meters. Practice and proficiency firing can be conducted on any range equipped with mechanical lifters and muzzle flash simulators. A small square of reflective material and a shielded low wattage flashing light (protected from bullet impact) may be used to facilitate target detection. (Figures 5-3, page 5-6) The light should be placed to highlight the center of the target with a flashing, faint glow (intended to represent a muzzle flash). The light should not be on constantly, when the target is not exposed, or on when the target is exposed but not being used in actual engagement. The light should provide the firer with a momentary indication that a target is presenting itself for engagement. It should not be attached to the target or provide the firer with a distinct aiming point, regardless of how dim it may be. Practice can also be accomplished by the use of MILES equipment and target interface devices.

When an automated record fire range (RETS) is used for this exercise, the two 50-meter mechanisms are used. Before training, one E-type silhouette target is replaced with an F-type silhouette target. The F-type silhouette target is engaged at 25 meters from the prone unsupported position. The soldier is issued one magazine of 15 rounds (5 rounds ball; 10 rounds tracer) and presented 15 ten-second exposures. The firing line is moved, and the soldier engages the E-type silhouette target at 50 meters. He is issued a second 15-round magazine (5 rounds ball; 10 rounds tracer) to engage 15 ten-second exposures.

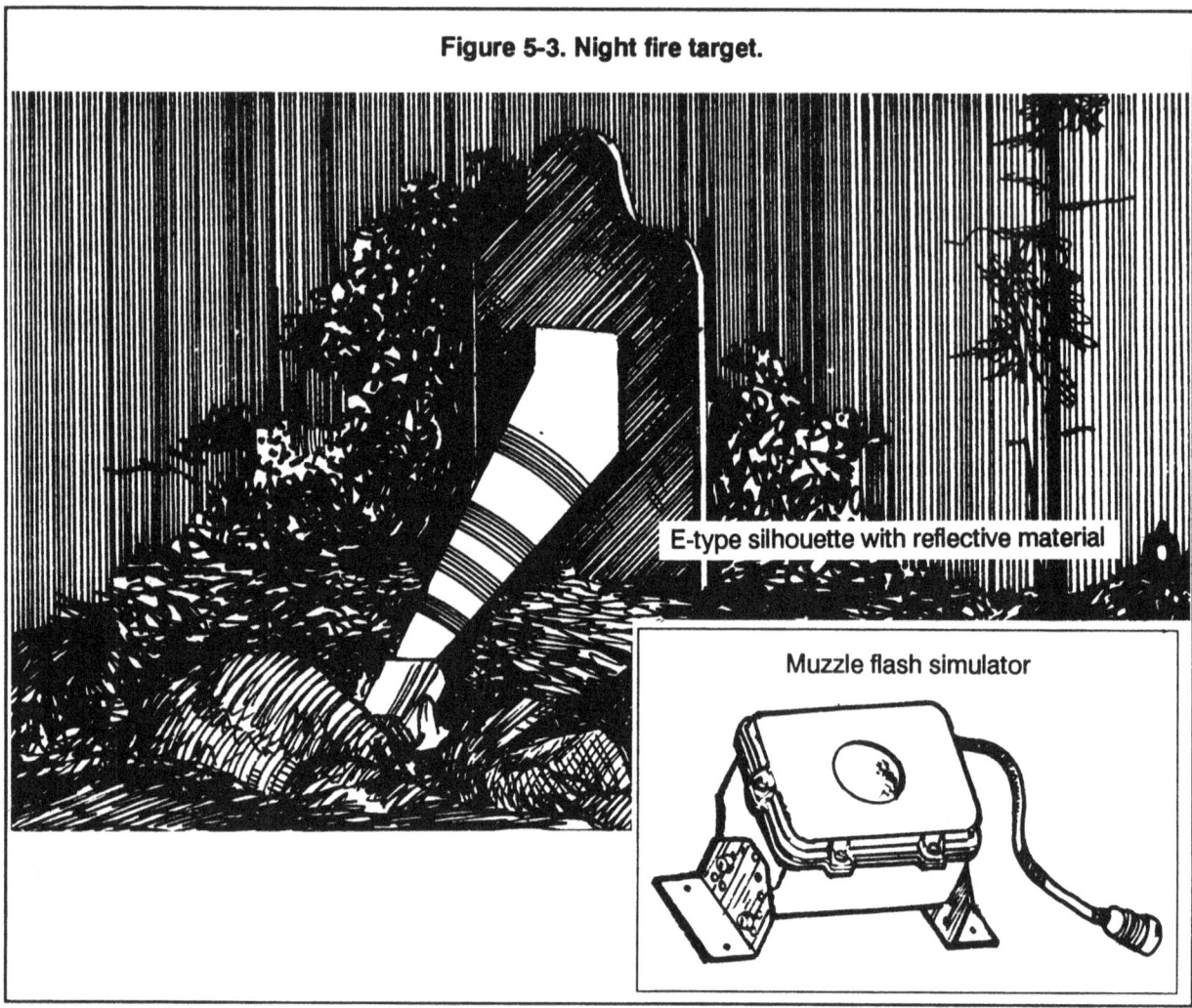

Figure 5-3. Night fire target.

When the automated range is used, the soldier's performance is recorded in the tower. If automatic scoring is not available, F-type and E-type silhouette paper facings are attached to the mechanical target, and bullet holes are counted. Facings may be repaired or replaced for each firer.

To meet the annual/semiannual minimum performance requirements, all soldiers must hit and kill seven separate targets out of 30 exposures. The results are annotated on the soldier's record fire scorecard.

- *Individual.* Application of immediate action, rapid magazine changes, and refinements of the modified quick-fire aiming point should be tested and evaluated for further training.

  - *Unassisted.* After soldiers exhibit proficiency of individual tasks, training and evaluation at ranges beyond those possible using only the rifle are indicated.

  - *Artificial illumination.* After mastering the unassisted night fire task and after repeated dry-fire training under artificial illumination, the soldier is ready to be tested and evaluated using live fire under illumination. Pop-up or stationary targets at ranges out to 175 meters (depending on light conditions, terrain features, and vegetation) may be used. Illumination is provided by flares, mortar/artillery, or floodlights. Once these tasks are mastered, further training and evaluation using NVDs is indicated. Multipurpose range complexes can be used for night firing by using artificial illumination. Automated field fire or record fire ranges can also be used by adding lighting. During this training, soldiers engage targets at 75 to 175 meters. Several target scenarios are possible. A typical training exercise would present 30 random exposures of the 75-meter and 175-meter targets (or optional 100-meter and 200-meter targets). Soldiers should be expected to hit at least 10 targets. Tracer ammunition can be used to enhance training.

- *Night vision devices.* Repeated training and dry-fire practice on the proper use of NVDs are essential to the successful conduct of any live-fire training using these devices. Firers must understand the equipment and skillfully employ it. NVDs can provide engagement capabilities out to 300 meters.

NOTE: Spotlights or floodlights can be modified through use of a rheostat to simulate the flickering, bright/dim nature of artificial illumination. Lights should not be used to continuously spotlight targets. Unanticipated artificial illumination may render NVDs difficult to see through or may shut the device off. Live-fire training should consider any problems incurred by such unexpected illumination.

- *Unit.* Parts of unit STXs, FTXs, and LFXs should be conducted at night. This training should include target detection, unassisted MILES and live fire, artificial illumination, and NVDs. Targets out to 300 meters may be used, depending on the existing conditions. Emphasis is on soldier knowledge and skills displayed.

NOTE: See FM 25-7 for a description of ranges available and recommended for live-fire training.

APPENDIX A

Year-Round Marksmanship Training

An effective unit marksmanship program reflects the priority, emphasis, and interest of commanders and trainers. This appendix proposes a rifle marksmanship training strategy as guidance in establishing and conducting an effective training program. The strategy consists of the individual and leader refresher training for maintaining the basic skills learned during IET. It progresses to training advanced and collective skills under near-combat conditions during live-fire STXs.

MARKSMANSHIP AND THE METL

Marksmanship proficiency is critical and basic to soldiering and is required for any unit deployed to a wartime theater. All commanders should develop a METL and organize a training program that devotes adequate time to marksmanship.

The unit's combat mission must be considered when establishing training priorities. This not only applies to the tasks selected for the unit's METL but also the conditions under which the tasks are to be performed. If a unit may be employed in a MOUT environment, the effects of range, gravity, and wind may not be too important, but automatic/burst fire, quick fire, and assault fire would be. The reverse may be true of a unit that expects to engage the enemy at long range with rifle fire.

ASSESSMENT OF MARKSMANSHIP STATUS

To conduct an effective marksmanship program, the unit commander must determine the current marksmanship proficiency of all assigned personnel. To check the effectiveness of a unit's marksmanship program, constant evaluation is required. Observing and accurately recording performance reveals the status of rifle and magazine maintenance, the quality of rifle zeros, and the ability of each soldier to hit targets. This also allows the commander to identify soldiers who need special assistance in order to reach required standards, and to recognize soldiers who exceed these standards. Based on this evaluation, marksmanship training programs can be developed and executed.

This assessment is continuous, and the program is modified as required. Spot-checks of individual marksmanship performance, such as interviews and evaluations of soldiers, provide valuable information as to whether the soldier knows how to zero, to use NVDs, and to perform other marksmanship tasks.

In addition to spot-checks and direct observation of training, assessment includes a review of past training, which provides valuable information for developing a training plan. The assessment should include how record fire was conducted, what course of fire was used, how often the units conducted collective NBC or night fire, and so on. The results are reviewed to determine unit weaknesses and which individuals require special attention.

Based on the commander's evaluation, goals, and missions, training events are identified that should be conducted quarterly, semiannually, or annually—rifle marksmanship programs must be continuous. While the unit may only qualify its soldiers annually or semiannually, test results show that sustainment training is required at least quarterly to maintain marksmanship skills.

TRAINING THE TRAINER

Knowledgeable small-unit leaders are the key to marksmanship training. This manual and other training publications provide the unit instructor with the required information for developing a good train-the-trainer program. The commander should identify unit personnel who have had assignments as marksmanship instructors. These individuals should be used to train other unit cadre by conducting preliminary rifle instruction and live-fire exercises for their soldiers.

Assistance and expertise from outside the unit may also be available such as the Army Marksmanship Unit, Fort Benning, Georgia. A suggested train-the-trainer program is outlined below:

- Marksmanship diagnostic test.
- Review of operation and function, immediate action, and safety of rifle and ammunition.
- Conduct of PRI; review of four fundamentals.
- Review of coaching techniques and device usage.
- Principles and execution of grouping and zeroing.
- Effects of wind and gravity on long-range firing out to 300 meters (then 400 to 500 meters).
- Purpose and conduct of practice fire, and scaled target, and at range out to 300 meters (then 400 to 500 meters).
- Range operations.
- Purpose and conduct of qualification/record firing.
- Diagnosis of firing problems

References.

- AR 350-41.
- DA Pamphlet 350-38.
- TRADOC BT/OSUT POI.
- FC 23-11.
- FM 25-1, -2, -3, and -4.
- Supporting television tapes that teach BRM parts TVT 7-1 and TVT 7-2.
- FM 25-7.

QUALIFICATION TRAINING

Although marksmanship is a continuous training requirement, units normally conduct a refresher program before qualification. Soldiers must be well-rounded in marksmanship fundamentals and have preparatory marksmanship training before qualification. This applies to qualification for the entire unit or for newly assigned personnel. All trainers must understand that rifle marksmanship is not a series of exercises to be trained in a planned sequence. The unit must prepare for training by —

- Issuing soldiers a serviceable and accurately firing rifle.

- Repairing and replacing bad magazines.

- Issuing and assigning each soldier his own rifle that only he zeros and fires.

- Considering available or required resources early such as targets, ranges, ammunition, training aids, devices, and publications.

Before the soldier can fire, he must know how to adjust rifle sights and should understand ballistics to include the effects of wind and gravity on a bullet strike. A refresher training program can prevent frustration and loss of confidence in the soldier, and also prevent wasting ammunition and training time. This program is conducted for all soldiers so they can meet the standards outlined in this manual and soldier's manuals.

NOTE: Many individual marksmanship tasks, such as operation and function checks, immediate action, target detection, and dry fire, **DO NOT** require live fire.

Feedback (precise knowledge of bullet strike) must be included in all live-fire training. Downrange feedback, known-distance (KD) firing, or scaled-silhouette target exercises provide the results at extended ranges. Feedback is not adequate when bullets from previous firings cannot be identified such as previous shot groups on a zero target that are not triangulated and clearly marked.

The initial live fire should be a grouping exercise, which allows soldiers to apply marksmanship fundamentals to obtain tight, consistent shot groups. Following a successful grouping exercise, zeroing is quick and simple using only a few rounds.

After zeroing, downrange feedback should be conducted. A series of scaled-silhouette targets provide unlimited situations for training on the 25-meter range if modified field-fire or KD ranges are not available. The timed-fire scaled-silhouette target can add to successful record fire performance since it represents targets at six different ranges, requires quick response, and allows precise feedback. It is another way to confirm zero and requires the application of the four fundamentals. This exercise can benefit units that have access only to 25-meter ranges. (See Appendix E for use of scaled targets.)

Field-fire training is a transitional phase that stresses the focusing on a certain area. Soldiers must detect the target as soon as it comes up and quickly fire with only hit-or-miss feedback; this is an important combat skill. Soldiers who are exposed to the field-fire range before they have refined their basic firing skills cannot benefit from the exercise. For example, if most 175- and 300-meter targets are missed, additional feedback or PRI training should be conducted.

The Army standard record fire course involves an element of surprise in that the soldier is not be familiar with the lane in which he qualifies. He must scan the sector and apply detection skills and range estimation skills. However, practice can be repeated on the record fire course when available. This course provides the best opportunity for practicing target detection skills and for engaging targets at ranges from 50 to 300 meters.

For poor firers, remedial training is conducted to include the use of the Weaponeer device discussed in Appendix C. Soldiers proficient in marksmanship skills can assist in the remedial training effort.

MARKSMANSHIP TRAINING TASKS

The following marksmanship training guide contains the current tasks that are trained in basic rifle marksmanship programs, during basic combat training at ATCs, and during infantry OSUT. It provides a basis for structuring unit sustainment programs. The unit normally trains by performing a diagnostic test of the tasks and conducts training only for soldiers who must improve their firing skills. Training is usually conducted in a shorter time frame than at IET.

Introduction to Rifle Marksmanship and Mechanical Training.

(4 hours)

TASK: Perform operator maintenance on M16A1/A2 rifle, magazine, and ammunition.

CONDITIONS: Given an M16A1/A2 rifle, magazine, 5.56-mm dummy ammunition, and small-arms maintenance equipment case.

STANDARDS:

    a. Clear and disassemble an M16A1/A2 rifle.

    b. Inspect, clean, and lubricate the rifle.

    c. Assemble the rifle and perform a functional check.

    d. Disassemble the magazine.

    e. Clean and lubricate the magazine.

    f. Assemble the magazine.

    g. Clean the ammunition.

TASK: Perform immediate-action procedures on the M16A1/A2 rifle to reduce a stoppage.

CONDITIONS: Given an M16A1/A2 rifle, 5 rounds of dummy 5.56-mm ammunition, and a 20-/30-round magazine.

STANDARDS: Soldiers must demonstrate the ability to reduce a stoppage by applying the six steps of immediate action.

NOTE: Care must be taken in teaching immediate action (SPORTS) to clear a weapon stoppage. This technique must not be confused with the procedure for correctly loading a magazine into the weapon due to the position of the bolt.

TASK: Load and unload an M16A1/A2 rifle magazine.

CONDITIONS: Given either a 20- or 30-round magazine and 5 rounds of dummy ammunition.

STANDARDS: Load and unload the magazine.

    a. **Loading the Magazine.** Cartridges are loaded into the magazine so that the projectile or the rounds point in the same direction as the raised portion of the follower.

    b. **Unloading the Magazine.** Hold the magazine open end away from the body and depress the center of the top round in the magazine, allowing the first round to be slipped out of the magazine. This process is repeated until all rounds have been removed.

TASK: Adjust front and rear sights on the M16A1/A2 rifle.

CONDITIONS: Given an M16A1/A2 rifle, dummy 5.56-mm bullet, nail or other suitable instrument, explanation, demonstration, and practical exercise.

STANDARDS: Demonstrate an understanding of sight adjustment procedures by moving the front and rear sights in relationship to the intended movement of the strike of a bullet.

Fundamentals of Rifle Marksmanship (Dry Fire).
(6 hours)

TASK: Apply the four basic fundamentals of marksmanship.

    a. Steady position.

    b. Aiming.

    c. Breath control.

    d. Trigger squeeze.

CONDITIONS: Day, in a suitable training area given, an M16A1 rifle from either the supported or prone unsupported position; and using the target-box exercise, dime (washer) exercise, M16 sighting device, and riddle sighting device while wearing a helmet and LBE.

STANDARDS: Correctly apply the basic fundamentals while dry firing from or using —

    a. The supported position at 250-meter (M16A1) or 300-meter (M16A2) zero targets.

    b. The prone unsupported position at 250-meter (M16A1) or 300-meter (M16A2) zero targets.

    c. The dime (washer) exercise, M16 sighting device, riddle device, immediate action, 250-meter (M16A1) or 300-meter (M16A2) zero target, and target-box exercise.

Fundamentals of Rifle Marksmanship and Diagnostic: (Dry Fire).
(6 hours)

TASK: Demonstrate the integrated act of firing while using the Weaponeer device.

CONDITIONS: Given a demonstration and practical application with 12 shots and a diagnostic of 9 shots in a suitable training area, and with a Weaponeer and simulated supported firing position while wearing a helmet and LBE.

STANDARDS: Each soldier demonstrates the integrated act of firing by the proper application of the four fundamentals of marksmanship and achieves 6 hits out of 9 shots on the 250-meter (M16A1) or 300-meter (M16A2) zero target displayed on the device.

NOTE: Soldiers who do not meet the standard will receive remedial training on the fundamentals of rifle marksmanship before subsequent instruction.

Shot Grouping: (Live Fire).
(4 hours)

TASK: Apply the four fundamentals of rifle marksmanship in the integrated act of firing with the M16A1/A2 rifle (live fire).

CONDITIONS: Day, on a 25-meter firing range, given a 250-meter (M16A1) or 300-meter (M16A2) zero target placed in the center of an E-type silhouette, M16A1/A2 rifle, and 27 rounds of ammunition while wearing a helmet and LBE.

STANDARDS: From the supported position, fire up to 27 rounds or less in 3-round shot groups and achieve two consecutive 3-round shot groups (measured separately) within the plastic target-box paddle template (DVC-T 7-86) 4-cm circle.

NOTE: Once the soldier has demonstrated a consistency in his point of aim and achieved the standard in less than 27 rounds, a bold sight adjustment can be made to bring the groups closer to the aim point in preparation for subsequent training.

25-meter Battlesight Zeroing.
(8 hours)

TASK: Battlesight zero an M16A1/A2 rifle.

CONDITIONS: On a 25-meter range, given an M16A1/A2 from the supported position, 18 rounds of 5.56-mm ammunition, 250-meter (M16A1) or 300-meter (M16A2) zeroing target placed on the standard E-type silhouette, sandbag for support, and an M16A1/A2 rifle while wearing a helmet and LBE.

STANDARDS: Each student must adjust the sights so that 5 out of 6 rounds in two consecutive shot groups strike within the 4-cm circle on the 25-meter zero target.

NOTE: Bullets that break the line of the 4-cm circle should be used in evaluating the soldier's overall performance of the standard. Soldiers not achieving an acceptable zero in 18 rounds will be diagnosed on the Weaponeer and given appropriate remedial training. Once the soldier's problem has been corrected, he returns to the firing line and is given up to 18 additional rounds for zeroing. A careful serviceability check of the weapon is made for all soldiers failing to zero on the second attempt.

Downrange Feedback Firing: (75, 175, 300 meters).
(8 hours)

TASK: Confirm zero at 175 meters.

CONDITIONS: Day, from the supported position, on a modified field or known-distance (KD) firing range with an E-type silhouette feedback target at 175 meters and given an M16A1/A2 rifle and 6 rounds of 5.56-mm ammunition while wearing a helmet and LBE.

STANDARDS: Demonstrate consistent application of the integrated act of firing and obtain 4 out of 6 hits within the 11-inch circle on the 175-meter downrange feedback target.

NOTES: 1. Soldiers are given preliminary instructions on the effects of wind and gravity, and proper holdoff at range before conducting live fire. Soldiers go downrange and record target hits after each 3-round shot group.

2. Instructors/cadre critique the soldier's performance after each 3-round shot group and have the soldier make the necessary sight or aiming adjustments to bring the shot groups within the 11-inch circle. If a KD range is used, feedback is provided by spotters in the pits, and soldiers do not leave the firing line.

3. Soldiers not confirming zero at range receive remedial training by instructors/cadre personnel and refire with the last firing order.

TASK: Obtain downrange feedback at 75, 175, and 300 meters.

CONDITIONS: Day, given an M16A1/A2 rifle on a modified field or known-distance firing range with F&E feedback targets while wearing a helmet and LBE, engage the 75-meter target with 5 rounds from the prone unsupported position and with 5 rounds from the supported position, engage the 175-meter target from the supported position with 10 rounds and with 10 rounds from the prone unsupported position, and engage the 300-meter target with 5 rounds from the supported position and with 5 rounds from the prone unsupported position.

STANDARDS: Demonstrate consistent application of the four fundamentals in the integrated act of firing and obtain 8 hits out of 10 shots on the 75-meter target, 14 hits out of 20 shots on the 175-meter target, and 5 hits out of 10 shots on the 300-meter target.

NOTES: 1. If a KD or location-of-misses-and-hits (LOMAH) range is available, training should be conducted there. If not, use a modified field-fire range.

2. Soldiers go downrange and record target hits after each 5-round shot group. Instructors/cadre critique the soldier's performance after each 5-round exercise. Sights or aiming point is adjusted as necessary. Soldiers should do not adjust sights to zero for each range versus using the correct hold-off or adjusted aiming point for range.

Field Fire I (Single Timed Targets) and Target Detection.
(3 hours)

TASK: Detect and engage single timed targets with the M16A1/A2 rifle.

CONDITIONS: Day, given an M16A1/A2 rifle on a field-fire range with timed single target exposures presented at 75, 175, and 300 meters; and given 18 rounds of 5.56-mm ammunition for an introduction to field fire and 36 rounds of ammunition, and a requirement to engage all targets within the time exposed while wearing a helmet and LBE.

STANDARDS: Detect and achieve a total of 22 target hits out of 36 timed target exposures.

TASK: Detect targets.

CONDITIONS: Given instruction on target detection during daylight hours on a target-detection range with 10 target exposures at ranges from 50 to 300 meters.

STANDARDS: Each soldier must detect 9 out of 10 targets to receive a GO.

Field Fire II (Single and Multiple Timed Targets).
(3 hours)

TASK: Detect and engage single and multiple timed targets with the M16A1/A2 rifle.

CONDITIONS: Day, given an M16A1/A2 rifle, while wearing a helmet and LBE, on a field-fire range with timed single and multiple target exposures presented at 75, 175, and 300 meters; and given 10 rounds of 5.56-mm ammunition for an introduction to field fire single and multiple targets and 44 rounds of ammunition, and a requirement to engage all targets within the time exposed.

STANDARDS: Detect and achieve 27 target hits out of the 44 timed target exposures.

NOTE: Using peer coaching for firing performance and for assisting in target detection on the firing line is permitted during Practice Record I and II.

Practice Record Fire I.
(4 hours)

TASK: Detect and engage timed targets with the M16A1/A2 rifle.

CONDITIONS: Day, given an M16A1/A2 rifle on a record fire range, 40 timed target exposures at ranges from 50 meters to 300 meters, and 40 rounds of ammunition. Engage 20 targets from the supported position and 20 targets from the prone unsupported position while wearing a helmet and LBE.

STANDARDS: Detect and obtain at least 23 target hits on the 40 exposed timed targets.

Practice Record Fire II.
(4 hours)

TASK: Detect and engage timed targets with the M16A1/A2 rifle.

CONDITIONS: Given an M16A1/A2 rifle on a record fire range, 40 target exposures at ranges from 50 to 300 meters, and 40 rounds of ammunition; engage 20 targets from the supported position and 20 targets from the prone unsupported position while wearing a helmet and LBE.

STANDARDS: Obtain at least 23 hits on the 40 targets exposed.

NOTE: Practice Record Fire I should not be fired on the same range as Practice Record Fire II. If scheduling requires use of the same practice range, soldiers fire from different positions or lanes. The same range is not used for Practice Record Fire II and Record Fire Qualification, which are not conducted on the same day. Coaching the firer on the firing lane is permitted during Practice Record Fire II.

Record Fire.
(4 hours)

TASK: Detect and engage timed targets with an M16A1/A2 rifle.

CONDITIONS: Given an M16A1/A2 rifle on a record fire range, 40 target exposures at ranges from 50 meters to 300 meters and 40 rounds of ammunition; engage 20 targets from the supported position and 20 targets from the prone unsupported position while wearing a helmet and LBE.

STANDARDS: Without assistance, the soldier detects and engages targets with the M16A1/A2 rifle, and achieves a minimum of 23 target hits out of 40 target exposures.

Qualification Table

- Personnel hitting 22 or fewer targets are unqualified.
- Personnel hitting 23 to 29 targets qualify as marksman.
- Personnel hitting 30 to 35 targets qualify as sharpshooter.
- Personnel hitting 36 to 40 targets qualify as expert.

NOTE: Verifiable inoperative targets are the only alibis allowed during record fire. Soldiers whose weapons fail to operate due to broken parts or bad ammunition are withdrawn from the firing line to have the weapon repaired or ammunition replaced. Then the soldier refires as a first-time firer. Failures to fire due to the soldier not applying immediate action or detecting a target are not considered alibis.

Automatic Firing (M16A1)/Burst Firing (M16A2).
(2 hours)

TASK: Apply the integrated act of automatic rifle/burst firing using automatic firing positions.

CONDITIONS: Given an explanation, demonstration, M16A1/A2 rifle with bipod, 21 rounds of ammunition, and using the automatic firing position from a supported position on a field fire range, engage target exposures at 75, 175, and 300 meters while wearing a helmet and LBE.

STANDARDS: Correctly fire the weapon in 3-round bursts and obtain target hits at each range while demonstrating control of the weapon in the automatic/burst mode (instructional requirement only).

Protective Mask Firing.
(3 hours)

TASK: Engage targets in an NBC environment with the M16A1/A2 rifle.

CONDITIONS: Given an explanation, demonstration, M16A1/A2 rifle, and 20 rounds of 5.56-mm ammunition, engage 20 five-second exposures of F-type silhouette targets at 25 meters, firing 10 from the the prone unsupported and 10 from the prone or supported position, using semiautomatic fire, while wearing helmet, mask with hood, gloves, and LBE.

STANDARDS: Obtain a total of 11 hits out of the 20 target exposures. This is a GO/NO-GO exercise.

Night Fire.
(3 hours)

TASK: Engage targets while applying night-fire techniques with the M16A1/A2 rifle.

CONDITIONS: Given an explanation, demonstration, and practical exercise with 5 rounds during the day, and with semiautomatic fire at night (EENT), engage F-type silhouette targets at 25 meters from the prone unsupported position and E-type silhouette targets at 50 meters from the prone supported position; given 35 rounds of tracer and ball mix ammunition and an M16A1/A2 rifle while wearing a helmet and LBE.

STANDARDS: Achieve 7 hits out of 30 target exposures. This is a GO/NO-GO exercise.

NOTE: Five rounds are fired during daylight hours, using night-fire techniques for a practice exercise.

FM 23-9

Advanced Rifle Marksmanship.
(8 hours)

TASK: Detect and engage moving and stationary targets with the M16A1/A2 rifle (practice).

CONDITIONS: Day, on a remoted target system range, given an M16A1/A2 rifle, 8 moving target exposures at ranges from 35 to 125 meters, 2 stationary target exposures at ranges of 175 and 300 meters, and 10 rounds of ammunition; engage targets from the semisupported firing position while wearing a helmet and LBE.

STANDARDS: Obtain at least 5 target hits on the 10 targets exposed.

NOTE: Soldiers who do not hit at least 5 out of 10 target exposures will perform remedial training before firing the 40-round scenario.

TASK: Explain the semisupported firing position.

CONDITIONS: Given an explanation and demonstration.

STANDARDS: Explain the following performance measures:

    a. Increase the grip of the nonfiring hand.

    b. Lift the nonfiring elbow to maintain a smooth track.

    c. Maintain rearward pressure with the firing hand.

    d. Lift the firing elbow to maintain a smooth track.

    e. Spread feet a comfortable distance apart.

    f. Do not assume chest-to-wall contact.

TASK: Apply lead guidance rules using the Aid to Improved Marksmanship (AIM) booklet.

CONDITIONS: Given an explanation and practical application using the AIM booklet.

STANDARDS: Each soldier must select the correct lead on a target in an AIM booklet exercise.

NOTE: Each soldier is tested by the instructor at the end of the exercise.

TASK: Perform rapid magazine change.

CONDITIONS: Given an M16A1/A2 rifle and two magazines.

STANDARDS: Perform a rapid magazine change within 5 seconds.

TASK: Engage the dry-fire mover.

CONDITIONS: Given an explanation, practical exercise, and an M16A1/A2 rifle, the dry-fire mover is engaged from the semisupported firing position.

STANDARDS: Correctly apply the moving target techniques of tracking and trapping.

TASK: Detect and engage moving and stationary targets with the M16A1/A2 rifle (record fire).

CONDITIONS: Day, on a remoted target system range, given an M16A1/A2 rifle, 30 moving target exposures at ranges from 35 to 125 meters, 10 stationary target exposures at ranges of 175 and 300 meters, and 40 rounds of ammunition; engage targets from the semisupported firing position.

STANDARDS: Obtain at least 15 target hits on the 40 targets exposed.

Advanced Rifle Marksmanship Qualification.
(8 hours)

TASK: Detect and engage moving and stationary targets with the M16A1/A2 rifle.

CONDITIONS: Day, on a remoted target system range, given an M16A1/A2 rifle, 25 moving target exposures at ranges from 35 to 185 meters, 25 stationary target exposures at ranges from 50 to 300 meters, and two magazines with 25 rounds each of 5.56-mm ball ammunition.

STANDARDS: Each soldier must achieve 18 target hits out of 50 target exposures.

NOTE: Firing scenario is engaged once for practice and then for qualification. Soldiers who fail to qualify on the initial day of qualification receive only one refire the same day.

Quick Fire.
(2 hours)

TASK: Engage F-type and E-type silhouettes at 15 and 25 meters, respectively, using quick-fire techniques.

CONDITIONS: Day, given an operational M16A1/A2 rifle, 20 rounds of 5.56-mm ball ammunition, and 10 target exposures of 2 seconds each at 15 meters and 25 meters.

STANDARDS: Achieve 7 target hits at 15 meters and 5 target hits at 25 meters for 10 target exposures.

NBC Fire.
(2 hours)

TASK: Engage personnel targets while wearing a protective mask.

CONDITIONS: Day, on a field-fire range, given an M16A1/A2 rifle, 30 rounds of 5.56-mm ball ammunition, 3 magazines, and E-type silhouette targets at 75, 175, and 300 meters.

STANDARDS: Each firer has 60 seconds at each target distance. The standard for each distance is as follows:

- 75 meters: 7 target hits of 10 rounds.
- 175 meters: 6 target hits of 10 rounds.
- 300 meters: 2 target hits of 10 rounds.

Night Fire.
(4 hours)

TASK: Engage targets under artificial illumination.

CONDITIONS: Night, on a field-fire range, given an M16A1/A2 rifle with bipod, 20 rounds of 5.56-mm ball ammunition, 10 rounds of 5.56-mm tracer ammunition, and 2 magazines of 15 rounds each; assuming a prone bipod supported position; using 75- and 175-meter E-type silhouettes; and given 90 seconds at each distance.

STANDARDS: Each soldier must obtain a total of 5 target hits out of 15 rounds at 75 meters and 3 target hits out of 15 rounds at 175 meters.

TASK: Engage targets using an AN/PVS-4 and M16A1/A2 rifle.

CONDITIONS: Night, on a field-fire range, given an M16A1/A2 rifle, 30 rounds of 5.56-mm ball ammunition, 3 magazines of 10 rounds each, and E-type silhouettes at 75, 175, and 300 meters.

STANDARDS: Each firer has 60 seconds at each distance. The standards for each distance are as follows:

- 75 meters: 7 target hits of 10 rounds.
- 175 meters: 5 target hits of 10 rounds.
- 300 meters: 2 target hits of 10 rounds.
-

TASK: Mount, dismount, and place into operation the AN/PVS-4.

CONDITIONS: During daylight or limited visibility, given an AN/PVS-4 with mounting knob assembly and M16A1/A2 rifle.

STANDARDS: Each firer performs the following to the standard:

    a. Mount the AN/PVS-4 to an M16A1/A2 rifle in three minutes.
    b. Dismount the AN/PVS-4 from an M16A1/A2 rifle in one minute.
    c. Place the AN/PVS-4 into operation in five minutes.

TASK: Determine the effects of wind and gravity on sight adjustments for engagements out to 400 meters (500 meters, M16A2).

CONDITIONS: Given an M16A1/A2 rifle, scaled targets that represent various ranges, scaled sight training aid, information on wind, paper and pencil, and an explanation and practical application.

STANDARDS: Each soldier demonstrates the correct sight adjustment technique, considering the effects of wind and gravity on a bullet at long ranges, and how to apply hold-off on a silhouette target.

TASK: Zero the M16A1/A2 rifle at ranges of 300, 400, 500, and 600 yards.

CONDITIONS: Given an E-type silhouette and 36 rounds of 5.56-mm ammunition fired in four 3-round shot groups at each range from the prone supported position.

STANDARDS: Each soldier correctly demonstrates the techniques of sight adjustment to zero the M16A1/A2 rifle for each range, and obtains a zero within 12 rounds at 300, 400, and 500 yards.

TASK: Engage targets at distances of 300, 400, 500, and 600 yards with the M16A1/A2 rifle.

CONDITIONS: Given an M16A1/A2 rifle and 46 rounds of 5.56-mm ammunition on a KD range from the prone supported position, fire at E-type silhouette targets with 2 sighter rounds fired at each range, 10 rounds at 300, 15 at 400, and 15 at 500 fired in succession.

STANDARDS: Within the time specified, one minute at 300, two minutes at 400 and two minutes at 500, each soldier is required to meet the minimum gate at each range and an overall score of 30 hits out of 40 rounds on an E-type silhouette. Times specified are: 300 yards: 1 minute, 10 rounds achieve 8 hits minimum; 400 yards: 2 minutes, 15 rounds achieve 12 hits minimum; and 500 yards: 2 minutes, 15 rounds achieve 10 hits minimum.

TASK: Engage scaled silhouette targets using rapid semiautomatic fire.

CONDITIONS: During daylight hours on a 25-meter range; given an M16A1 rifle, one 20-round magazine of 5.56-mm ball ammunition, and a 25-meter scaled silhouette timed-fire target; using the prone supported firing position; fire 1 round at each of the 10 silhouettes on the timed-fire target using rapid semiautomatic fire.

NOTE: This exercise is fired twice: the first iteration is fired in a time limit of 40 seconds, and the second iteration is fired in a time limit of 30 seconds. Targets are inspected and posted after each iteration.

STANDARDS: Each soldier must obtain 14 silhouette target hits.

TASK: Engage a scaled landscape target using suppressive fire.

CONDITIONS: During daylight hours on a 25-meter range, given an M16A1/A2 rifle, 25-meter scaled landscape suppressive fire range, one 15-round magazine, one 10-round magazine, and three 5-round magazines of 5.56-mm ball ammunition; using the supported firing position. Using rapid semiautomatic fire, fire 9 rounds at the "open window" area of the target and 12 rounds at the "fence/hedgerow" area of the target. Using automatic fire, fire three 3-round bursts at the "tank turret" area of the target.

STANDARDS: Each soldier must obtain 10 hits inside the dotted lines surrounding the "fence/hedgerow" area within 24 seconds, 5 hits inside the "open window" area within 18 seconds, and 3 hits inside the "tank turret" area (no time limit).

TASK: Apply the integrated act of automatic rifle firing using the automatic firing position.

CONDITIONS: Given the M16A1/A2 rifle with bipod, 36 rounds of ammunition, and scaled silhouette targets at 25 meters; assume the M60 modified prone position or M60 modified supported position.

STANDARDS: Fire the weapon in rapid 3-round bursts and obtain 5 target hits while demonstrating control of the weapon in the automatic mode.

UNIT LIVE-FIRE EXERCISES

Unit live-fire exercises are planned, prepared, and performed as outlined in the mission training plan for the infantry platoon and squad. It is within the framework of these exercises that the soldier performs marksmanship tasks under realistic combat conditions.

During training, the fundamentals must apply to combat as well as to the range. Too often soldiers disregard the fundamentals while under the pressure of combat. Therefore, it is imperative the soldier receives feedback regarding his firing results and his use of the fundamentals during collective live-fire exercises. This training should also discuss target acquisition, area fire, quick fire, assuming firing positions, responding to oral fire commands, and safety. Dry fire or MILES rehearsals at crawl, walk, and run paces are required to learn SOPs and proper procedures.

Enough evaluators must be present during training to observe each soldier to provide performance feedback. The evaluator must know the scenario, the location of targets, the friendly plan, and SOPs. He must watch to determine if the soldier identifies targets in his sector and successfully engages them. The evaluator must also know the fundamentals of marksmanship to detect soldier mistakes and review them during the AAR.

UNIT SUSTAINMENT TRAINING

A unit cannot sustain marksmanship proficiency based only on qualification preceded by preliminary training. Soldier skills and marksmanship fundamentals deteriorate within two months, and mechanical skills weaken even sooner. Dry-fire training can strengthen marksmanship skills and assess their deterioration.

Each new soldier should be assigned a rifle and must perform shot grouping, zero firing, and diagnostic testing to assess training status. Many of the nonfiring tasks may be accomplished as opportunity training or concurrent training.

FM 23-9

> NOTE: Mechanical training can be considered as part of the normal weapons maintenance. Using details to clean weapons deprives soldiers of refresher training.

Problem firers must be given special attention. They must participate in bimonthly training on the Weaponeer or dry-fire exercises. When fielded, the basic/advanced rifle marksmanship trainer will provide leaders with a better method of sustaining marksmanship skills since it provides accurate feedback.

SAMPLE EVALUATION GUIDE

The following questions are provided as an example of a self-evaluation guide for commanders/leaders to help evaluate unit marksmanship training and small-arms readiness. Other areas of interest should be developed locally, based on the unit's missions and state of training.

COMMANDER'S/LEADER'S RIFLE MARKSMANSHIP EVALUATION GUIDE

1. Have you clearly stated the priority of rifle (small-arms) proficiency in your unit? What is it? Is this priority supported by the staff and subordinates? Is it based on your METL and an understanding of FM 25-100?

2. Have you clearly stated the intent of record fire? Are leaders accurately evaluating firing performance, based on accurately recorded data and results?

3. Have you clearly stated that weapons qualification or record fire is one of the commander's opportunities to assess several skills relating to small-arms readiness?

4. What qualification course will be used to evaluate your unit's rifle (small-arms) readiness?
    a. Is the standard combat course 300-yard KD or 25-meter scaled target used?
    b. How will it be conducted? Will the prescribed procedures be followed?
    c. Who will collect the data?

5. Have you clearly stated the purpose and intent of PRI?
    a. What skills will PRI address?
    b. Will PRI be performance-oriented? Are tasks integrated?

6. Do soldiers maintain their assigned weapons and magazines IAW the technical manual? Do they have a manual?

7. Do soldiers conduct serviceability checks of weapons and magazines before training? Were maintenance deficiencies corrected?

8. Do soldiers demonstrate an understanding of the weapon's operation, functioning, and capabilities?

9. Can your soldiers correctly apply immediate-action procedures to reduce weapon stoppages and then quickly continue to fire? Have they demonstrated this during dry fire?

10. Are your soldiers firing their assigned weapons?
    a. How often are weapons reassigned between individuals?
    b. What is the value of a recorded zero?

11. Can your soldiers precisely and consistently apply the four fundamentals of rifle marksmanship? To what standard? Have they demonstrated their mastery on a device —
    a. During dry fire?
    b. During live fire?
    c. During firing on the 25-meter course?
    d. During KD firing?

12. Can your soldiers accurately battlesight zero their assigned rifle to standards?
    a. Do they understand sight adjustment procedures?
    b. Do they record rifle zeros? How is it done? Why?
    c. Do they record the date the specific soldier last zeroed his specific rifle? What is the specific sight setting? Are these linked? How do you check this?

13. Do your soldiers demonstrate their knowledge of the effects of wind and gravity while firing out to 300 meters? What feedback was provided? How?

14. Can your soldiers scan a designated area or sector of fire and detect all targets out to 300 meters? If not, why?

15. Can your soldiers quickly engage timed single and multiple targets from both supported and unsupported firing positions out to 300 meters? If not, which targets were not engaged? Which were missed? Why?

16. During individual and collective training, do soldiers demonstrate their ability to manage allocated ammunition and to service all targets? Do they fire several rounds at one target? Which targets? Why?

17. Based on an analysis of individual qualification scores, what is the distribution?
    a. Are most soldiers just meeting the minimum acceptable performance (marksman)?
    b. Are most soldiers distributed in the upper half of the performance spectrum (sharpshooter, expert)?
    c. What is the hit distribution during collective LFXs?

18. Do your soldiers demonstrate proficiency during night-fire, target detection and acquisition, and night fire engagement techniques? Use of night vision devices?

19. Do your soldiers demonstrate individual marksmanship proficiency during MOPP firing conditions? During collective exercises?

20. Do your soldiers demonstrate proficiency in moving target engagements? Do they demonstrate proficiency collectively at the multipurpose range complex by hitting moving targets? If not, do you conduct moving target training?

21. Do you integrate marksmanship skills into tactical exercises and unit live-fire exercises? If so, do you conduct suppressive fire, rapid-semiautomatic fire, and automatic/burst fire? What tasks in the mission training plan are evaluated?

22. Based on your on-site observations and analysis of training and firing performance, what skills or tasks show a readiness deficiency?
    a. What skills need training emphasis? Individual emphasis? Leader emphasis?
    b. What are your performance goals?

23. Who has trained or will train the trainers?
    a. What is the subject matter expertise of the cadre?
    b. Are they actually training the critical skills?
    c. Have they addressed the nonfiring skills first?
    d. What aids and devices are being used?

24. What administrative constraints or training distractors can you overcome for the junior officer and NCO?
    a. At what level are the recourses necessary to train marksmanship controlled — time, aids, weapons, ammunition, ranges?
    b. Do the sergeants do the job they are charged with?

FM 23-9

APPENDIX B

Target Detection and Exercises

The material contained in this appendix provides the detailed information on the skills required for proper target detection and identification. The exercises contained herein will train and sustain those skills.

Section I. TARGET DETECTION AND ENGAGEMENT

Target detection is the process of locating, marking, and determining the range to combat targets. For most soldiers, finding the target can be a greater problem than hitting it. Target detection must be conducted as part of individual training and tactical exercises and must be integrated into night live-fire exercises. The observation lines on target detection ranges approximate the location a soldier might occupy when in combat. Provisions must be made to incorporate night target detection.

TARGET LOCATION

The ability to locate a combat target depends on the observer's position and skill in searching and maintaining observation of an area, and the target indications of the "enemy" during day or night.

Selection of a Position. A good position is one that offers maximum visibility of the area while affording cover and concealment. Position has two considerations—the observer's tactical position in a location and his body position at that location.

Usually, the firer is told where to prepare his defensive position. However, some situations (such as the attack and reorganization on the objective) require him to choose his own defensive position.

Although target training courses prescribe conferences and demonstrations on choice of steady firing positions, the instruction does not normally include applying this skill. Therefore, instructors/trainers must emphasize the importance of the observer's position when conducting practical exercises in other target-detection techniques.

Observation of an Area. When a soldier moves into a new area, he quickly checks for enemy activity that could be an immediate danger. This search entails quick glances at specific points throughout the area rather than just sweeping the eyes across the terrain. The eyes are sensitive to slight movements occurring within the arc on which they are focused. However, they must be focused on a certain point to have this sensitivity.

If the soldier fails to locate the enemy during the initial search, he then begins a systematic examination known as the 50-meter overlapping strip technique of search (Figure B-1). Normally, the area nearest the soldier offers the greatest danger to him. Therefore, the search begins with the terrain nearest the observer's position.

Beginning at either flank, the soldier searches the terrain to his front in a 180-degree arc that is 50 meters deep. After reaching the opposite flank, the soldier searches over a second 50-meter strip farther out but overlapping the first strip by about 10 meters. The soldier continues until the entire area has been searched.

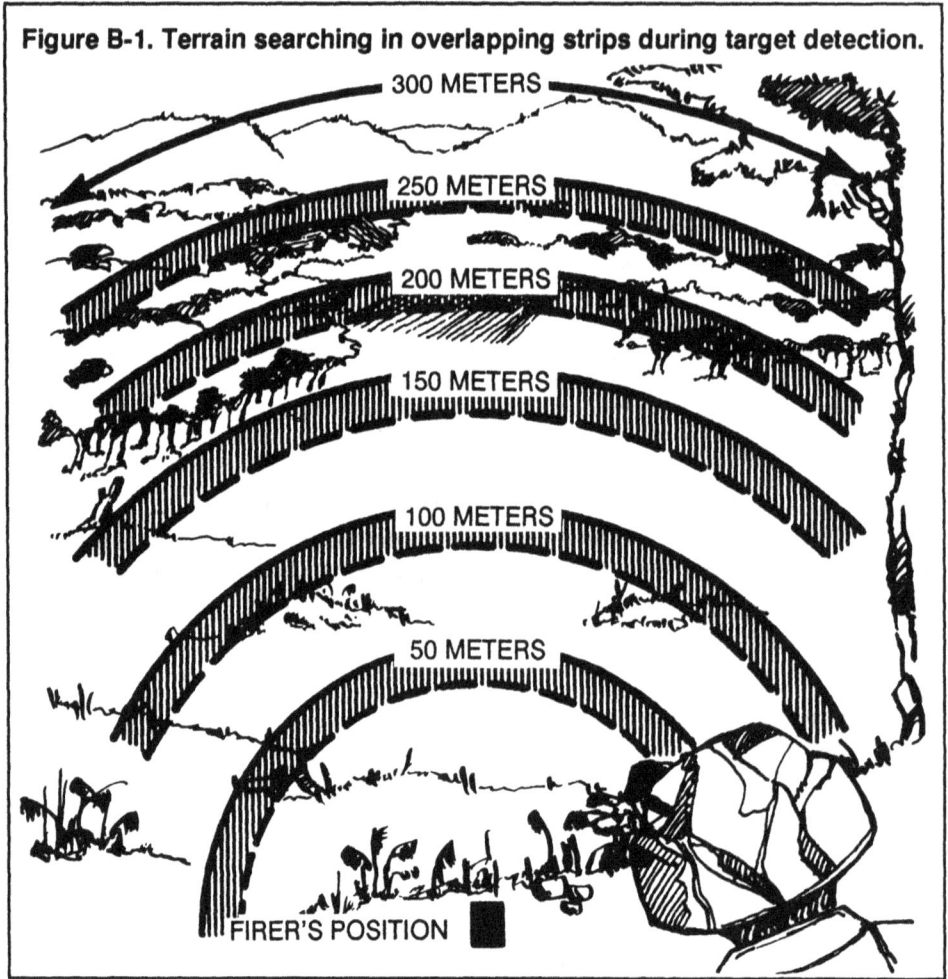

Figure B-1. Terrain searching in overlapping strips during target detection.

To benefit from his side vision, the soldier looks at certain points as he searches from one flank to the other. He remembers prominent terrain features and areas that offer cover and concealment to the enemy, learning the terrain as he searches it.

After completing his detailed search, the soldier maintains observation of the area. He should use a technique the same as his initial quick search of the area. He uses quick glances at various points throughout the entire area, focusing his eyes on certain features as he conducts this search. He devises a set sequence of searching the area to ensure complete coverage of all terrain. Since this quick search could fail to detect the initial enemy movement, the soldier routinely repeats a systematic search of the area. This systematic search is conducted anytime the attention of the soldier has been distracted from his area of responsibility.

Target Indicators. A target indicator is anything a soldier (friendly or enemy) does or fails to do that reveals his position. Since these indicators apply to both sides of the battlefield, a soldier learns target indicators from the standpoint of locating the enemy. At the same time, he must prevent the enemy from using the same indicators to locate him. These indicators can be grouped into three areas for instruction: sound, movement, and identifiable shapes.

Sound. Potential targets indicated by sounds (footsteps, coughing, or equipment noises) provide only a direction and general location. It is difficult to pin-point a specific target's location only by sound. However, the fact that an observer was alerted by a sound greatly increases the chances that he will locate the target through subsequent target indicators.

Movement. The problem in locating moving targets depends mainly on the speed of movement. Slow, deliberate movements are much harder to notice than those that are quick and jerky. The techniques previously outlined are the best procedures for locating moving targets.

Identifiable shapes. The lack of or poor use of camouflage and concealment are indicators that reveal most of the targets on the battlefield. Light reflecting from shiny surfaces or a contrast that presents a clearly defined outline are indicators easily noticed by an alert observer. For instruction, camouflage indicators are divided into three groups:

- Shine. Items such as belt buckles or other metal objects reflect light and act as a beacon to the wearer's position; therefore, such objects should be camouflaged. This is true during the day or night.

- Outlines. The human body and most types of military equipment are outlines known to all soldiers. The reliability of this indicator depends upon the visibility and experience of the observer. On a clear day, most soldiers can identify the enemy or equipment if there is a distinct outline. During poor visibility, it is not only harder to see outlines, but inexperienced troops often mistake stumps and rocks for enemy soldiers. Therefore, the soldier should learn the terrain during good visibility.

- Contrasts. If a soldier wearing a dark uniform moves in front of a snowbank, the contrast between the white snow and dark uniform makes him clearly visible. However, if he wears a white (or light-colored) uniform, he is harder to see. Contrast with the background is one of the hardest target indicators to avoid. During operations in which the soldier is moving, he is usually exposed to many types of color backgrounds. No camouflage uniform exists that can blend into all backgrounds. Therefore, a moving soldier must always be aware of the surrounding terrain and vegetation. A parapet of freshly dug earth around a fighting position is noticeable. Even if camouflaged, the position can still be located due to the materials used for concealing. Camouflage materials are usually cut from vegetation close by but eventually wilt and change color. An observer, seeing an area that has been stripped of natural growth, can assume there are close camouflaged emplacements.

TARGET MARKING

When a target has been located, the soldier should mark its location in relation to visible terrain or a man-made feature. If the soldier observes several targets at one time, he can fire on only one of them; therefore, he must mark the locations of the others for later engagement. To mark the location of a target, the soldier uses an aiming or reference point. An aiming point is a feature directly on line between the soldier and target, such as a tree trunk, which is usually the most effective means of delivering accurate fire. Using a reference point or aiming point to mark targets moving from one location to another depends on the following factors:

Number of Targets. If several targets appear and disappear at the same time, the point of disappearance of each is hard to determine.

Exposure Time. Usually, moving targets are exposed for only a short period; therefore, the observer must be alert to see the point of disappearance for most of the targets.

Target Spacing. The greater the distance between targets, the harder it is to see the movements of each. When there is a great distance between targets, the observer should carefully locate and mark the one nearest his position first.

Aiming Points. Aiming points can be either good or poor. Good aiming points are easily determined in the nearby terrain. Targets disappearing behind good aiming points, such as man-made objects and large terrain features, can be easily marked for future engagement. Poor aiming points are not easily distinguishable within the surrounding terrain. Targets disappearing behind poor aiming points are hard to mark and are easily lost, and they should be engaged first.

RANGE DETERMINATION

Range determination is the process of finding the distance between two points — one point is usually the observer's own position and the other a target or prominent feature. Range determination is an important skill in completing several types of missions since it affects combat marksmanship proficiency. It is needed in reporting information, and in adjusting artillery and mortar fires.

Many techniques are used to determine range: measuring distances on maps, pacing the distance between two points, using an optical range finder. However, the soldier does not usually have a map, and he rarely has access to an optical range finder. Pacing the distance between two points is one technique a soldier can use, as long as the enemy is not near. A sector sketch is a rough schematic map of an observer's area of responsibility (Figure B-2). It shows the range and direction from the soldier's position to recognizable objects, terrain features, avenues of approach, and possible enemy positions. The soldier paces the distance between his position and reference points to reduce range errors. By referring to the sector sketch, the soldier can quickly find the range to a target appearing near a reference point.

The 100-Meter Unit-of-Measure Technique. To use this technique, the soldier must visualize a distance of 100 meters on the ground. For ranges up to 500 meters, he determines the number of 100-meter increments between the two points (Figure B-3). Beyond 500 meters, the soldier must be select a point halfway to the target, determine the number of 100-meter increments to the halfway point, and then double it to find the range to the target (Figure B-4). During training exercises, the soldier must aware of the effect that sloping ground has on the appearance of a 100-meter increment. Ground that slopes upward gives the illusion of greater distance and soldiers have a tendency to overestimate a 100-meter increment. Conversely, ground that slopes downward gives the illusion of a shorter distance; therefore, the soldier tends to underestimate.

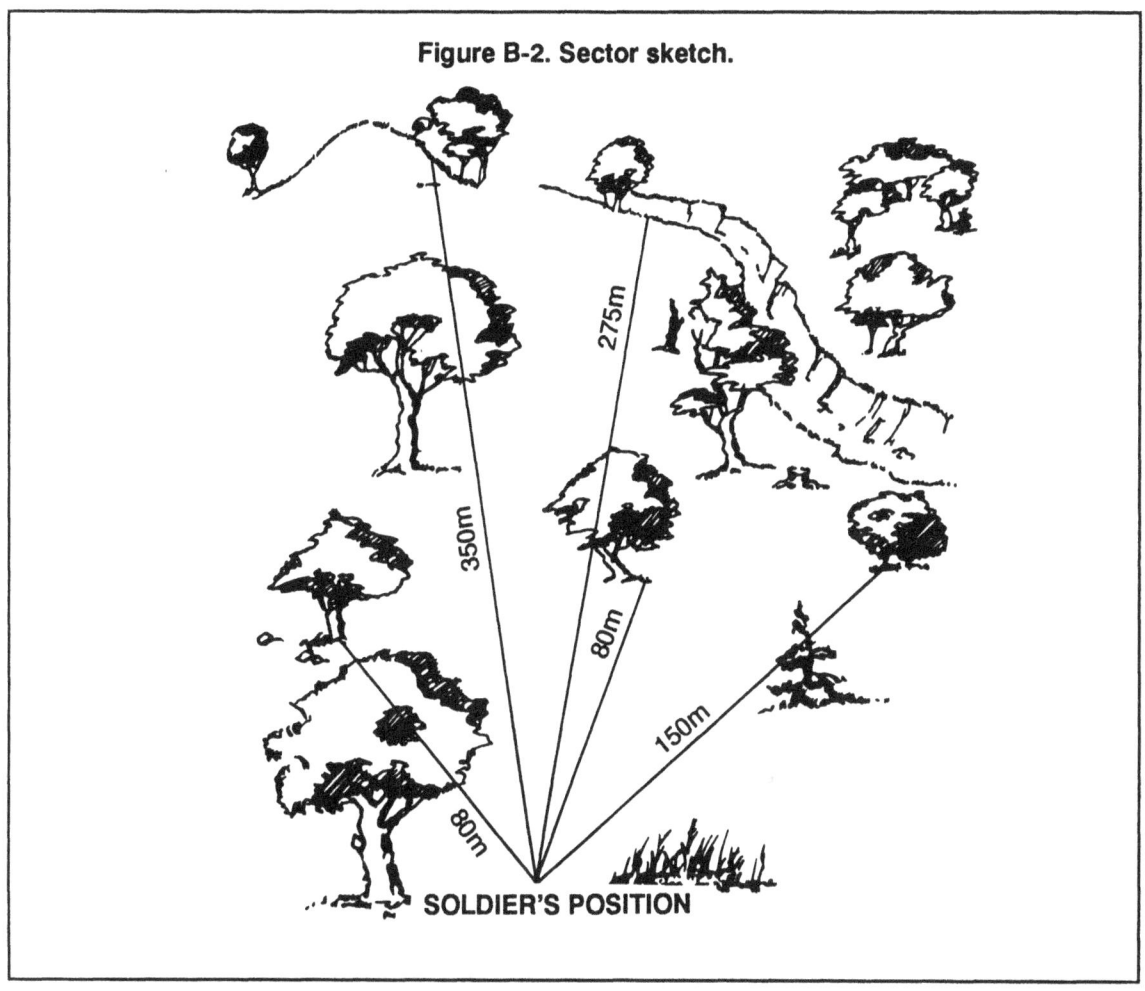

Figure B-2. Sector sketch.

To obtain proficiency in the 100-meter unit-of-measure technique requires dedicated practice. Throughout training, the soldier should compare his estimated range to the actual range determined by pacing or other reliable means. The best training technique is to require the soldier to pace the range after he has made a visual estimation, realizing the actual range for himself. This teaches him more than being told by the instructor/trainer.

One shortcoming of the 100-meter unit-of-measure technique is that its accuracy is depends upon the amount of visible terrain for ranges up to 500 meters. If a target appears at a range greater than 500 meters, and the soldier can see only a portion of the ground between himself and the target, it is hard to accurately use the 100-meter unit-of-measure technique.

The Appearance-of-Objects Technique. This technique determines range by the size of the object observed. This is a common technique of determining distances and is used by most people in their everyday living. For example, a motorist trying to pass another car must judge the distance of an oncoming vehicle. He does this based on his knowledge of how vehicles appear at various distances. Suppose the motorist knows that at a distance of 1 mile an oncoming vehicle seems to be 1 inch wide and 2 inches high. Then, anytime he sees another oncoming vehicle that fits this dimension, he knows it is about 1 mile away. This same technique can be used by the firer to determine ranges on the battlefield. If he knows the size and detail of personnel and equipment at known ranges, then he can compare these traits to like objects at unknown ranges — when the traits match, so do the ranges.

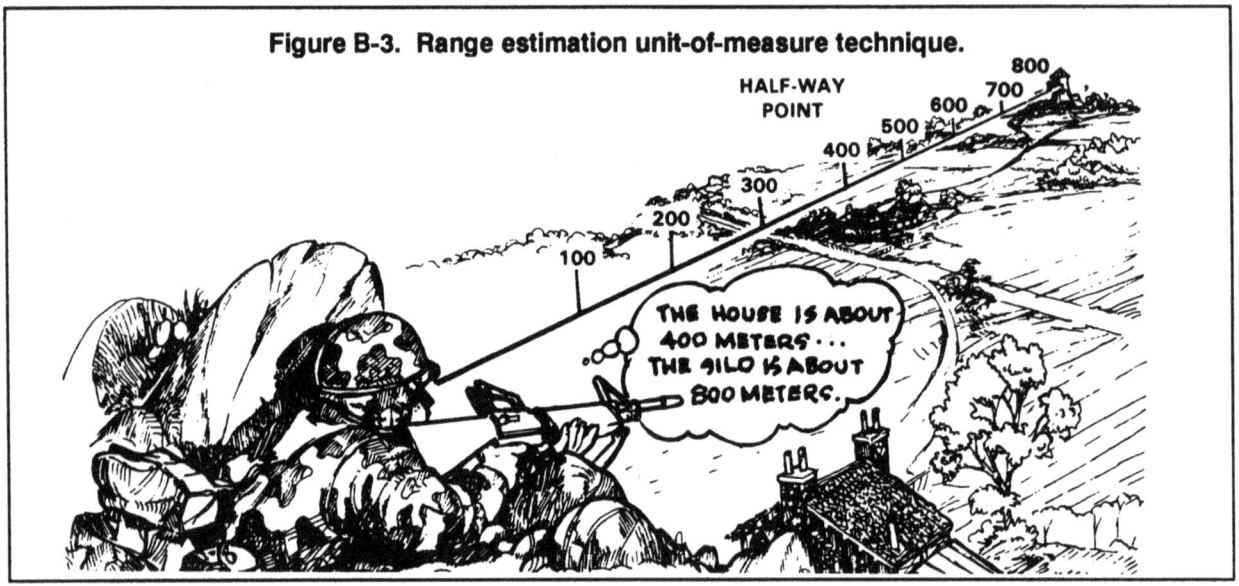

Figure B-3. Range estimation unit-of-measure technique.

The Front Sight Post Estimation. The front sight post can be used to estimate range. The targets in Figure B-4 show the soldier perceives the front sight post to be the same width as a man-size target when the target is located at a distance of 175 meters. A man can be covered using half of the front sight post when the range to the target is doubled to 350 meters. An easy rule to remember: if the target is bigger than the front sight post, the target must be within 175 meters; when the target is less than the full width of the front sight post, the target is beyond 175 meters. The silhouette zeroing target provides the same perception to the firer as a man-sized target at 250 meters. The various scaled-silhouette targets provide a means for soldiers to practice range estimation with the front sight post. This is a method of dry-fire training, and soldiers should be aware of the importance of range estimation during all of their marksmanship training.

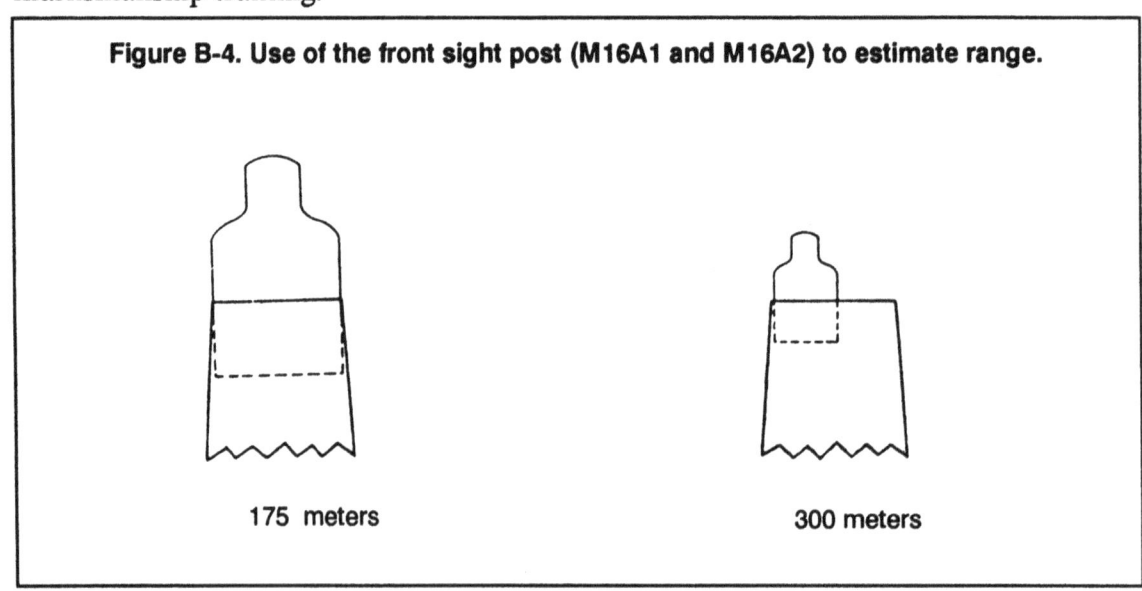

Figure B-4. Use of the front sight post (M16A1 and M16A2) to estimate range.

Section II. TARGET DETECTION RANGES

The following information gives specific instructions on target detection range design. Additional information should be reviewed in FM 25-7.

CONSTRUCTION

Target detection ranges must be located in areas having good natural vegetation and close to the firing range. (Figure B-5.)

The observation line should be the first area of the target detection range to be built. The location of all downrange panels, sound systems, and foliage depends on the degree of visibility from the defensive position (observation) line. The observation line should be wide enough to accommodate 50 points. The distance between observation points should be no closer than two meters. An observation line of this size can accommodate half of a 200-man unit (50 two-man teams).

NOTE: An initial rehearsal should be conducted for each target detection exercise. More rehearsals are needed only if target men are changed. A presenta-tion refers to each time one exercise is conducted. A rehearsal is counted as a presentation.

The observation fan should cover an area between 30 degrees left of the left flank point of the observation line to 30 degrees right of the right flank point. To provide maximum flexibility in conducting exercises in range determination, the target detection range should be deeper than 500 meters. Installations having limited training space can conduct effective training on ranges at least 300 meters deep.

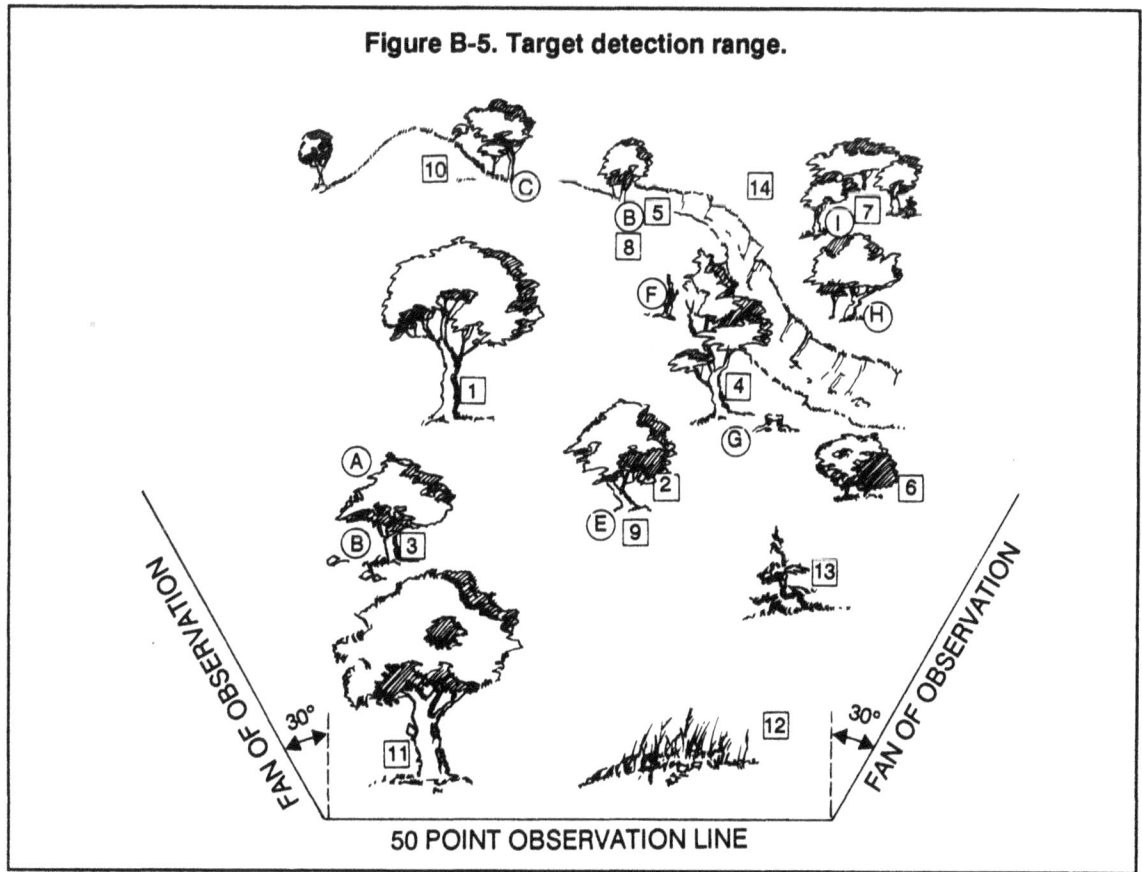

Figure B-5. Target detection range.

Both lettered and numbered panels are placed throughout the observation area. The lettered panels serve two purposes: first, they divide the range into sectors defining a firer's area of responsibility; and second, they serve as reference points for marking targets. The numbered panels are used during exercises to locate sound targets. These panels are built so they can be raised or lowered, as needed.

The number of panels needed depends on the size of the range. For a range having a 50-point observation line and 300 meters deep, about 7 lettered panels and 14 numbered panels will be needed.

Numbered stakes are placed downrange. These stakes are not visible from the observation line since they are used only by instructors and target men in presenting target situations. The number of stakes needed depends on the depth of the range. As a guide, a range 300 meters deep has about 150 stakes. When placing numbered stakes, one technique is to divide the range into three sectors: A, B, and C. This allows for easy reference. Stakes are then numbered beginning at the maximum depth of the range and proceeding forward to the observation line. All stakes in one sector would have the sector letter following the number. For example, if the right sector is designated A, all numbers on stakes in that sector are followed by the letter A. Stakes in the center and left sectors have the letter B or C, respectively, after the number. The location of all panels and stakes must be recorded on the master trial sheets (see Figure B-6).

Figure B-6. Example master trial sheet.

Trial No	Range (meters)	Target man	Description of requirements
1	200	1	Standing exposed by tree. Down to kneeling, exposed. Slow movement to out-of-sight position. Out-of-sight fire round for smoke indications. Stake 1A.
2	150	2	Same as above with a poor aiming point. Stake 3B.
3	175	3	Start standing. Disappear on command. Reappear in same position. Make five 4-second rushes with a good aiming point. Fire one round from last position. Stakes 2C-3C-4C-5C-6C-7C.
4	300	1	Start from kneeling position behind bush. Make five 4- to 5-second rushes. Disappear where there is a poor aiming point. Reappear from same position. 5-3-3-5-5-seconds. Fire round from last position. Stakes 1A-2A-3A-4A-5A-6A.
5	300	2	Start prone. Make five 4-5-8-second rushes. Disappear after each rush and roll or crouch to new position. Three-second rush, crawl left. Six-second rush, rush, crawl right. Cross small draw. Appear and make 8-second rush, crawl left. Three-second rush, crawl right. Fire one round from last position. Stakes 3B-4B-5B-6B-7B-8B.
6	175	3	Start prone. Make three 4-second and two 6-second lateral rushes to new concealment. Reappear at same point of disappearance. Vary time between rushes. Fire one round from last position. Stakes 7L-8L-9L-10C-11C-12C.
7	175	3	Do same in reverse. Crawl or roll to new position after disappearing. Fire round from last position. Stakes 12C-11C-10C-9C-8C-7C.
8	200	1	Run 200 yards from tree to position with a poor aiming point. Fire two blanks 1 minute after disappearance. Stakes 6A-11C.
9	300	2	Start prone. Three-second rush, crawl left. Five-second rush, crawl right. 5-L-3-3-6-R-4-5, through draw. Fire round from last position. (Numbers indicate duration of rush; letters L and R indicate direction of roll or crawl after each rush.) Stakes 8B-9B-10B-11B.
10	300	3	Start behind bush. 6-8-R-3-R-4-3. Fire round from last position. Stakes 7L-9L-10L-11L-12C.

Note. Target trial sheets should be prepared from a master trial sheet similar to the one above containing only the trials and target indications performed by a specific target.

For proper control of target men, sound equipment is used throughout the observation area. Since problems of adequate sound vary according to location, it is best that a sound survey be conducted of each target detection range before the equipment is installed.

The exact positioning of panels, stakes, and sound equipment is checked from the observation line. Sound equipment should be concealed from the observation line.

FIELD-EXPEDIENT AREA FOR TRAINING

If standard detection ranges are not available, the principles can be applied to parks, wooded fields, vegetated areas, and urban areas. The following checklist provides suggestions for adapting areas for target detection training.

The range should have more depth than a standard target detection range. The fan of observation should be increased, depending on the degree of camouflage in the area.

Target men should be spaced farther apart in areas having little natural vegetation. Logs, man-made objects, and piles of logs may be needed to add to the number of concealed positions.

For MOUT training, a MOUT site should be used. If one is not available, buildings in a company area could be used with target men or silhouettes.

PERSONNEL AND EQUIPMENT

The following personnel are needed to conduct and supervise target-detection training:

- OIC or principal instructor.
- Four assistant instructors (based on a 50-point observation line).
- Target men for the period of instruction.

The following equipment is needed to conduct target-detection training:

- One aiming device (see Figure B-7) for each observation point for the period of instruction.

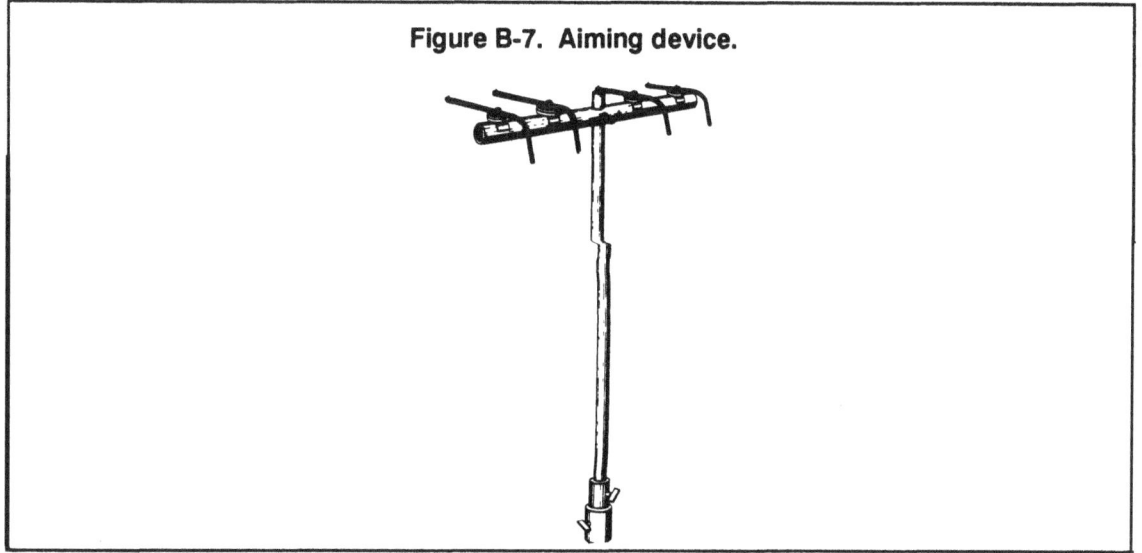

Figure B-7. Aiming device.

- One master trial sheet (Figure B-8) for each instructor and assistant instructor.
- One answer sheet for each observer.
- One target trial card (Figure B-9) for each target man.
- Camouflage paint tubes.
- One rifle for each observation point (for exercises in which observers simulate firing on target men).
- Combat field equipment including helmet and rifle for target men.

Section III. CONDUCT OF TRAINING

Demonstrators for target detection training wear combat field equipment. This increases their value as target men simulating the movements and appearance of "enemy" soldiers. Target detection is taught in three phases: first, how to locate a target; second, how to mark the location of the target; and third, how to determine the range to the target.

MASTER TRIAL SHEETS AND TARGET TRIAL CARDS

Trial sheets and trial cards contain the following information:

Master Trial Sheets. The master trial sheet (Figure B-6) shows the number of target men needed for an exercise, the actions to be performed by the target men, the duration of the actions, and the panel or stake locations where the actions occur. A master trial sheet is made for each period of instruction.

Target Trial Cards. A target trial card (Figure B-8) is issued to each man who acts as an enemy "target" in the area of observation. These men, called *target men*, use the target trial cards as a basis for their location and actions throughout an exercise. All actions performed by a target man, which lead to his eventual disclosure, are termed trials.

CONDUCT OF TRIALS

Before a trial is conducted, soldiers face away from the range area so that target men can assume their positions unobserved. When target men are in position, soldiers are told to face downrange. There are four types of trials conducted during target detection training: stationary trials, moving trials, stationary sound trials, and multiple moving and sound target trials. Certain factors can affect the appearance of objects and should be known by the firer (Table B-1, page B-12).

Stationary Trials. Normally, there are four phases in each stationary trial. The first three phases last 30 seconds each.

PHASE ONE: The target man remains motionless in a slightly exposed position. This allows him to observe the heads and chests of soldiers along the observation line.

PHASE TWO: The same target man slowly raises his head and shoulders until he can observe the soldiers on the observation line from the ground up.

FM 23-9

Figure B-8. Example of target trial card (locally fabricated).

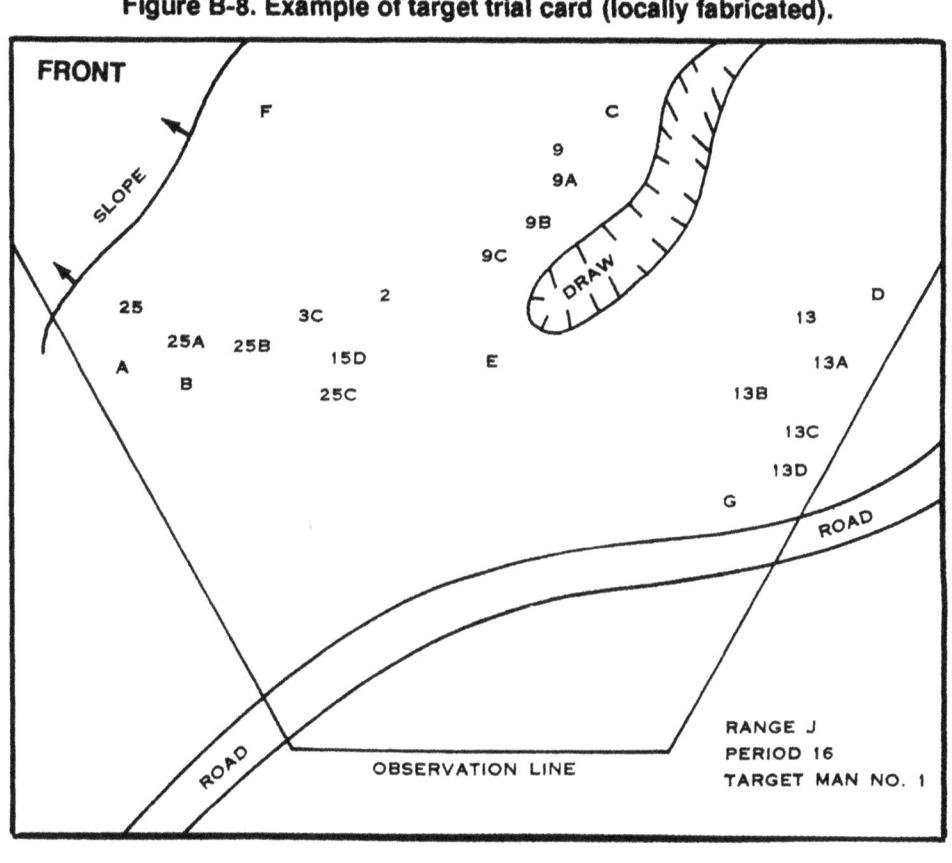

B-11

PHASE THREE: The same target man makes quick, jerky movements constantly for 30 seconds.

PHASE FOUR: The same target man fires one or two blank rounds toward the observation line (safety permitting). The command to begin a stationary target trial is TRIAL ONE, PHASE ONE, OBSERVE.

Table B-1. Factors affecting the appearance of objects.		
Factors in determining range by eye	Objects appear nearer than they are—	Objects appear more distant than they are—
The target—its clearness of out line and details.	When most of the target is visible and offers a clear outline.	When only a small part of the target may be seen or is small in relation to its surroundings.
Nature of the terrain or position of the observer.	When looking across a depression, most of which is hidden from view.	When looking across a depression, all of which is visible.
	When looking downward from high ground.	When looking from low ground toward high ground
	When looking down a straight, open road or along a railroad track.	When field of vision is narrowly confined as in twisted streets, draws, or forest trails.
Light and atmosphere.	When looking over uniform surfaces like water, snow, desert, or grain fields or when the sun is shining from behind the observer.	In poor light such as dawn and dusk in rain, snow, or fog, or when the sun is in the observer's eyes.
		When the target blends into the background or terrain.
	When the target is in sharp contrast with the background or is silhouetted by reason of size, shape, or color.	
	When seen in the clear atmosphere of high altitudes.in bright light	

If during the first phase the observer thinks he has located the target, he notes the letter of the panel nearest the target and determines the range from his position to the target. He enters this information on his answer sheet, and an assistant instructor checks his answer. A range error of not more than 10 percent is satisfactory. If the observer has chosen the wrong panel or the error in range exceeds 10 percent, he is told his answer is incorrect and to continue his observation.

If the answer is correct, the soldier continues his observation of the area, recording the required information on his answer sheet for the subsequent phases. This procedure is followed throughout the four phases of stationary trials.

NOTE: For more detailed information, see Exercise 1 and 2.

Moving Trials. The target trial cards for moving trials indicate the certain trials in which the target man engages, the stake location to which he moves, and the type of movement or other actions he performs. For example, the target trial card for target man No. 1 might indicate that he engages in trials 1, 5, 6, and 8. In trial 1, the instructions state that he perform four phases of a stationary target exercise. In trial 5, he is told to make five short rushes from stake 25 to stake 25C.

To check the accuracy of soldiers, aiming devices are used to mark the points of disappearance of multiple moving targets.

The observer aligns the two sight knobs on the aiming device where he thinks the targets are located. Normally, two soldiers are assigned to an aiming device: one to act as the observer and one to check the observer's work.

To begin a moving trial, the command is MOVING TARGET(S) STAND UP; DISAPPEAR AND BEGIN YOUR MOVEMENTS. On these commands, the applicable target men reveal themselves to the observers, move back into their concealed positions, and begin the movements as directed on their target trial cards. During some exercises, the target men may fire blank rounds after reaching a new location. Observers are allowed 30 seconds to mark the point(s) of disappearance with the aiming device. The instructor then commands, TARGETS STAND UP, ALTERNATE OBSERVERS CHECK ALIGNMENT. The alternate observer then checks the accuracy of the observer's work. This procedure continues until all of the trials have been conducted.

NOTE: For more detailed information, see Exercises 3, 4, and 6.

Stationary Sound Trials. Before the trials begin, the observers should draw a sector sketch of the area.

All of the numbered panels should then be raised for stationary sound trials. Each target man occupies a concealed position near one of the numbered panels. The instructor then informs the observer that a shot will be fired from one of the numbered panels. The observers must determine the panel location nearest the sound and record the information on their answer sheets. The commands to conduct the exercise are TRIAL NUMBER (ONE): READY, AIM, FIRE. OBSERVERS RECORD YOUR ANSWERS.

Should it be necessary to reposition target men for subsequent trials, the observers should face away from the range while the movement is taking place. In some trials, two target men should fire at the same time to demonstrate how hard it is to locate similar sounds coming from two directions at the same time.

NOTE: For more detailed information, see Exercise 5.

Multiple Moving and Sound Targets. To conduct multiple moving and sound target trials, eight target men are needed (two 4-man teams).

Soldiers are divided into two groups with each pair having one aiming device. The command to begin the exercise is MOVING TARGETS STAND UP; DISAPPEAR AND BEGIN YOUR MOVEMENT. The moving target men expose themselves, resume their concealed positions, and begin their rushes forward. After making their move, some of the target men should fire one or more blank rounds. The observer uses the aiming device to mark the point of disappearance of as many moving targets as possible.

Upon completing a trial, the instructor commands, TARGETS STAND UP, CHECK ALIGNMENT. The target men stand up and the alternate observer checks the accuracy of the observer's work. In the next trial, the alternate observer and observer change places.

NOTE: For more detailed information, see Exercise 7.

TESTS

In the final stage of target detection training, soldiers are tested on their ability to detect and determine ranges to single stationary targets. They mark the points of disappearance of single and multiple moving targets and locate targets by sound.

Test Number One—Stationary Targets. Test number one is conducted using the same four phases prescribed for the target detection trials of stationary targets. The soldier receives points in proportion to the number of phases needed to detect the target. If the soldier detects the target in phase one, he receives four points; in phase two, three points; and so on down to zero points if he fails to detect the target after four phases. To be considered correct, the soldier (again) selects the lettered panel nearest the target and determines the range from his position to the target. A range error of 10 percent or less is satisfactory. Master trial sheets, target cards, and range procedures are the same as prescribed for the practical exercises in detecting stationary targets. Each soldier is given 16 trials involving detection of stationary targets to provide enough informa- tion to judge his ability.

NOTE: For more detailed information, see Exercise 8.

Test Number Two—Moving Targets. Target detection test number two requires the soldier to mark the points of disappearance of multiple moving targets. These tests are conducted the same as the practical exercises for moving targets. After the target men have completed their movements, soldiers are allowed 30 seconds to mark the points of disappearance using the aiming device. Assistant instructors check the results and award one point for each correctly marked target location.

NOTE: For more detailed information, see Exercise 9.

Test Number Three — Sound Targets. Test number three involves sound targets. The test is conducted the same as practical exercises for locating sound targets. On command, one or two target men fire their rifles, and the observer tries to locate the sound using the numbered panels as reference points. One point is awarded for each correct answer.

NOTE: For more detailed information see Exercis 9. All reproducible forms are contained in Appendix H.

EXERCISES

The following exercises teach soldiers the skills and techniques of detecting, marking, and determining the range to realistic battlefield targets.

Exercise 1: Introduction to Target Detection (Two Hours).

Range facilities. Two target detection ranges.
Personnel.
- Two principal instructors (one for each range).
- Six target men (three for each range).

NOTE: One principal instructor is needed at each range. He sets up the range, trains target men, and conducts the class. Four assistant instructors are needed for each range. They control the observers, assist in scoring, and are familiar with the position of the targets. The six target men (three for each range) must be trained to perform the duties of "targets." Each is assigned a number of target placements within a certain area. All target men are given a target trial card containing only the trial numbers and the indications he is to perform.

Blank ammunition requirements.
- For each presentation:

    First hour: Five rounds for demonstration.

    Second hour: Fifteen rounds for practice exercise.
- For each rehearsal:

    First hour: Five rounds.

    Second hour: Fifteen rounds.

Master trial sheet: Exercises 1, 2 and 8. (Table B-2).

Table B-2. Exercises 1, 2, and 8.

TRIAL NO.	TARGET MEN	DESCRIPTION OF REQUIREMENTS	LOCATION	RANGE (METERS)	STAKE
1	1	1. Slightly exposed, motionless. 2. Raise and lower head, slowly. 3. Repeat 2 (above), rapidly. 4. Fire one blank round.	A-B	22	41A
2	2	1. Slightly exposed, motionless. 2. Slowly move head from side to side. 3. Raise head slowly; drop quickly. 4. Fire one blank round.	B-D	66	31B

FM 23-9

Table B-2. Exercises 1, 2, and 8 (continued).

TRIAL NO.	TARGET MEN	DESCRIPTION OF REQUIREMENTS	LOCATION	RANGE (METERS)	STAKE
3	3	1. Slightly exposed, motionless. 2. Move forward and back every ten seconds. 3. Step out and back rapidly. 4. Fire two blank rounds.	E-F	161	11C
		Change Target Locations			
4	1	1. Slightly exposed, motionless. 2. Shake bush every five seconds. 3. Raise and lower head slowly while wearing shiny helmet liner. 4. Fire two blank rounds.	G	119	13B
5	2	1. Kneeling partly exposed, motionless. 2. Slowly move head and shoulders from side to side. 3. Jump out and back every five seconds. 4. Fire one blank round.	C	44	32A
6	3	1. Slightly exposed, motionless. 2. Slowly assume kneeling position. 3. Raise up slowly; drop quickly. 4. Fire two blank rounds.	D	95	19B
		Change Target Locations			
7	1	1. Slightly exposed, kneeling. 2. Raise head slowly; drop quickly. 3. Repeat 2 (above), rapidly. 4. Fire two blank rounds.	B	91	29B
8	2	1. Slightly exposed, motionless. 2. Slowly move up and down. 3. Make rapid, jerky movements. 4. Fire one blank round.	D	51	27
9	3	1. Slightly exposed, motionless. 2. Slowly move head and shoulders from side to side. 3. Same as 2 (above) with shiny helmet liner. 4. Fire one blank round.	A-D	41	29A
		Change Target Locations			
10	1	1. Slightly exposed. 2. Move slowly. 3. Move quickly. 4. Fire two blank rounds.	D-E	88	28B

FM 23-9

NOTE: An individual target trial card (Figure B-8) is prepared for each target man. It contains only those trials in which he engages, the location (stake number) used, and the action performed in each trial. Although each target man has rehearsed, the target trial cards ensure that no mistakes are made.

Answer sheet. See DA Form 3009-R (Target Detection Exercise Answer Sheet — Periods 1, 2, and 8) (Figure B-9).

Figure B-9. Example of completed DA Form 3009-R (Exercises 1, 2, and 8).

TARGET DETECTION EXERCISE ANSWER SHEET PERIODS 1, 2, AND 8

For use of this form, see FM 23-8 and FM 23-9; proponent agency is TRADOC.

NAME	(LAST) JORDAN	(FIRST) FRANK		PLATOON 3D	SQUAD 3D	DATE 4 APR 89
TRIAL NO.	\multicolumn{4}{c	}{PHASE NUMBER}	WHERE (LETTER OF NEAREST LANDMARK)	RANGE (METERS)

TRIAL NO.	1	2	3	4	WHERE	RANGE (METERS)
1	X	✓			C	50
2	X	X	✓		D	75
3	X				B	50
4	X	X	X	✓	F	278
5	X	✓			A	83
6	X	X	✓		E	115
7	X				C	196
8	X	✓			B	58
9	X	X	X	X		
10	X	X	✓		F	280
11	X	✓			D	89
12	X	✓			B	63
13	X	✓			A	90
14	X	X	✓		B	70
15	✓				E	120
16	✓				C	230
TOTAL	16	18	18	1		

DA FORM 3009-R, 1 Nov 73   REPLACES DA FORM 3009-R, 1 JUN 65, WHICH IS OBSOLETE.

B-17

FM 23-9

Exercise 2: Detection of Realistic Battlefield Targets (Two Hours). If possible, this exercise is conducted the same as Exercise 1, but on a different range. Range facilities, personnel, organization, ammunition requirements, master trial sheet, and answer sheet are the same as outlined for Exercise 1.

Exercise 3: Detection of Single Moving Targets (Two Hours). This exercise gives the soldier practice in detecting and simulating the engagement of single, combat-- type moving targets.

Range facilities. One target detection range.
Personnel.
- One principal instructor.
- Three target men.

Blank ammunition requirements.
- Rounds for each presentation – 10.
- Rounds for each rehearsal – 10.

Master trial sheet: Exercise 3. (Table B-3).

Table B-3. Exercise 3.

TRIAL NO.	TARGET MEN	DESCRIPTION OF REQUIREMENTS	RANGE (METERS)	STAKE
1	1	1. Standing by tree. 2. Kneel slowly. 3. Move slowly from view.	200	6A
2	2	4. Fire two blank rounds. Same as trial 1 (above). Poor aiming point location.	150	28B
3	3	5. Start standing. Disappear; on command, reappear in same position. Make four 4-second rushes to good aiming points.	175	27C-28C-29C-30C-32C
		Change Target Locations		
4	1	1. Start from kneeling position behind bush. Make five rushes. Disappear at poor aiming points. Reappear from same position; 5-3-3-5-5 seconds. Fire one blank round from last position.	300	1A-2A-3A-4A-5A-6A
5	3	2. Start prone. Make five rushes. Disappear after each rush and roll or crawl to a new location before reappearing; 5-3-2-8-8 seconds. Fire one blank round from last position.	300	1C-2C-3C-4C-5C-6C
6	2	3. Start prone. Make five lateral rushes. Reappear at same location; 4-5-4-4-3 seconds. Fire one blank round from last position.	175	28B-29B-30B-31B-32B-33B

Table B-3. Exercise 3 (continued)

TRIAL NO.	TARGET MEN	DESCRIPTION OF REQUIREMENTS	RANGE (METERS)	STAKE
		Change Target Location		
7	2	1. Run 6 meters in reverse	175	33B-32B-31B-30B-29B-28B
8	3	2. Run 100 meters from tree to position with poor aiming point.	200	25C-39C
9	1	3. Start prone; 5-L-3-3-6-R-4-5 seconds; through draw. Fire one blank round from last position. (Numbers indicate duration of rush; letters indicate direction of roll or crawl after each rush.)	260	6A-7A-8A-9A 10A-11A
10	2	4. Start behind bush; 6-8-R-3-R-4-3 seconds. Fire one blank round from last position.	300	1B-2B-3B-4B-5B-6B

Answer sheet: See DA Form 3010-R (Target Detection Exercise Answer Sheet — Period 3) (Figure B-10).

Figure B-10. Example of completed DA Form 3010-R (Exercise 3).

TARGET DETECTION EXERCISE
ANSWER SHEET
PERIOD 3
For use of this form, see FM 23-8 and FM 23-9, proponent agency is TRADOC.

NAME	PLATOON	SQUAD	DATE
THIELE, STEVE	3D	3D	30 JULY 1989

TRIAL NO	WHERE (LETTER OF NEAREST LANDMARK)	RANGE (METERS)
1	✓ G	✓ 86
2	X A	X 90
3	✓ E	✓ 110
4	✓ B	✓ 225
5	✓ C	✓ 50
6	X F	✓ 120
7	✓ G	✓ 86
8	✓ D	✓ 105
9	✓ H	✓ 75
10	✓ A	✓ 125

DA FORM 3010-R, 1 Nov 73   REPLACES DA FORM 3010-R, 1 JUN 65, WHICH IS OBSOLETE

Exercise 4: Detection of Multiple Moving Targets (Two Hours). This exercise gives the soldier practice in detecting and aiming at multiple, combat-type moving targets.

Range facilities. Two target detection ranges.

Personnel.

- Two principal instructors (one for each range).
- Eight assistant instructors (four for each range).
- Sixteen target men (eight for each range).

Organization. One order of soldiers is assigned to each range.

Blank ammunition requirements.

- Rounds for each presentation—47.
- Rounds for each rehearsal—47.

Master trial sheet: Exercise 4. Soldiers use target aiming device to mark the points of disappearance of moving targets (Table B-4).

Table B-4. Exercise 4.

TRIAL NO.	TARGET MEN	DESCRIPTION OF REQUIREMENTS	RANGE (METERS)	STAKE
1	1	1. Kneeling exposed. Crawl to new position in five 5-meter crawling movements. Fire one blank round from each new position. (Five rounds for each target man.) Good aiming point positions.	75	42A-43A-44A-45A-46A
	2			40B-41B-42B-43B-44B
2	3	2. Same as above. Poor aiming point positions, but reference points available. Reference points increase in difficulty each time.	100	39A-40A-41A-42A-43A
	4			37B-38B-39B-40B-41B
3	5	3. Start with target men walking through woods or other partial concealment. Disappear when fired on. Make five 4-second rushes to positions with good aiming points. Fire one blank round from last position.	300	9A-10A-11A-12A-13A-14A
	6			11B-12B-13B-14B-15B-16B- 7C-8C-9C-10C-11C-12C- 9C-10C-11C-12C-13C-14C
4	7	4. Start from kneeling position behind bush. Make five rushes. Disappear where there is a poor aiming point.	200	12C-13C-14C-15C-16C-17C 14C-15C-16C-17C-18C-19C
	8	Reference points available but not easy; 4-2-4-4 seconds. Fire one blank round from last position.		

Table B-4. Exercise 4 (continued).

TRIAL NO.	TARGET MEN	DESCRIPTION OF REQUIREMENTS	RANGE (METERS)	STAKE
		Change Target Locations		
5	4	1. Start prone. Make five rushes, good and poor aiming point positions; 2-4-6-2 seconds. Fire one blank round from last position.	300	8A-9A-10A-11A-12A-13A
	5			11A-12A-13A-15A-16A
	6			10B-11B-12B-13B-14B-15B
6	1	2. Start at tree. Make five rushes to new positions affording good and poor aiming points; 2-3-4-2 seconds. Fire one blank round from last position.	175	14A-15A-16A-17A-18A-19A
				18B-19B-20B-21B-22B-23B
				17C-18C-19C-20C-21C-22C
				19C-20C-21C-22C-23C-24C
7	3	3. Start at different distances. Make five approach rushes; 4-2-4-3-4 seconds. Varied good and poor aiming point positions. Fire one blank from last position.	175	17A-18A-19A-20A-21A-22A
				13A-14A-15A-16A-17A-18A
				16A-17A-18A-19A 20A-21A
		Change Target Locations		
8	1	1. Make five 5-meter crawling movements to positions with good aiming points. Reference points increase in difficulty each time. Fire one blank round from last position.	200	11A-12A-13A-14A-15A-16A
				12B-13B-14B-15B-16B
				12C-13C-14C-15C-16C 9A-10A-11A-12A-13A-14A
9	3	2. Start walking in woods. Make five rushes; 4-2-2-6-4 seconds. Fire one blank round from last position.	300	10B-11B-12B-13B-14B-15B
	4			10C-11C-12C-13C-14C-15C
	5			
10	3	3. Make five rushes; 4-3-6-2-3 seconds. Fire one blank round from last position.	150	20A-21A-22A-23A-24A-25A
	4			21B-22B-23B-24B-25B-26B
	5			18B-19B-20B-21B-22B23B
	6			19C-20C-21C-22C-23C-24C

Exercise 5: Locating Target by Sound (Two Hours). This exercise gives the soldier practice in locating targets by the sound of firing from hostile firing position.
Range facilities. One target detection range.
Personnel.
- One principal instructor.
- One assistant instructor for every 10 soldiers.
- Five target men.

Organization. One order of soldiers on the range at a time.
Blank ammunition requirements.
- Rounds for each presentation — 46.
- Rounds for each rehearsal — 46.

Master trial sheet: Exercise 5 (Table B-5).

Table B-5. Exercise 5.

TRIAL NO.	TARGET MEN	PANEL LOCATION
1	1	14
2	2	11
	3	4
3	4	7
	5	9
4	1	14
	2	11
5	3	4
6	4	7
7	5	9
	2	11
Change Target Locations		
8	1	2
	3	8
9	4	12
10	2	13
	5	6
11	3	8
	4	12
12	1	2
	2	13
13	4	12
	5	6
Change Target Locations		
15	4	1
	5	3
16	1	10
17	3	5
	2	9
18	4	1
	1	10
19	3	5

Table B-5. Exercise 5 (continued).

TRIAL NO.	TARGET MEN	PANEL LOCATION
20	5	3
	2	9
21	4	1
	3	5

Change Target Locations

TRIAL NO.	TARGET MEN	PANEL LOCATION
22	1	3
23	2	9
24	3	14
	4	10
25	5	8
	1	3
26	2	9
	3	14
27	4	10
	5	8
28	4	8

Answer sheet. See DA Form 3011-R (Target Detection Answer Sheet—Period 5) (Figure B-11).

Figure B-11. Example of completed DA Form 3011-R (Exercise 5).

TARGET DECTION EXERCISE
ANSWER SHEET
PERIOD 5

For use of this form, see FM 23-8 and FM 23-9, the proponent agency is TRADOC.

OBSERVER'S NAME	(LAST)	(FIRST)	PLATOON
	CAPRARO	LEE	3D

OBSERVATION POINT	DATE
23	30 JULY 1989

TRIAL NO	SOUND POSITION	TRIAL NO	SOUND POSITION
1	4-3	15	6-13
2	6 X	16	14-4
3	8	17	8
4	12-14	18	13
5	1-3 X	19	6-5
6	7	20	7-1 X
7	4	21	3
8	2-5	22	7
9	9	23	4
10	10-12	24	7-12
11	4	25	10-1
12	8 X	26	8-4 X
13	10	27	3
14	1	28	6-4 X

TOTAL ---------- RIGHT 35   WRONG 6

DA FORM 3011-R, 1 Nov 73   REPLACES DA FORM 3011-R, 1 JUN 66, WHICH IS OBSOLETE.

Exercise 6: Detection of Movement by Opposing Teams, Personal Camouflage. This exercise gives soldiers practical work in target detection and movement as target teams. Demonstrations and practical work in personal camouflage are conducted.

Range facilities. Two target detection ranges.

Personnel.

- Two principal instructors (one for each range).
- Four assistant instructors (two for each range).
- Four demonstrators (two for each range).

Organization. One order of observers is assigned to each range.

Blank ammunition requirements. None.

Master trial sheet, Exercise 6. (Table B-6.)

Table B-6. Exercise 6.

Trial NO.	(1) 6-sec rush	(2) 2-sec rush	(3) 5-m low crawl	(4) 4-sec rush	(5) 5-m low crawl	(6) 4-sec rush	(7) 10-m high crawl	(8) 2-sec rush	(9) 6-sec rush	(10) 50-m bound
1		x			x	x			x	x
2	x		x			x		x		x
3		x				x	x		x	x
4	x		x	x				x		x
5				x	x			x	x	x
6	x		x			x		x		x
7	x	x		x			x			x
8				x			x	x	x	x
9		x			x	x		x		x
10	x			x	x			x		x
11		x				x	x	x		x
12		x		x	x				x	x
13	x		x			x		x		x
14		x		x	x				x	x

This master trial sheet reflects 10 trials for 14 soldiers acting as targets. Units may revise the master trial sheet to include additional target requirements to ensure maximum participation when larger squads are used.

Exercise 7: Combination of Sound and Multiple Moving Targets (Two Hours). This exercise gives soldiers practice in locating, marking, aiming, and firing at moving combat-type targets.

Range facilities. Two target detection ranges.

Personnel.

- Two principal instructors (one for each range).
- Ten assistant instructors (five for each range).
- Sixteen target men (eight for each range).

Organization. One order of soldiers assigned to each range.
Blank ammunition requirements.

- Rounds for each presentation — 75.
- Rounds for each rehearsal — 75.

Master trial sheet: Exercise 7. (Table B-7.)

\<td colspan="5"\>				
TRIAL NO.	TARGET MEN	DESCRIPTION OF REQUIREMENTS	RANGE (METERS)	STAKE
1	1 2 3 4	**1.** Two targets make clumsy 5-meter crawls; two target men fire four flank rounds each toward the observation line. All located at good aiming point positions.	150	28A-31A 28B-31B P14 P3
2	5 6 7 8	**2.** Three target men make skilled 5-meter crawls; one target man fires two blank rounds. All positions lack good aiming points. Distance between targets is 25 meters.	200	22A-23A 20B-22B 21C-23C P4
		Change Target Locations		
3	1 2 3 4	**1.** Two target men are walking through woods, disappear on command, and make one 4-second rush. Two target men fire two blank rounds each.	300	1A-2A 2B-3B 1C-2C P12
4	5 6 7 8	**2.** Two target men make one 3-second rush and disappear at poor aiming point positions. Two target men fire three blank rounds each.	250	14B-15B 13C-14C P13 P10
		Change Target Locations		
5	1 2 3 4	**1.** One target man makes a 5-meter crawl and stops at a good aiming point position. One target man makes a 5-meter rush and stops at a poor aiming point position. One target makes a 10-meter rush. One target man fires five blanks.	100	31A-32A 31B-32B 29C-31C P9
6	5 6 7 8	**2.** Three target men make a 5-meter clumsy crawl. One target man man fires two blank rounds. All positions at poor aiming points. Good reference points available.	150	27B-28B 26C-27C 25A-26A P2

Table B-7. Exercise 7.

Table B-7. Exercise 7 (continued).

TRIAL NO.	TARGET MEN	DESCRIPTION OF REQUIREMENTS	RANGE (METERS)	STAKE
		Change Target Locations		
7	1	1. Two target men make a skilled 5-meter crawl. Two target men fire two blank rounds each. All positions have good aiming points.	225	P11
	2			P1
	3			13C-14C
	4			14B-15B
8	5	2. Three target men walk through woods until fired on from observation line. They disappear and make a 5-meter crawl. One target man fires four blank rounds.	300	P6
	6			2C-3C
	7			2A-3A
	8			2B-3B
		Change Target Locations		
9	1	1. Same as above except all target men stop at poor aiming point positions.	150	24A-25A
	2			25B-24B
	3			P6-26C-27C
10	4	2. Two target men make a five-second rush. Two target men fire one blank round each.	225	15C-17C
	5			P8
	6			19B-20B
		Change Target Locations		
11	1	1. Four target men fire one blank round each.	250	P11
	2			P13
	3			P9
	4			P12
12	5	2. One target man makes a 10-meter rush to a poor aiming point position. Three target men fire two blank rounds each.	100	P6
	6			P10
	7			P4
	8			39B-42B
		Change Target Locations		
13	1	1. Two target men fire one blank round each. Two target men make a 5-meter crawl. Varied good and poor aiming point positions.	200	11A-12A
	2			9B-10B
	3			P2
	4			P5
14	5	2. Same as above except all positions are at poor aiming points and require the use of reference points.	75	P3
	6			P1
	7			44B-45B
	8			44C-45C
		Change Target Locations		
15	1	1. Two target men, spaced far apart, make a three-second rush. Two target men, close together, fire two blank rounds each.	225	8A-9A
	2			7C-8C
	3			P14
	4			P11

Table B-7. Exercise 7 (continued).

TRIAL NO.	TARGET MEN	DESCRIPTION OF REQUIREMENTS	RANGE (METERS)	STAKE
16	5	**2.** Four target men make one-, two-, three-, and four second rushes after being fired on from the observation line. Each target man fires one blank round two seconds after disappearing. Varied, good and poor aiming point positions.	275	15A-16A
	6			14B-13B
	7			16B-15B
	8			14C-15C
17	1	**3.** Four target men alternately fire one round each; varied good and poor aiming point positions.	125	36A
	2			35A
	3			31B
	4			32C
18	5	**4.** Two target men make a one-second rush; two target men fire one blank round each. Varied, good and poor aiming points.	300	P9
	6			P13
	7			1A-2A
	8			2C-3C
Change Target Locations				
19	1	**1.** Three target men make a three-second rush, and one target man fires one blank round.	125	35A-34A
	2			36A-37A
	3			31B-32B
	4			P9
20	5	**2.** One target man makes a 5-meter crawl and three target men fire one round each. Varied, good and poor aiming points.	175	23B-24B
	6			P14
	7			P11
	8			P1

Answer sheet. See DA Form 5791-R (Target Detection Exercise Answer Sheet—Period 7) (Figure B-12).

FM 23-9

Figure B-12. Example of completed DA Form 5791-R (Exercise 7).

TARGET DETECTION EXERCISE
ANSWER SHEET
PERIOD 7
For use of this form, see FM 23-8 and FM 23-9; proponent agency is TRADOC.

COMBINATION OF SOUND LOCALIZATION AND MULTIPLE MOVING TARGETS (OBSERVERS CHECK EACH OTHER'S ALINEMENT AND PLACE NUMBER OF TARGETS CORRECTLY ALINED IN SPACE OPPOSITE APPROPRIATE TRIAL NUMBER.)

NAME	PLATOON	SQUAD	DATE
WILSON, PETE	3D	3D	2 FEB 89

TRIAL NO.	NO. CORRECT	TRIAL NO.	NO. CORRECT
1	1	12	4
2	3	13	3
3	2	14	2
4	1	15	2
5	3	16	3
6	1	17	1
7	0	18	0
8	2	19	3
9	2	20	1
10	3	TOTAL CORRECT	38
11	1		

TRIAL NO.	NO. CORRECT	TRIAL NO.	NO. CORRECT
1	0	12	2
2	2	13	0
3	2	14	1
4	1	15	1
5	3	16	2
6	3	17	1
7	1	18	3
8	4	19	2
9	2	20	1
10	2	TOTAL CORRECT	34
11	1		

DA FORM 5791-R, JUN 89

Exercise 8: Target Detection Test One (One Hour). This exercise tests the soldier's ability to locate and deter- mine ranges to single, stationary battlefield targets.

Range facilities. One target detection range.
Personnel.

- One principal instructor.
- Seven assistant instructors.
- Three target men.

Blank ammunition requirements.

- Rounds for each presentation – 15.
- Rounds for each rehearsal – 15.

Master trial sheet. Same as for Exercise 1, except locations of target men should be changed.

Answer sheet. Same as for Exercise 1 (Figure B-9).

Exercise 9: Target Detection Tests Two and Three (One Hour). This exercise tests the soldier's ability to locate and mark the points of disappearance of single and multiple moving targets (test two) and his ability to locate sound targets (test three).

Range facilities. One target detection range.
Personnel.

- One principal instructor.
- Four assistant instructors.
- Four target men.

Blank ammunition requirements.

- Rounds for each presentation – 30.
- Rounds for each rehearsal – 30.

Master trial sheet, target detection, test two. (See Table B-8.)
Master trial sheet, sound detection, test three. (See Table B-9.)

		Table B-8. Test two, Exercise 9.		
TRIAL NO.	TARGET MEN	DESCRIPTION OF REQUIREMENTS	RANGE (METERS)	STAKE
1	1	1. Kneeling by tree, up on command. Ten-meter bound to a poor aiming point.	300	1A-2A
2	2	2. Rush 15 meters to a poor aiming point.	75	40A-41A
	3			39B-40B
				40C-41C
3	1	1. Ten-meter rush to poor aiming points.	200	11A-12A
	2			13A-14A
	3			11B-12B
	4			11C-12C
4	1	2. Ten-meter lateral rush. Good aiming points.	150	12A-12B
	2			14A-14B
	3			12B-12C
	4			12C-11B

FM 23-9

Table B-8. Test two, Exercise 9.

TRIAL NO.	TARGET MEN	DESCRIPTION OF REQUIREMENTS	RANGE (METERS)	STAKE
		Change Target Locations		
5	1	1. One man rush right, the other	200	13A-12B
	2	left; both stop at poor aiming points.		12C-11B
6	3	2. Ten-meter rush. One to a good	75	35C-36C
	4	aiming point and the other	100	29B-30B
	1	to poor aiming points.	125	27A-28A
7	2	1. Five-meter rush to poor aiming	75	34B-35C
	3	points.		
	4			35A-35B
8	1	2. Twenty-meter rush to a poor aiming point.	200	16A-17A
9	2	3. One rush 5 meters to a good	300	1B-2B
	3	aiming point and the other 10 meters to a poor aiming point.		1C-3C
10	4	4. Five-meter lateral rush to	150	25A-26A
	1	poor aiming points.		24B-25B

Table B-9. Test three, Exercise 9.

TRIAL NO.	TARGET MEN	PANEL LOCATION
1	1	6
2	2, 3	4, 12
3	4, 1	7, 9
4	2	8
5	3	14
6	4, 1	14, 3
7	2	1
8	3	6
9	4	7
10	1, 2	13, 8
11	3, 4	10, 1
12	1, 2	2, 6
13	3, 4	7, 3
14	1, 2	11, 6
15	3	7
16	4	3
17	1, 2	6, 14
18	3, 4	12, 1
19	1	8
20	2	3

Answer sheet. See DA Form 3014-R (Target Detection Exercise Answer Sheets Tests No. 2 and 3--Period 9) (see Figure B-15).

Figure B-15. Example of completed DA Form 3014-R (Exercise 9).

TARGET DETECTION EXERCISE
ANSWER SHEETS TESTS NO. 2 AND 3
PERIOD 9
For use of this form, see FM 23-8 and FM 23-9; the proponent agency is TRADOC.

OBSERVER'S NAME	(LAST) HUBBARD,	(FIRST) JOHN	PLATOON
OBSERVATION POINT	23		DATE 3 MAR 89

TRIAL NUMBER	NO. OF TARGETS PRESENTED	RIGHT	WRONG
1	1	1	0
2	3	2	1
3	4	1	3
4	4	3	1
5	2	2	0
6	3	2	1
7	3	2	1
8	1	1	0
9	2	1	1
10	2	2	0
TOTAL 25	RIGHT 17		WRONG 8

DA FORM 3014-R, JUN 89

FM 23-9

APPENDIX C

Training Aids and Devices

Training aids and devices must be included in a marksmanship program. This appendix lists those that are available and provides information on how to obtain them for marksmanship training.

Section I. TRAINING RESOURCES

This section provides the classification and nomenclature for training aids, devices, and targets.

CLASSIFICATIONS

Information on the classification of various training resources are listed below along with a general description and source publication.

TYPE	DISCRIPTION	SOURCE PUBLICATION
Graphic Training Aids	Charts, handout cards, diagrams, posters, overhead transparencies, 35-mm slides, and small plastic aids	DA Pam 108-1 Index of Army Motion Picture and related autovisual aids.
Devices	Three-dimensional training aids such as scale models and simulators	Da Pam 310-12 Index Description of Army Training Devices.
Training Extension Course Leassons	Audiovisual 8-mm film cartridge audiocassette and student instruction sheet.	Extension training material status list published by the US Army Training Support Center, Fort Eustis, Va 23604, Catalog of Training Extension Course Lessons.

TRAINING AND AUDIOVISUAL SUPPORT CENTER

TASCs are located throughout the world and are the POCs for obtaining all training aids and devices. Each TASC provides training aid services to customers in their geographic area of support to include Active Army units and schools, Reserve Components, and ROTC units.

NOTE: For more information concerning TASC operations, write Commander, United States Army Training Support Center, ATTN: ATIC-DM, Fort Eustis, VA 23604.

TRAINING DEVICES AND EXERCISES

Several marksmanship training devices are available to aid in sustainment training when used with the appropriate training strategies. They are beneficial when ammunition is limited for training or practice exercises. Some training devices are complex, costly, and in limited supply, while others are relatively simple, cheap, and in large supply. Devices and aids can be used alone or in combinations. Individuals or squads can sustain/practice basic marksmanship skills and fundamentals with devices/aids.

Aiming Card. The M15A1 card (Figure C-1) determines if the soldier understands how to aim at target center of mass. The card is misaligned, and the soldier is instructed to establish the correct point of aim. It is checked by a trainer. Several aimings provide an understanding of center of mass. Also, this card is used to ensure the soldier understands adjustment of the aiming point, how to allow for gravity, and how to engage a moving target. The sight-target relationship on the card is the same visual perception the soldier should have when he is zeroing on a standard silhouette target.

Figure C-1. The M15A1 aiming card (NSN 6910-00-716-0930).

Riddle Sighting Device. The Riddle sighting device (Figure C-2) indicates to the trainer if the soldier understands the aiming process while using the rifle. It is a small plastic plate with a magnet and a drawing of an E-type silhouette target. A two-man team is needed. The soldier assumes a supported or prone firing position. The assistant places the Riddle device on the front sight assembly and adjusts the plastic plate at the direction of the firer until he reports the proper sight picture. Without disturbing the plastic plate, the trainer or coach must aim through the sights to determine if the soldier has aligned the target and sight properly. Many sightings are conducted, and the trainer may include variations to ensure the soldier understands the process.

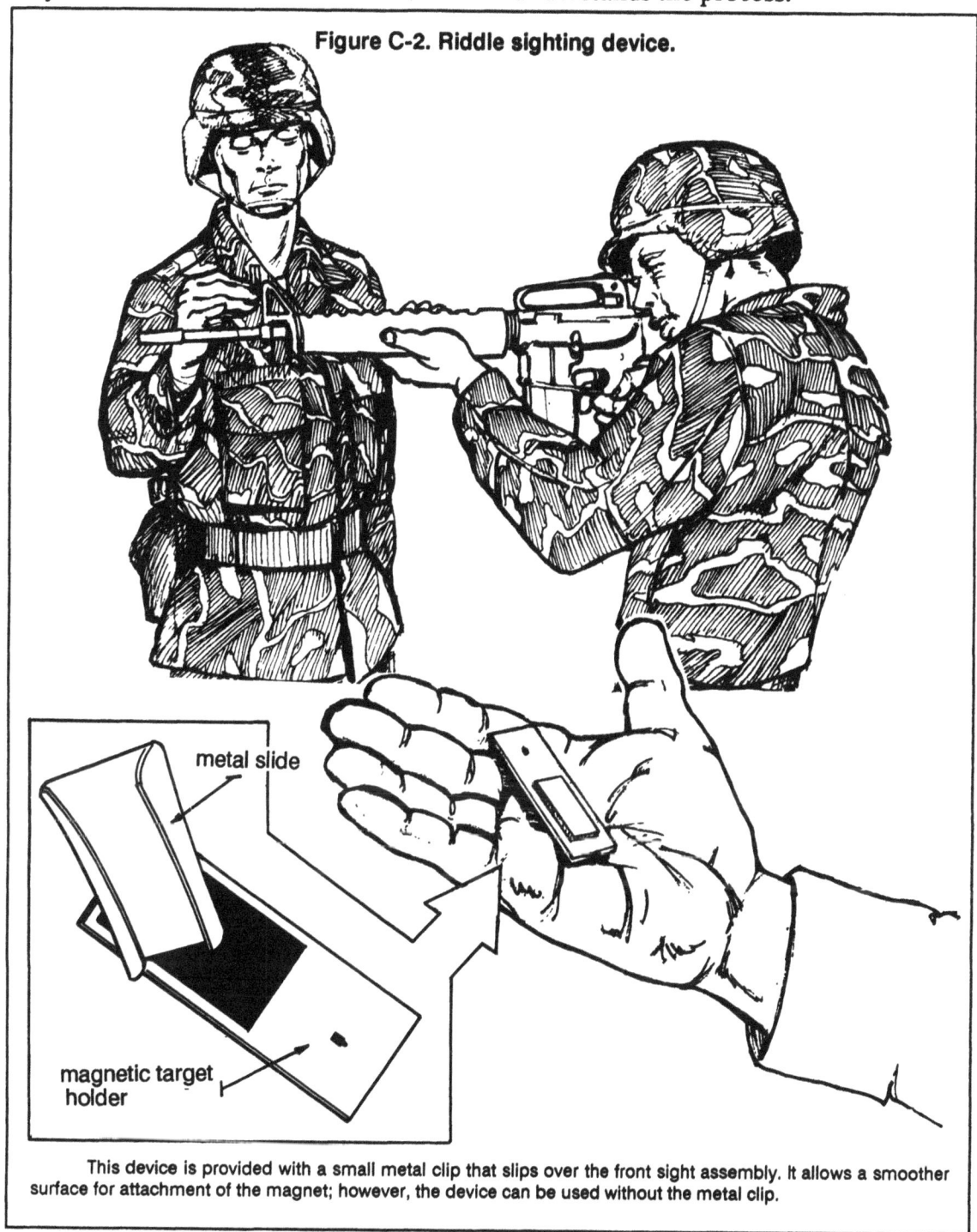

Figure C-2. Riddle sighting device.

This device is provided with a small metal clip that slips over the front sight assembly. It allows a smoother surface for attachment of the magnet; however, the device can be used without the metal clip.

M16 Sighting Device. The M16 sighting device (Figure C-3) is made of metal with a tinted square of glass placed at an angle.

When the device is attached to the rear of the M16A1 carrying handle, an observer can look through the sight to see what the firer sees. The M16 sighting device can be mounted on the M16A2 rifle. The charging handle must be pulled to the rear first. Then, the M16 sighting device is mounted on the rear of the carrying handle, and the charging handle is returned.

The M16 sighting device can be used in a dry-fire or live-fire environment, but a brass cartridge deflector must be used during live fire. The observer must practice with the sight for it to be effective. For example, the observer looks at a reflected image; if the soldier is aiming to the right, it appears left to the observer. Also, the device must be precisely positioned on the rifle (it may need to be bent to stay on). The observer's position must remain constant. At the same time, the observer talks with the firer to ensure a correct analysis of the aiming procedures.

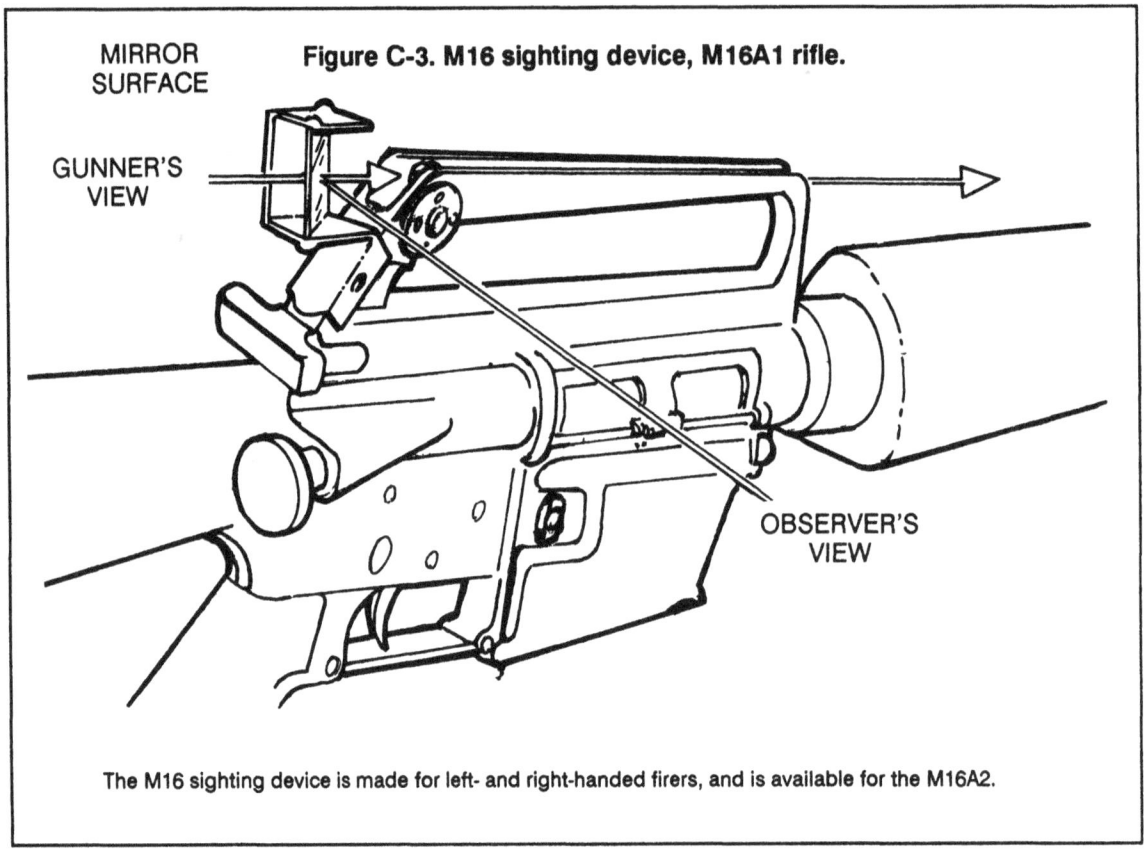

Figure C-3. M16 sighting device, M16A1 rifle.

The M16 sighting device is made for left- and right-handed firers, and is available for the M16A2.

Blank Firing Attachment (BFA), M15A2. This device (Figure C-4) is attached to the muzzle of the M16A1 or M16A2 rifle. It is designed to keep sufficient gas in the barrel of the weapon to allow semiautomatic, automatic, or burst firing with blank ammunition (M200). After firing 50 rounds, the attachment should be checked for a tight fit. Continuous blank firing results in a carbon buildup in the bore, gas tube, and carrier key. When this occurs, the cleaning procedures in TM 9-1005-249-10 for TM 9-1005-249-34 should be followed.

Target-Box Exercise. The target-box exercise checks the consistency of aiming and placement of three-round shot groups in a dry-fire environment (Figure C-5).

To conduct the exercise, the target man places the silhouette anywhere on the plain sheet of paper and moves the silhouette target as directed by the firer. The two positions must have already been established so that the rifle is pointed at some place on the paper. The positions are separated by 15 yards or 25 meters. When the firer establishes proper aiming, he signals the target man to "Mark." Only hand signals are

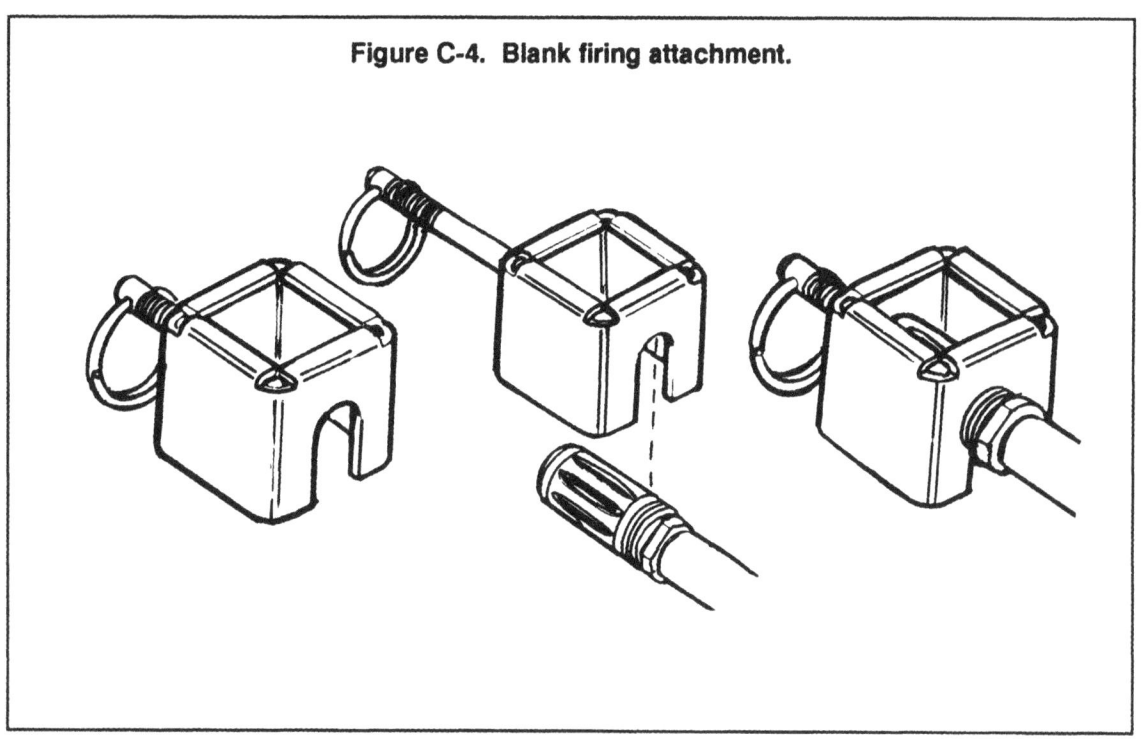

Figure C-4. Blank firing attachment.

Figure C-5. Target-box exercise.

used since voice commands would be impractical when training several pairs of soldiers at one time.

The target man then places the pencil through the hole in the silhouette target and makes a dot on the paper. Then he moves the silhouette to another spot on the paper and indicates to the firer that he is ready for another shot. When the three shots are completed, the target man triangulates the three shots and labels it shot group number one. The firer and instructor view the shot group.

A simulated shot group covered within a 1-cm (diameter) circle indicates consistent aiming. Since no rifle or ammunition variability is involved and since there is no requirement to place the shot group in a certain location, a 1-cm standard may be compared to obtaining a 4-cm shot group on the 25-meter live-fire zero range. The soldier fires several shot groups. After two or three shot groups are completed in one location, the rifle, paper holder, or paper is moved so shots fall on a clean section of the paper.

Any movement of the rifle or paper between the first and third shots of a group voids the exercise. Two devices are available to hold the rifle (Figures C-6 and Figure C-7). The rifle holding device and rifle holding box are positioned on level ground, or are secured by sandbags or stakes to ensure there is no rifle movement during the firing of the three shots. Movement of the paper is eased by using a solid backing (Figure C-8). Any movement of either is reflected in the size of the shot group. Several varieties of wooden target boxes have been locally fabricated. A new rifle holder has been developed and should be used (Figure C-7).

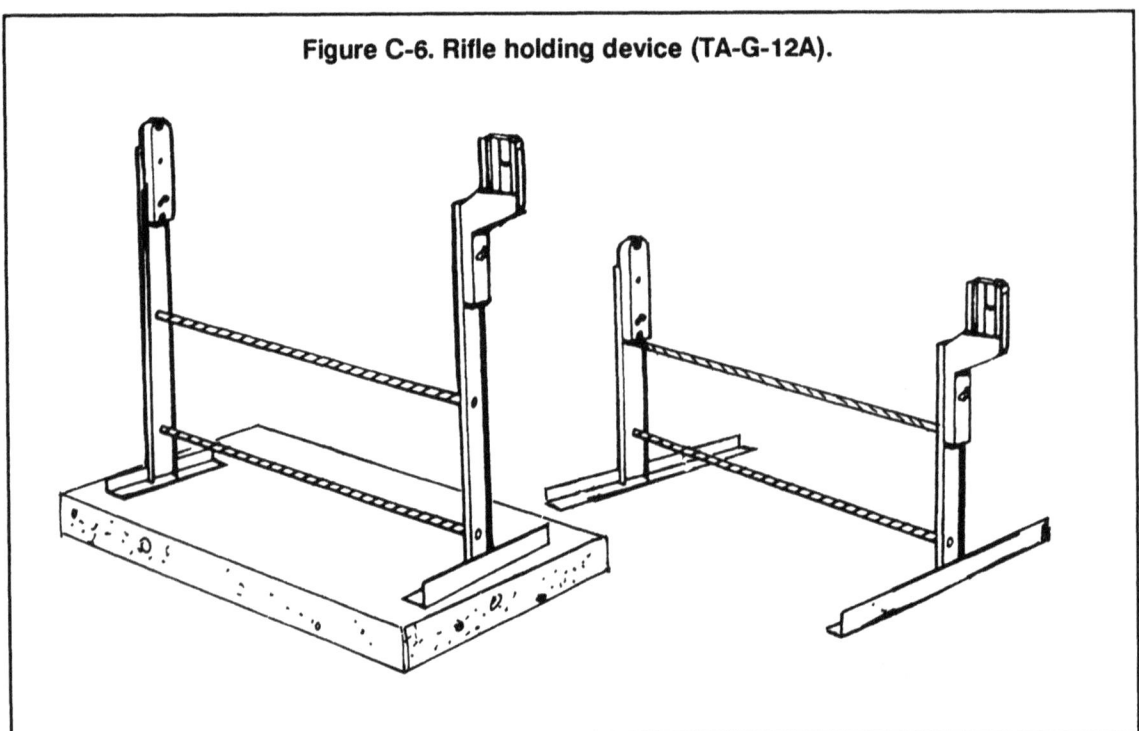

Figure C-6. Rifle holding device (TA-G-12A).

The silhouettes on the plastic paddle (Figure C-9) are scaled to represent an E-type silhouette target at 250 meters. The visual perception during the target-box exercise is similar to what a soldier sees while zeroing on a standard zeroing target. The small E-type silhouette is the same scale at 15 yards as the larger silhouette is at the 25-meter range (some training areas are set up at 15 yards; others are set up at 25 meters). While there are some benefits to representing a 250-meter target, the main benefit of this exercise can be obtained at any distance. A standard zero target can be used at 25 meters in place of the paddle by placing a small hole in the center (dot), moving the target sheet over the paper, and marking as previously outlined.

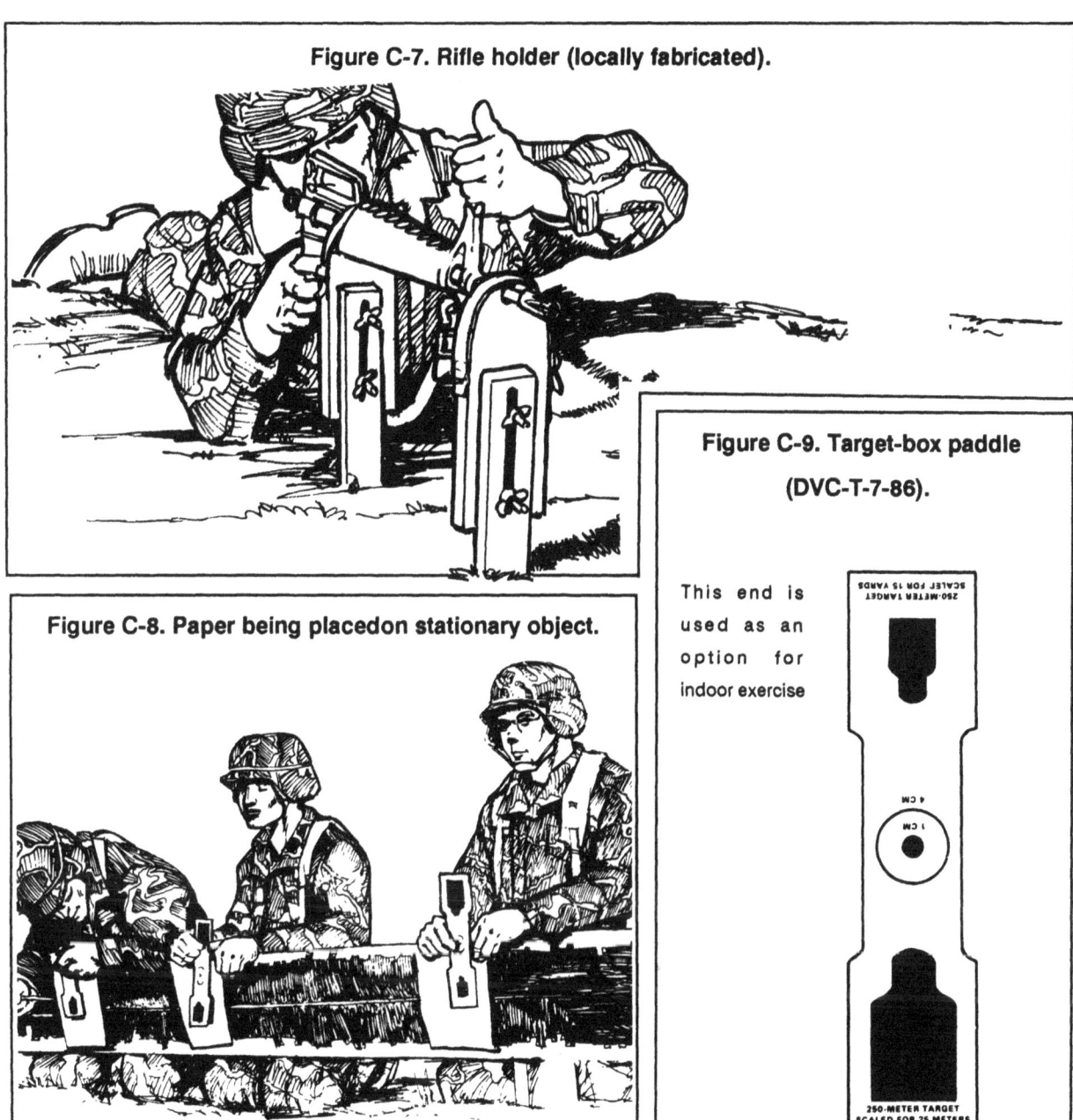

Figure C-7. Rifle holder (locally fabricated).

Figure C-8. Paper being placed on stationary object.

Figure C-9. Target-box paddle (DVC-T-7-86).

The shot-group exercise provides a chance for the trainer to critique the soldier on his aiming procedures, aiming consistency, and placement of shot groups. Assuming that the rifle and paper remain stationary and that the target man properly marks the three shots, the only factor to cause separation of the dots on the paper is error in the soldier's aiming procedures. When the soldier can consistently direct the target into alignment with the sights on this exercise, he should be able to aim at the same center-of-mass point on the zero range or on targets at actual range.

Ball-and-Dummy Exercise. This exercise is conducted on a live-fire range. The coach or designated assistant inserts a dummy round into a magazine of live rounds. In this way, the coach can detect if the firer knows when the rifle is going to fire. The firer must not know when a dummy round is in the magazine. When the hammer falls on a dummy round, which the firer thought was live, the firer and his coach may see movement. This is caused by the firer anticipating the shot or using improper trigger squeeze. Proper trigger squeeze results in no movement when the hammer falls. If the firer knows when the hammer is going to fall, movement can often be observed at that moment.

Dime (Washer) Exercise. This dry-fire technique is used to teach or evaluate the skill of trigger squeeze and is effective when conducted from an unsupported position. When using the M16A1 rifle for this exercise, the soldier must cock the weapon, assume an unsupported firing position, and aim at the target. An assistant places a dime (washer) on the rifle's barrel between the flash suppressor and front sight post assembly. The soldier then tries to squeeze the trigger naturally without causing the dime (washer) to fall off. Several repetitions of this exercise must be conducted to determine if the soldier has problems with trigger squeeze.

If the dime (washer) is allowed to touch the sight assembly or flash suppressor, it may fall off due to the jolt of the hammer. Also, the strength of the hammer spring on some rifles can make this a difficult exercise to perform. Instructors should allow soldiers to use another rifle when the one they are using is defective or needs repair.

When using the M16A2 rifle, the dime (washer) exercise is conducted the same except that a locally fabricated device must be attached to the weapon. A piece of 3/4-inch bonding material is folded into a clothes-pin shape and inserted into the compensator of the weapon so that the dime (washer) can be placed on top of it.

NOTE: The Weaponeer is discussed in Section II.

SELECTION OF TRAINING AIDS AND DEVICES

After training requirements have been established, appropriate training aids and devices can be selected from the TASC. To help in selecting these aids and devices, many of those available and their identification numbers are listed here.

TYPE/NOMENCLATURE	DENTIFICATION NUMBER
Training Aids and Devices	
Weaponeer — Remedial Rifle Marksmanship Trainer	DVC 7-57
M15A2 Blank Firing Aattachment	Supply Item (see TM)
Chamber Block (M16A1/A2)	Local TASC Expendable Item
M16 Sighting Device (A1 or A2)(Left and Right)	DVC-T 7-84
Target Box Paddle	DVC-T 7-86
Riddle Device	DVC-T 7-87
M16 Rifle Brass Deflector	DVC-T 7-87

M15A1 Aiming Card	DVC-T-07-26
M16A1 Disassembly Mat (Paper)	GTA 09-06-43
M16A1 Display Mat (Canvas)	TAD-0034 (locally)
Rifle Rest (for target-box exercise)	TAD-12 (locally)
Rifle Rest (Portable)	TAD-12 (locally)
Front and Rear Sight, M16 Rifle	TAD-26 (locally)
Front and Rear Sight, M16A2 Rifle	TAD-0026A
Rotating Panel Chalkboard Holder for GTA Charts	TAD-4 (locally)

Graphic Training Aids (GTAs)

Rifle, 5.56-mm, M16A1 Mechanical Training (1973)	GTA 7-1-26
Rifle, M16 Disassembly (M16A1)	GTA 9-6-43
M16A1 Rifle Malfunction	GTA 9-6-44
M16A1 Rifle Maintenance Card	GTA 21-1-3

Training Films

*Rifle, M16A1--Part I, Care, Cleaning, Lubrication	TF 21-3907
*Rifle, M16A1--Part II, Field Expedients	TF 21-3908

*Also available in video tape.

TYPE/NOMENCLATURE	IDENTIFICATION NUMBER
Video Tapes	
Engagement of Moving Personnel Targets with the M16A1 Rifle Team from the Foxhole Position	2E/010-071-1271-B
Cycle of Functioning M16A1 Rifle	2E/010-071-0444-B
Overview of BRM Training	010-071-0086-B
7-13 (Feb 87)	2E/010-071-0725-B
TVT 7-1 Teaching Rifle Marksmanship: Part One.	
TVT 7-2 Teaching Rifle Marksmanship: Part Two.	
TEC Lessons	
Loading and Unloading the M16A1 Rifle	1-939-071-0009-F
Disassembly and Assembly M16A1 Rifle	1-939-071-0010-F

Maintaining the M16A1 Rifle	1-939-071-0011-F
Preventing and Correcting Common Malfunctions	1-939-071-0012-F
Zero the M16A1 Rifle	1-939-071-0213-J
Zero M16A1 Aim and Fire Techniques	1-939-071-0214-F
Zero M16A1 Analyze and Correct Errors	1-939-071-0215-F
Perform Operator Maintenance on an M16A1 Rifle, Magazine, and Ammunition (Plus hands-on test)	071-311-2001
Load, Reduce a Stoppage, and Clear an M16A1 Rifle (Plus hands-on test)	071-311-2003
Battlesight Zero an M16A1 Rifle	071-311-2004

TARGET ORDERING NUMBERS

The following numbers can be used when ordering marksmanship targets.

Designation	Description	NSN
D prone	Full-length face with V through two scoring areas	6920-00-922-7450
D prone	Repair center with V through two scoring areas	6920-00-922-7451
Disignation	**Discription**	**NSN**
E-Silhouette	Full-length face, solid-color paper	6920-00-600-6874
E-Silhouette	Full-length, pop-up, solid-color plastic	6920-00-071-4780
E-Silhouette	Full-length face, cardboard, kneeling	6920-00-795-1806
F-Silhouette	Short-length face, solid-color paper	6920-00-610-9086
F-Silhouette	Short-length, pop-up, solid-color plastic	6920-00-071-4589

Designation	Description	NSN
F-Silhouette	Short-length face pasteboard	6920-00-795-1807
25-Meter Alternate-Course Scaled Qualification Target	50- to 300-meter scaled-silhouette target	6920-01-167-1398
15-meter Battlesight-Zero Target (.22-Caliber RFA)	250-meter scaled-silhouette target (50-foot indoor range)	6920-01-167-1393
15-Meter Alternate Course C (.22-Caliber RFA)	50- to 300-meter scaled-silhouette (50-foot in door range)	6920-01-167-1396
25-Meter M16A1 Zero Target	250-meter scaled silhouette-target	6920-01-167-1392
DESIGNATION	**DESCRIPTION**	**NSN**
25-Meter M16A2 Zero Target	300-meter scaled-silhouette target	6920-01-253-4005
25-meter M16A1 Slow-Fire Target	75- to 300-meter scaled-silhouette target	6920-01-167-1391
25-Meter M16A1 Timed-Fire Target	50- to 300-meter scaled-silhouette target	6920-01-167-1397
75-Meter M16A1 Feedback Target	75-meter scaled F-type silhouette	6920-01-169-6921
75-Meter M16A2 Feedback Target	75-meter scaled F-type silhouette	6920-01-253-4006
175-Meter M16A Feedback Target	1175-meter scaled E-type silhouette	6920-01-167-1395
175-Meter M16A2 Feedback Target	175-meter scaled E-type silhouette	6920-01-167-1395
Pasters, Black		6920-00-165-6354

Pasters, Buff	6920-00-172-3572
Landscape Target	6920-00-713-8253
Spindle, Target Spotter Wood	6920-00-713-8257
Spotters, 1 1/2 inches in diameter	6920-00-789-0864
Spotters, 3 inches in diameter	6920-00-713-8255

LOCATION OF MISS AND HIT (LOMAH) SYSTEM

LOMAH is a range aid used during downrange feedback exercises. The device uses acoustical triangulation to compute the exact location of a supersonic bullet as it passes through a target. The bullet impact is displayed instantly on a video monitor at the firing line. Of more importance, it shows the location of a bullet miss, thereby, allowing the firer to make either a sight adjustment or a holdoff for subsequent shots.

LOMAH, like other devices, is only an aid. Understanding the weapon and firing techniques, and having a coach/instructor are required when the soldier uses LOMAH.

LOMAH ranges have been fielded in USAREUR and Korea. In locations where known distance (KD) ranges are not available and restrictions prohibit walking downrange, LOMAH is a practical alternative to essential downrange feedback. Requests for LOMAH devices should be sent to: Commander, US Army Training Support Center, ATIC-DM, Fort Eustis VA 23604.

CALIBER .22 RIMFIRE ADAPTER, M261

The RFA can contribute to a unit's marksmanship program when 5.56-mm ammunition is not available or when ranges that allow the firing of 5.56-mm ammunition are not available. The RFA can be useful for marksmanship training such as night fire, quick fire, and assault fire. It is not recommended for primary marksmanship training.

Training Considerations. When service ammunition is in short supply, the RFA can be used to complement a unit's training program.

Rifle performance. The RFA/.22-caliber rimfire ammunition cannot replicate the exact ballistics of the 5.56-mm ammunition. Efforts to match RFAs with specific rifles can result in reasonable replication. Under ideal training conditions, the RFA should be used with dedicated rifles. Some variability can be eliminated by finding the right match of RFA and rifle. A trial-and-error technique can match RFAs to rifles, which results in good firing weapons. The RFA cannot be depended on to fire in the same place as 5.56-mm ammunition; therefore, it is not necessary for the soldier to use his own weapon during RFA training.

Rifle zero. The RFA will not usually group in the same location as 5.56-mm ammunition at 25 meters; therefore, it cannot be used for weapon zero. It normally fires a slightly larger shot group than 5.56-mm ammunition. When a soldier uses an RFA in his rifle, he must be careful not to lose his 5.56-mm zero. This can be accomplished by using hold-off while firing .22-caliber ammunition or keeping a record of sight changes so the sights can be moved back. The .22-caliber round approximates the 5.56-mm trajectory out to 25 meters. The correct zeroing target or appropriate scaled-silhouette targets can be used for practice firing exercises at 15 meters (50 feet) or 25 meters.

Advantages and Disadvantages. If the RFA is selected as a training aid, the advantages and disadvantages of the service must be considered during training.

Advantages. The .22-caliber ammunition is cheaper and, therefore, may be available in larger quantities than 5.56-mm ammunition. It can be fired on all approved indoor ranges and in other close-in ranges where 5.56-mm ammunition is prohibited. RFA training can be used to sustain marksmanship skills between periods when full caliber 5.56-mm ammunition training cannot be conducted.

Disadvantages. Some negative training aspects exist because of differences in the weapon's functioning when using the RFA. These differences include the forward assist not working, and the bolt not locking to the rear after the last round is fired. More malfunctions can occur with the RFA than with 5.56-mm ammunition, and immediate-action procedures are different.

MULTIPURPOSE ARCADE COMBAT SIMULATOR

MACS has been developed by the US Army as an inexpensive part-task marksmanship trainer (Figure C-10).

The system consists of a Commodore 64 microcomputer, 13-inch color monitor, specially designed long-distance light pen, and mount, which attaches to the M16 rifle. (Some versions use a permanent mount on a demilitarized rifle.) The system is activated by a program cartridge, which contains several training exercises.

Figure C-10. Multipurpose arcade combat simulator.

MACS was designed to enhance other training techniques and existing training aids and devices that are used to train and sustain marksmanship skills. It is not designed to replace live-fire training or to eliminate the need for knowledgeable instructors. MACS provides additional practice for those units that do not have access to adequate range facilities or have other resource constraints.

Section II. M2 BOLT and SHORT-RANGE TRAINING AMMUNITION

To augment the use of the 5.56-mm ammunition, a new generation of short-range training ammunition (SRTA), M862, is available.

SHORT-RANGE TRAINING AMMUNITION

The SRTA is about .06-inch shorter than the corresponding service cartridge, and has a blunt nose and projectile. The case and base that enclose the primer cap are made of brass, which is lined with plastic. The projectile is made of light-blue plastic. SRTA has an overall weight of about 114.6 grains. The muzzle velocity of the projectile is about 4,750 feet per second. Because of the light weight of the projectile, the velocity The projectile has a usable range of 25 meters and a maximum range of less than 300 meters.

NOTE: The SRTA can be used in an unmodified rifle; however, the rifle only functions as a single-shot weapon.

WARNING

THE PROJECTILE CAUSES CONSIDERABLE DAMAGE OUT TO 50 METERS AND IS CONSIDERED DANGEROUS OUT TO ITS MAXIMUM RANGE.

M2 BOLT

The M2 bolt is required to cycle the rifle when firing the SRTA (Figure C-11), and is interchangeable with the M16A1 and M16A2 rifles. Use of the M2 bolt converts the rifle from gas-operated weapons to blowback-operated weapons. The modified rifle functions in both semiautomatic and automatic modes. The bolts are stored and issued by the installation TASC. Issue of the M2 bolt is based on the percentage of troop population.

FM 23-9

> **WARNING**
>
> THE RECESS IN THE M2 BOLT FACE AND THE SRTA CARTRIDGE RIM ARE SMALLER THAN THE REGULAR M16A1/A2 AMMUNITION AND BOLT FACE. DO NOT USE REGULAR AMMUNITION WITH THE M2 BOLT INSTALLED.

SRTA AND M2 BOLT USE IN TRAINING

The US Army is constantly faced with training constraints such as lack of suitable real estate, safety restrictions, and cost of transporting troops to live-fire ranges. Short-range training ammunition allows training in small local training areas without fixed training facilities, in MOUT facilities, and in combat training theaters. With its 25-meter usable range, the SRTA can train on grouping, zeroing, 25-meter scaled silhouette firing,

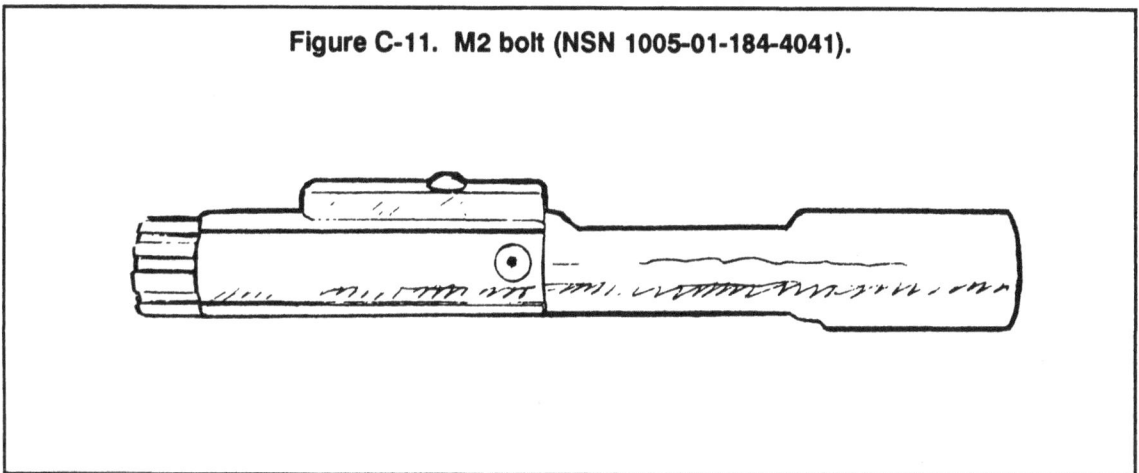

Figure C-11. M2 bolt (NSN 1005-01-184-4041).

Section III. WEAPONEER

The Weaponeer is an effective rifle marksmanship training device that simulates the life firing of the M16A1/A2 rifle. The system can be used for developing and sustaining marksmanship skills, diagnosing and correcting problems, and assessing basic skills.

CHARACTERISTICS

The Weaponeer operates on 110 to 130 volts AC, 10 amperes, 50 or 60 Hz, grounded electrical power. (A stand-alone voltage transformer is provided for oversea units.) The recommended training area for Weaponeer is 10 by 23 by 8 feet. The operational temperature range is 40 degrees to 100 degrees F. The Weaponeer must be protected from the elements, and it should not be subjected to excessive vibration, high dust levels, or condensing humidity. The M16A1/A2 attached to the Weaponeer is demilitarized and does not require the usual weapon security.

EQUIPMENT DATA

Major Components (unpacked)	Weight (pounds)	Length (inches)	Width (inches)	Height (inches)
Range assembly	119	99	30	8*/60
Target assembly	64	20	30	24
Operator's console	94	24	29	31
Firing pads -				
(stacked)	120	93	26	6
(prone layout)	120	93	74	2
(supported position)	120	93	52	46
Elevator ladder	20	3	24	61

*Prone position or when closed for transit.

DESCRIPTION

The Weaponeer is shown in Figure C-12 in the standing supported firing position. The rifle, with the exception of smoke and cartridge ejection, operates normally, and has the same weight and balance as the standard weapon. An infrared aiming sensor simulates round trajectory and hit point to an accuracy of better than one-minute-of-angle. Recoil is provided by the recoil rod that attaches at the muzzle end of the rifle. Recoil is provided in both semiautomatic and automatic modes of fire, and is adjustable from no-net force to 30 percent more than that of a live M16. Sound is provided through headphones and is adjustable from 115 to 135 decibels. Special magazines are used. One magazine simulates a continuous load; the other (used to train rapid magazine change) can be loaded with 1 to 30 simulated rounds. Selectable misfire can be used to detect gun shyness and drill immediate action. The front and rear sights are zeroed the same as standard rifles.

The Weaponeer range can be raised or lowered to accommodate all firing positions. The target assembly contains four targets: a scaled 25-meter zero target and three pop-up targets are standard. E-type and F-type silhouettes at ranges from 75 meters can be used on the Weaponeer. Also, known-distance and various other types of targets can be used, and can be displayed in fixed or random sequences. Target exposure times may be set to unlimited or from 1 to 30 seconds. The fall-when-hit mode can be selected with the KILL button.

The operator's console contains the system control buttons, graphics printer, and video feedback monitor. The back of the console has counters that total rounds and hours, and a storage bin for storing magazines, printer paper and ribbon, headphones, two wrenches for assembling the Weaponeer, and a small allen wrench for aligning the rifle sensor. A remote control, which attaches to the back of the console, enables a trainer or firer to operate select functions away from the console.

Figure C-12. Weaponeer set up in the standing supported position.

FEEDBACK

The Weaponeer provides feedback to help trainers to teach and soldiers to learn marksmanship skills.

Fall-When-Hit Mode. The fall-when-hit mode is enabled by lighting the KILL button. When the button is activated, targets fall when hit. This feedback provides the same hit/miss information as a train-fire (RETS) range.

Real-Time Aiming Point Display. When a firer aims on or near a target, his aiming point relative to the target is continuously displayed on the video screen. The aiming point display allows the trainer to teach and verify aiming techniques, and to continuously monitor the firer's steadiness, techniques, time on target, trigger squeeze, and recovery from recoil.

Immediate-Shot-Impact Display. When a shot is fired, its impact relative to the target is immediately displayed on the video screen as a blinking white dot. (The left target in Figure C-13.)

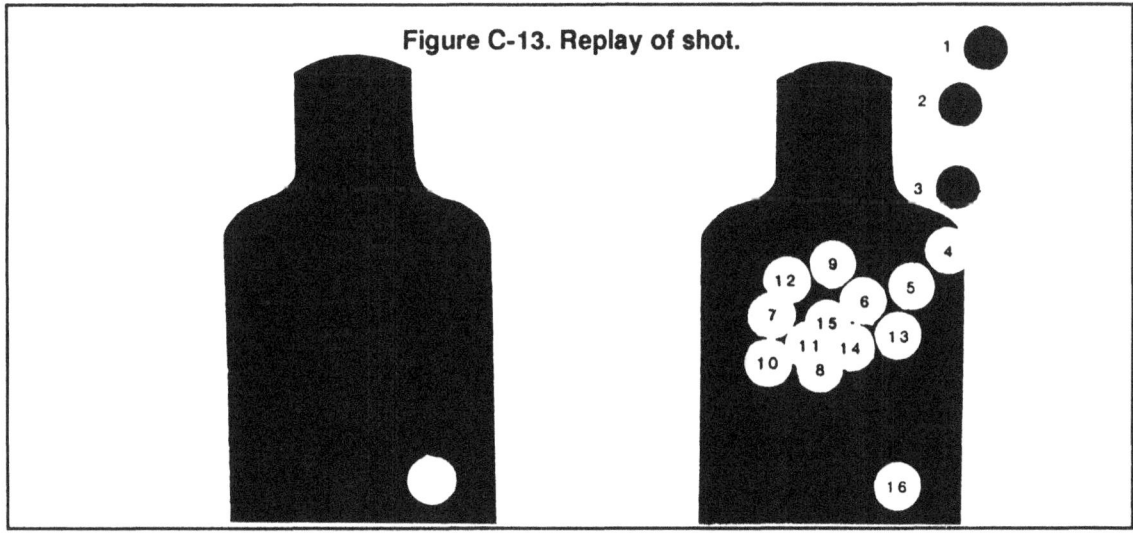

Figure C-13. Replay of shot.

FM 23-9

Replay. After a shot is fired, a real-rate display of how the firer engaged the target can be replayed on the video screen.

The target to the right in Figure C-14 shows the type of information that can be replayed on the video screen after a series of shots are fired. To show the sequence, the dots have been numbered.

To show a replay, the firer first selects the shot he wishes to replay by operating the EACH SHOT button. Then he presses the REPLAY button. Some Weaponeers record and store replays for just the first three shots.

Shot Groups. The impact location of up to 32 shots is automatically stored in the Weaponeer memory and displayed on the video screen. Each impact is indicated by a white dot, which blinks when indicating the last shot. All 32 shots can be fired at and displayed on a single target, or split among any combination of targets. The CLEAR button erases all shots from the Weaponeer memory.

Printer. A hard-copy printer is provided for postfiring analysis, for firer progress tracking, and for record keeping. Pressing the PRINT button causes the target displayed on the video to print. (Sample printouts are shown in Figure C-14.) Some Weaponeers can print the three pop-up targets at the same time by holding in the REPLAY button and then pressing the PRINT button.

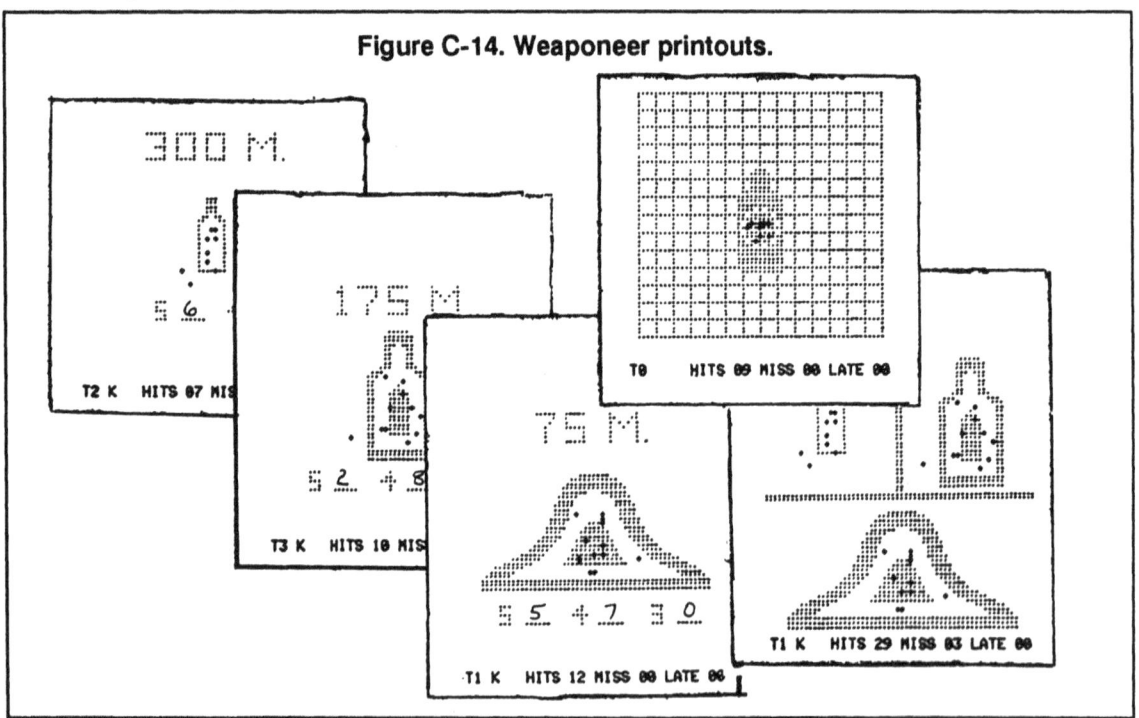

Figure C-14. Weaponeer printouts.

USE OF THE WEAPONEER

In BRM, the Weaponeer is used early in the program to evaluate firer's ability to apply the four fundamentals. It is used throughout the program to help diagnose and remediate problems. In the unit, the Weaponeer should be used much like it is used in BRM. Concurrent use of the Weaponeer at the rifle range provides valuable remedial training.

The preferred training configuration for the Weaponeer is shown in Figure C-14. One trainer operates the system while three to six soldiers observe the training. Soldiers should rotate, each receiving several short turns on the system. Where high throughput is required, consolidation of available Weaponeers may be considered.

When training soldiers on the Weaponeer:

- Proceed at a relaxed pace, and emphasize accuracy before speed.

- If possible, train with small groups, allowing each soldier several 10- to 15-minute turns on the device.

- For remedial training, try to relax the soldier; a nervous soldier will have trouble learning and gaining confidence in his marksmanship skills. For sustainment training, encourage competition between individuals or units.

In Figure C-15, five soldiers are being trained. One is firing and four are observing, awaiting their turns on the device. The video screen is carefully positioned just outside the vision of the firer, but the firer can easily turn his head to see replays and hit points. The position of the trainer is also important so he can see both the firer and video screen. This is a good position for detecting and correcting firing faults. When the firer is in the standing supported firing position, the console should be placed on a table so the trainer can see the video screen above the firer's rifle. (Figure C-16.) Observers can see the targets, firer, and video screen to help them perform during their turn. Observers learn procedures that speed up training and help avoid firing faults.

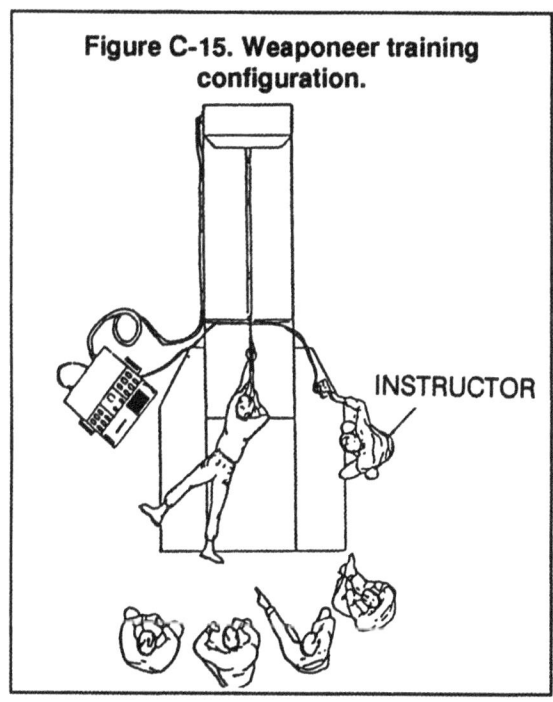

Figure C-15. Weaponeer training configuration.

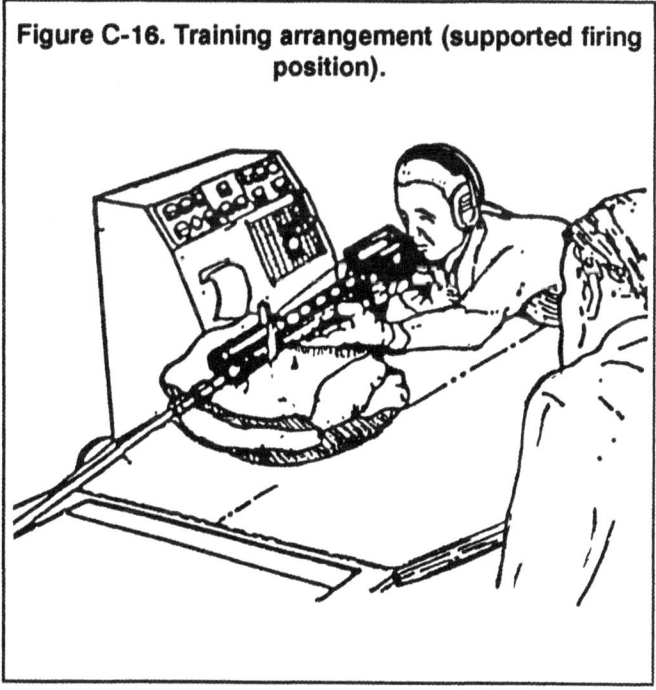

Figure C-16. Training arrangement (supported firing position).

MOBILE CONFIGURATION

To use the Weaponeer in a mobile confirguration, it must be shock mounted. (The manufacturer's conceptural mobile training unit is shown in Figure C-17.) The Training Audiovisual Support Center, Fort Benning, Georgia, has adopted a mobile

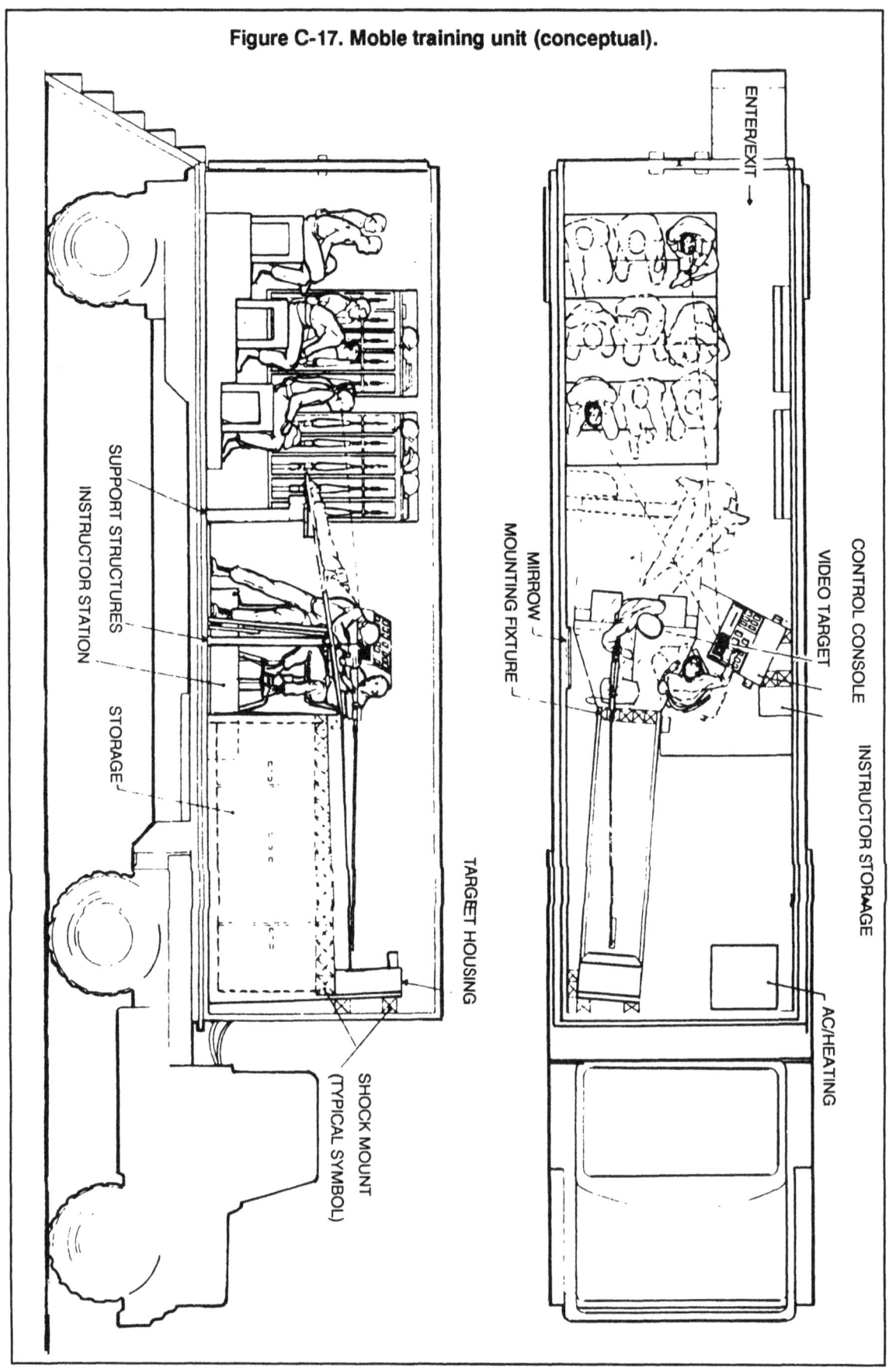

Figure C-17. Mobile training unit (conceptual).

Figure C-18. Mobile mounting stand.

mounting stand for supporting Weaponeer's range assembly and computer console (Figure C-18)

DIAGNOSIS OF FIRING PROBLEMS

Diagnosis of firing problems is the main purpose of the Weaponeer. The following seven-step program is recommended as a guide. Depending on the extent of the firer's problems and time constraints, the number of shots may be increased.

STEP 1. Tell the soldier to assume a good firing position, to aim at a target, and to hold steady (supported and prone unsupported positions).

STEP 2. Visually check the firer's firing position and correct any gross errors.

STEP 3. Observe the video screen. If there is no aiming dot on the video screen or if the aiming dot is far from target center, teach sight picture to the firer. If excessive movement is shown by the light dot, check and correct the techniques of the steady position and natural point of aim.

STEP 4. Tell the soldier to fire a three-round shot group aimed at the target's center of mass. Watch the video screen and firer as he fires. Note violations of the four fundamentals.

STEP 5. Replay each shot to show the firer his aim, steadiness, and trigger squeeze. In Figure C-13, page C-17, the target on the right shows a numbered series of 16 shots. Dots 1 through 4 indicate that the firer approached the target from high right. Dots 5 through 15 show that he is aiming near the center of the target but does not have a steady position. The sudden shift from dot 15 to 16 (dot 16 is the hit point of the shot) indicates that gun-shyness or improper trigger squeeze caused the firer to pull his aiming point down and to the right just before firing. Replay helps the firer understand and correct his firing errors.

STEP 6. Confirm and refine the diagnosis by allowing the soldier to fire additional three-round shot groups. Use replay to show the firer his firing faults.

STEP 7. Summarize and record the soldier's basic firing problems.

These seven steps are designed to efficiently diagnose and to show the soldier his firing errors. This could be enough to immediately correct the error. However, diagnosis should be followed up with remedial exercises either with the Weaponeer, target-box exercise, or dime (washer) exercise.

UNIT SUSTAINMENT TRAINING

Sustainment training and prequalification refresher training can be conducted, depending on the availability of the Weaponeers.

Direct the soldier to zero the Weaponeer rifle (sandbag supported position). Emphasize tight, consistently placed shot groups. Starting with the closest target and working out to the most distant, direct the soldier to practice slow precision fire at each target (supported and prone unsupported positions).

Direct the soldier to slow fire at random pop-up targets (both firing positions). Emphasize speed and precision. Direct him to slow fire at random pop-up targets with short exposure times (both firing positions).

OPTION: Direct the soldier to practice windage hold-off, rapid magazine change, and immediate action (both firing positions).

OPTION: Direct the soldier to practice night fire, automatic fire, and gas-mask fire.

ASSESSMENT OF SKILLS

The Weaponeer can aid in the objective assessment of basic marksmanship; therefore, periodic Weaponeer diagnosis should be conducted and recorded. Each soldier fires until zeroed on the Weaponeer. If unable to zero in 9 to 15 rounds, he should be withdrawn from testing and given remedial training. The soldier fires a surrogate record-fire scenario according to the following:

Scenario of target presentation. Presentation of the targets is controlled by the operator who uses the target buttons.

Order of target presentation. The scaled 100-meter and 250-meter targets (or 75 meters, 175 meters, and 300 meters) are presented in a mixed order according to a planned schedule.

Ratio of target presentation. Targets are presented in a ratio of three 250-meter targets to one 100-meter target (or three 300 meters, two 175 meters to one 75 meters). A 64-target scenario consisting of two 32-target scenarios (the first engaged from the supported position; the second from the prone unsupported position) is conducted with a short break.

Target exposure time. Exposure time is four seconds for the scaled 250-meter targets (or 175 meters and 300 meters) and two seconds for the scaled 100-meter target (or 75 meters).

Intertarget interval. The time between target exposures should be varied from one to eight seconds.

Target mode. The kill mode is used so that targets fall when they are hit.

A score of 41 hits out of the 64 targets on Weaponeer indicates that soldiers canproceed to actual record fire. Soldiers who score lower than 41 should receive remedial training.

FM 23-9

APPENDIX D

Rifle Range Safety Briefing/ Range Operations Checklist

All personnel training on a rifle range should be briefed on the safety and local requirements for that range. The briefing fulfills the minimum requirements for a rifle range safety briefing. Information may be added to conform to local requirements and safety regulation AR 210-21 should be reviewed.

RECOMMENDED BRIEFING

The first priority of this range is training, but safety must be at the forefront of the training program. The safety program will be corrected immediately. The safety program should include the following:

- The two red and white candy-striped poles, located on the far right and far left (point to them), are the range left and right limits. Firers never fire or point their rifles outside the limits of these poles.

- When not on the firing line, the selector lever is on SAFE, the bolt is locked to the rear, and the dust cover is open. (Demonstrate.)

- Firers will always enter and exit the firing line at the base of the tower. (Point.)

- Before occupying the firing positions, inspect the positions for harmful animals and insects.

- Firers will always point the muzzle of the rifle downrange whenever on the firing line. The firing finger is not placed within the trigger housing area. (Demonstrate.)

- Lock the bolt to the rear and place the rifle on SAFE on the firing line. Chamber blocks may be used.

- Smoke only in designated smoking areas. (Inform soldiers where.)

- You will not eat or drink on the firing line unless the tower operator permits you to drink from your canteen. Drink water often to prevent heat injuries.

- Never touch your rifle while personnel are downrange or in front of the firing line.

- Load the magazine into the rifle only on the command of the tower.

- Never fire without wearing hearing protection when within 25 meters of the firing line.

- Before leaving the firing line, the safety NCO must rod the rifle.

- Consider the rifle loaded at all times, even in the break areas. Never point the rifle at anyone.

- Left-handed firers will fire the M16A1 with left-handed brass deflectors attached to the rifle.

- Anyone observing an unsafe act will immediately call CEASE FIRE, place his rifle on SAFE, place it in the V-notch stake or on the sandbags, and then give both the vocal alarm and visual signal of cease fire. (Demonstrate and have soldiers demonstrate.)

- Once cleared off the firing line, firers report immediately to the ammunition point and turn in all ammunition and expended brass.

- No one will leave the range until he has been inspected for ammunition and brass.

NOTE: When an electrical storm occurs, the safety NCO will direct the tower operator to prepare to disperse soldiers. At that time, the tower gives the command LOCK AND CLEAR ALL WEAPONS AND GROUND ALL EQUIPMENT (except for wet-weather gear). Then the tower operator directs soldiers to a safe area.

The instructor/trainer should ask the firers if there are any questions concerning the safety procedures of the range.

PERSONNEL AND DUTIES

To provide both a safe and efficient range operation and effective instruction, the following is an example of personnel and duties that may be required.

OIC. He is responsible for the overall operation of the range before, during, and after live firing.

Range Safety Officer. He is responsible for the safe operation of the range to include conducting a safety orientation before each scheduled live-fire exercise. He ensures that a brass and ammunition check is made before the unit leaves the range. He ensures that all personnel comply with the safety regulations and procedures prescribed for the conduct of a live-fire exercise. He should ensure that left-handed firing devices are used by all left-handed firers. This officer should not be assigned other duties.

NCOIC. He assists the OIC and safety officer, as required—for example, by supervising enlisted personnel who are supporting the live-fire exercise.

Ammunition Detail. This detail is composed of one or more ammunition handlers whose responsibilities are to break down, issue, receive, account for, and safeguard live ammunition. The detail also collects expended ammunition casings and other residue.

Unit Armorer. He repairs the rifles to include replacing parts, as required.

Assistant Instructor. One assistant instructor is assigned for each one to ten firing points. Each assistant ensures that all firers observe safety regulations and procedures, and he assists firers having problems.

Medical Personnel. They provide medical support as required by regulations governing live-fire exercises.

Control Tower Operators. They raise and lower the targets, time the exposures, sound the audible signal, and give the fire commands. If possible, two men should be chosen to perform these functions.

Maintenance Detail. This detail should be composed of two segments: one to conduct small-arms repair and one to perform minor maintenance on the target-holding mechanisms.

RANGE PROCEDURES

Before beginning live-fire exercises, all personnel must receive an orientation on range operations. The orientation should outline the procedures for conducting the exercise to include the duties of the nonfiring orders. Scorers maintain the score of the firer. They may assist the firers by indicating the impact of the bullet in relation to the target—for example, short, right or high, left.

Regardless of unit size, soldiers are organized into four orders: one firing, one coaching, and two pulling targets. One order may be shifted from pulling targets to concurrent training. However, if a unit must use five or more orders due to range limitations, concurrent training must be conducted to effectively use waiting periods; no more than two orders should be sent to pull targets.

Practice Record Fire and Record Fire

Simple, standard fire commands are needed to avoid confusion and misunderstanding during practice record fire and record fire.

Practice record fire.

FIRERS, ASSUME A GOOD SUPPORTED (PRONE UNSUPPORTED) POSITION.

SCORERS, POINT OUT THE LIMITS OF YOUR LANE.

FIRERS, LOCK YOUR FIRST MAGAZINE, LOAD.

WATCH YOUR LANE.

CEASE FIRE.

CLEAR ALL WEAPONS.

Record fire.

    FIRERS, ASSUME A GOOD SUPPORTED (PRONE UNSUPPORTED) POSITION.

    SCORERS, POINT OUT THE LIMITS OF YOUR LANE.

    FIRERS, LOCK YOUR FIRST 20-ROUND MAGAZINE, LOAD.

    WATCH YOUR LANE.

    CEASE FIRE.

    CLEAR ALL WEAPONS.

Field Firing Exercises
    Simple, standard fire commands are needed to avoid confusion during field firing exercises. Commands for exercises from stationary positions are as follows:

    FIRERS, ASSUME A GOOD (_____) POSITION.

    LOCK ONE MAGAZINE OF (____) ROUNDS, LOAD.

    READY ON THE RIGHT?

    READY ON THE LEFT?

    THE FIRING LINE IS READY.

    PLACE YOUR SELECTOR LEVER ON SEMIAUTOMATIC.

    WATCH YOUR LANES.

    CEASE FIRE. LOCK YOUR WEAPONS.
    (Place the selector lever in the SAFE position.)

    Repeat the first seven commands above, or give the following commands:
    Commands for conduct of firing are minimal and standard. The proper commands are listed in the following paragraphs. Preliminary commands to describe the particular exercise may be used.
    The range officer relays his commands either by radio or telephone to the pit NCOIC so he can keep abreast of the conduct of fire. Before each firing exercise, the

range officer informs the pit NCOIC of the next exercise and any special instructions for target operation—for example, "The next firing will be for zero. Mark targets after each three-round group shot, or for slow fire, The next firing will be ten rounds, slow fire. Mark targets after each shot."

RATELOs relay commands to the pits and pass on special instructions to target operators as requested by the assistant instructors. RATELOs never identify a firer on a certain firing point. The command MARK TARGET NUMBER___ indicates that the target has been fired upon but has not been withdrawn for marking.

General Commands

The following are general commands and may be altered when necessary.

FIRERS, ASSUME THE _____ POSITION.
(Issue the firer _____ rounds of ammunition.)

COACH, SECURE _____ ROUNDS OF AMMUNITION.

LOCK ONE ROUND, LOAD.

READY ON THE RIGHT?

READY ON THE LEFT?

READY ON THE FIRING LINE?

COMMENCE FIRING WHEN YOUR TARGET APPEARS.

CEASE FIRING.

Rapid Fire Exercises

The following commands are used for rapid-fire exercises:

LOCK AND CLEAR ALL WEAPONS.

CLEAR ON THE RIGHT?

CLEAR ON THE LEFT?

THE FIRING LINE IS CLEAR.

FIRERS, ASSUME THE _____ POSITION.

ASSISTANTS, SECURE TWO MAGAZINES OF FIVE ROUNDS EACH.
(Issue the firer one magazine of five rounds.)

LOCK ONE MAGAZINE, LOAD.

READY ON THE RIGHT?

READY ON THE LEFT?

READY ON THE FIRING LINE?

WATCH YOUR TARGETS!

(Firers assume the appropriate firing position and commence firing when the targets are presented.)

When all the targets are withdrawn, the range officer checks for slow firers or malfunctions and then allows them to fire.

The pit NCOIC organizes, orients, and provides safety for the pit detail. The success of KD firing depends on efficient operation of the targets and the close coordination between the pit NCOIC and range officer. All operators must know the proper procedure for operating and marking the target.

Marking targets for zeroing and slow fire. Targets are marked quickly after each shot or group of shots without command. During slow fire, the firer has a time limit of one minute for each shot. Twenty seconds is considered the maximum time limit for marking. A marker (spotter) is placed in the hit regardless of its location on the target. Each time the target is marked, the marker is removed from the previous hit, and the hole is pasted. (Three-inch markers are used for 100, 200, and 300 meters; five-inch markers are used for 500 meters.)

Using disk markers. The target markers are painted black on one side and white on the opposite side. They may be procured in three dimensions: 1 1/2 inches (NSN 6920-00-789-0864), 3 inches (NSN 6920-00-713-8255), and 5 inches (NSN 6920-00-713-8254). The disk spindle may also be procured through supply channels (NSN 6920-00-713-8257).

RANGE OPERATIONS CHECKLIST

This checklist consists of nine sections, each covering a different topic relating to range operations.

The checklist should be modified to include local policy changes to the regulations or SOPs. The person responsible for the training must answer the questions in each section. Ask yourself each question in order. Record each "Yes" answer by placing a check in the GO column. Record a "No" or "Don't know" by checking the NO-GO column. Refer to the checklist to find the GO and NO-GO columns.

When all the questions in a section are asked, look back over the NO-GOs. Contact the people who reported them and ask if they have corrected each problem. If so, change the answer to GO. If any NO-GO remains, analyze it and implement a countermeasure for the shortfall. Afterwards, check to ensure the countermeasures work. Before range operations start, be sure a workable countermeasure is implemented for each safety hazard presented by a NO-GO answer.

FM 23-9

Section 1 — Mission Analysis

1. Who will be firing on the range?_____

   Number of personnel_____ Units _____

2. What weapons and course will be used?

   Weapon_____ Course_____

3. Where will the training be conducted?
   Range _____

4. When is the range scheduled for operations?
   Date _____ Opens_____ Closes _____

Section 2 — Double-Check

	go	no-go	REMARKS
1. Has sufficient ammunition been requested for the number of personnel?			
2. Are the range facilities adequate for the type of training to be conducted?			
3. Has enough time been scheduled to complete the training?			
4. Have conflicts that surfaced been resolved?			

Section 3 — Become an Expert

	go	no-go	REMARKS
1. Review TMs and FMs on the weapons to be fired			
2. Talk with the armorers and other personnel experienced with the weapons to be fired.			
3. Review AR 385-63			
4. Visit range control and read installation range instructions.			

go	no-go	REMARKS

5. Reconnoiter the range (preferably while it is in use).

6. Check ARTEPs and SQT manuals to see if training tasks can be integrated into the range training plan.

Section 4 — Determine Requirements

PERSONNEL
1. OIC.
2. Safety officer.
3. Assistant safety officer.
4. NCOIC.
5. Ammunition NCO.
6. Ammunition personnel (determined by type of range).
7. Target detail and target operators.
8. Tower operator.
9. Concurrent training instructors.
10. Assistant instructors.
11. RATELO.
12. Guards (range requirements).
13. Medic(s).
14. Air guard.
15. Armorer.
16. Truck driver (range personnel and equipment).
17. Mechanic for vehicles.
18. Have you overstaffed your range?

EQUIPMENT:
1. Range packet and clearance form.
2. Safety fan and diagram if applicable
3. Other safety equipment (aiming circle, compass).
4. Appropriate publications pertaining to the training that will be conducted.
5. Lesson plans, status reports, and reporting folder.
6. Range flag and light (night firing).
7. Radios.
8. Field telephone and wire.
9. 292 antenna, if necessary.
10. PA set with backup bullhorn(s).
11. Concurrent training markers.
12. Training aids for concurrent training stations.
13. Sandbags.
14. Tentage (briefing tent, warm-up tent).
15. Space heaters, if needed.

16. Colored helmets for control personnel.
17. Safety paddles and vehicle flag sets or lights.
18. Ambulance or designated vehicle.
19. Earplugs.
20. Water for drinking and cleaning.
21. Scorecards.
22. Master score sheet.
23. Armorers' tools and cleaning equipment for weapons.
24. Brooms, shovels, and other cleaning supplies and equipment.
25. Tables and chairs, if needed.
26. Target accessories.
27. Fire extinguishers.
28. Tarp, stakes, and rope to cover the ammunition.
29. Toilet paper
30. Spare weapons and repair parts as needed.
31. Tow bar and slave cables for vehicles.
32. Fuel and oil for vehicles and target mechanisms.

Section 5 — Determine Available Resources

1. Fill personnel spaces.
2. Keep unit integrity.
3. Utilize NCOs.
4. Effect coordination with supporting organizations:

- Ammunition.
- Transportation.
- Training Aids.
- Medics.
- Weapons.
- Other equipment.

Section 6 — Foolproofing

1. Write an overall lesson plan for the range.
2. Organize a plan for firing:

- Determine range organization.
- Outline courses of fire to be used.
- Have fire commands typed for use on range.
- Set rotation of stations.

3. Rehearse concurrent training instructors and assistants.
4. Brief RATELO on unique range control radio procedures.
5. Brief and rehearse reporting NCO on range operation and all his duties.

6. Collect and concentrate equipment for use on the range in one location.
7. Obtain training aids.
8. Pick up targets from range warehouse, if required.
9. Report to range control for safety briefing (if required) and sign for any special items.
10. Publish LOI:

- Uniform of range and firing personnel (helmets and earplugs).
- Mode of transportation, departure times and places.

Methods of messing to be used.
Any special requirements being placed on units.

Section 7 — Occupying the Range and Conducting Training

OCCUPY THE RANGE
1. Request permission to occupy the range.
2. Establish good communications.
3. Have designated areas prepared:

- Parking.
- Ammunition point.
- Medical station.
- Water point.
- Concurrent training.
- Mess.
- Helipad.
- Armorer.

4. Inspect range for operational condition.
5. Raise flag when occupying or firing according to the local SOP.
6. Check ammunition to ensure it is correct type and quantity.
7. Ensure that range personnel are in proper uniform and the equipment is in position.
8. Receive firing units.
9. Conduct safety checks on weapons.
10. Check for clean, fully operational weapons.
11. Conduct safety briefing (to include administrative personnel on range)
12. Organize personnel into firing orders (keep unit integrity if possible).
13. Request permission to commence firing from range control.

CONDUCT OF FIRING:
1. Are communications to range control satisfactory?
2. Commands from tower clear and concise?
3. Range areas policed?
4. Ammunition accountability maintained?
5. Master score sheet updated?
6. Personnel accountability maintained?
7. Vehicles parked in appropriate areas?
8. Air guard on duty and alert?
9. Personnel in proper uniform?
10. Earplugs in use?
11. Troops responding properly to commands?
12. On-the-spot corrections being made when troops use poor techniques or fail to hit the target?
13. Conservation of ammunition enforced?
14. Weapons cleared before they are taken from the firing line?
15. Personnel checked for brass or ammunition before they leave the range?
16. Anyone standing around not involved in training or support?

Section 8 — Closing of Range
1. Close downrange control according to the local SOP.
2. Remove all equipment and ammunition from range.
3. Police range.
4. Repaste and resurface targets as required by range instructions.
5. Perform other maintenance tasks as required by local SOP.
6. Request a range inspector from range control when ready to be cleared.
7. Submit after-action report to headquarters.
8. Report any noted safety hazards to proper authorities.

Section 9 — Known Distance Range
PERSONNEL: In addition to those identified in Section 4.
1. NCOIC of pit detail.
2. Assistant safety officer for pit area.

EQUIPMENT: In addition to equipment identified in Section 4.
1. Sound set for pit area.
2. Positive communication from the firing line to the pit area.
3. Pasters.
4. Glue and brushes for resurfacing targets.
5. Lubricant for target frames.
6. Proper targets mounted in target frames.
7. Briefing on how to operate a KD range.
8. Procedure for marking targets.
9. Procedure for pit safety.

APPENDIX E

Scaled Silhouette Targets

Scaled silhouette targets were developed in conjunction with the zero target to expand the use of widely available 25-meter ranges beyond that of just zeroing weapons. They are designed to provide an alternative to fill a significant training void. In the past, soldiers went directly from the zero range to the field fire environment. They fired at pop-up targets located at ranges of 75 to 300 meters and received only hit or miss feedback as to performance. The pop-up silhouette target represents an important skill that all soldiers should perform well. But hit and miss information does not constitute a good learning environment for the development and refinement of good marksmanship skills. The bad firer will miss most targets and never know what he is doing wrong. The good firer will hit most targets but will be unable to refine his skills to hit near target center.

BACKGROUND

The scaled silhouette target provides the same visual perception at 25 meters as the firer would see when firing targets at actual range. The use of this target at 25 meters allows the firer to practice aiming at various range targets and to see precisely where his bullets hit, whether they are target hits or misses. Scaled silhouette targets fit well into a marksmanship training program between zeroing and field fire training, but they provide for excellent training whenever a 25-meter range and ammunition are available. They are also appropriate targets for use during dry-fire training and for teaching adjusted aiming points, for making allowances for wind and range estimation, and when authorized to conduct record for qualification. Two scaled silhouette targets are appropriate for use during unit sustainment:

Slow-Fire Target. The slow-fire target shown in Figure E-1 may be used immediately following the zeroing period of instruction. This target is used to provide a smooth transition from firing at 25 meters to the engagement of KD feedback targets or pop-up targets on the field fire range. This target provides the same visual perception as the three targets the soldier will be required to engage on the field fire range. It allows the soldier to practice aiming at various range targets while seeing precisely where his bullets hit or miss the targets.

The soldier learns that a 75-meter target can be hit easily but that a good steady position, precise aiming, and smooth trigger squeeze are necessary to consistently hit the 300-meter target. The 4-cm circle is drawn at the center of each scaled silhouette so the soldier can relate his performance from the zeroing period to his performance on the scaled silhouettes. (All scaled targets in this manual have been reduced in size. The actual target sheet is about 18 by 23 inches.) During training, 18 rounds are fired on the slow-fire target. From the supported fighting position, the soldier fires three rounds at each of the three targets on the left side of the target sheet. He then moves downrange to inspect his target and to discuss his performance with his sergeant or instructor/trainer. Training the instructor/trainer may find that the soldier does not understand how to aim at target center of mass, that he aims too low at the 300-meter target, that his rifle does not have a good zero, or that he is not applying good firing fundamentals.

Moving back to the firing line, the soldier fires three rounds at each of the three targets on the right side of the target sheet from the prone unsupported position. The second inspection of the target allows the soldier to compare his firing ability in an unsupported position to that in a supported position. His sergeant or instructor/trainer has another opportunity to observe and critique the soldier's performance. For example, the soldier who fired the target in Figure E-2 is not doing a good job of applying the four firing fundamentals. Some soldiers will begin to lose precision after they complete zeroing. This soldier needs to work on the supported position, but particularly on the unsupported position. Target number 3 indicates he may have aimed too low on the 300-meter target. The 300-meter target is so small that when the rifle is aimed at center of mass, there is little black visible above the front sight. Target Number 6 indicates he may have overcompensated for this error with at least two shots.

Placement of all 18 bullets also indicates that the firer's rifle zero may be too far to the right. With a better firing performance, sight adjustments can be made while using this target. The analysis procedures used on the zeroing target may be applied to shot groups on these silhouette targets.

The shot-group center on each target indicates aiming techniques. The comparison between the targets fired supported and those fired unsupported also provides useful information. A good performance on this target indicates that the soldier is ready to engage targets at actual range. The soldier who does not hit these scaled silhouette targets at 25 meters should receive remedial training before being allowed to continue to a downrange feedback or the hit and miss portion of the program.

Timed-Fire Target. Following unit field fire training, the soldier may be brought back to the 25-meter range to confirm his zero (he should fire better now than at the start of the program) and to fire the timed-fire target shown in Figure E-3. This target sheet includes a scaled silhouette for each range target represented on the record fire course—50 to 300 meters. While the slow-fire target was fired at the soldier's own speed, the timed-fire target is designed to stress the soldier, emphasizing the task requirements of rapid target engagement under time pressure. The soldier must rapidly shift his point of aim from silhouette to silhouette in the absence of a pattern of clearly defined silhouettes on the target.

The initial firing of the timed-fire target is from the supported fighting position. The soldier is given 45 seconds and 10 rounds of ammunition with instructions to fire one round at each silhouette target. Upon inspection of the target, the soldier can review his own performance, which provides an excellent diagnostic checkpoint for his sergeant/instructor/trainer. The soldier who hits most of these targets will probably qualify with a good score. After the target has been critiqued and bullet holes have been marked or pasted, the soldier repeats the exercise from the prone unsupported position, having 55 seconds to engage the 10 targets.

The target shown in Figure E-4 represents a good firing performance from the fighting and prone positions. A good firer is aware of the bullet that was pulled (right side—middle target). Only with the worst firers are there any problems telling which bullet was fired at which target. Although there is room for improvement, this soldier

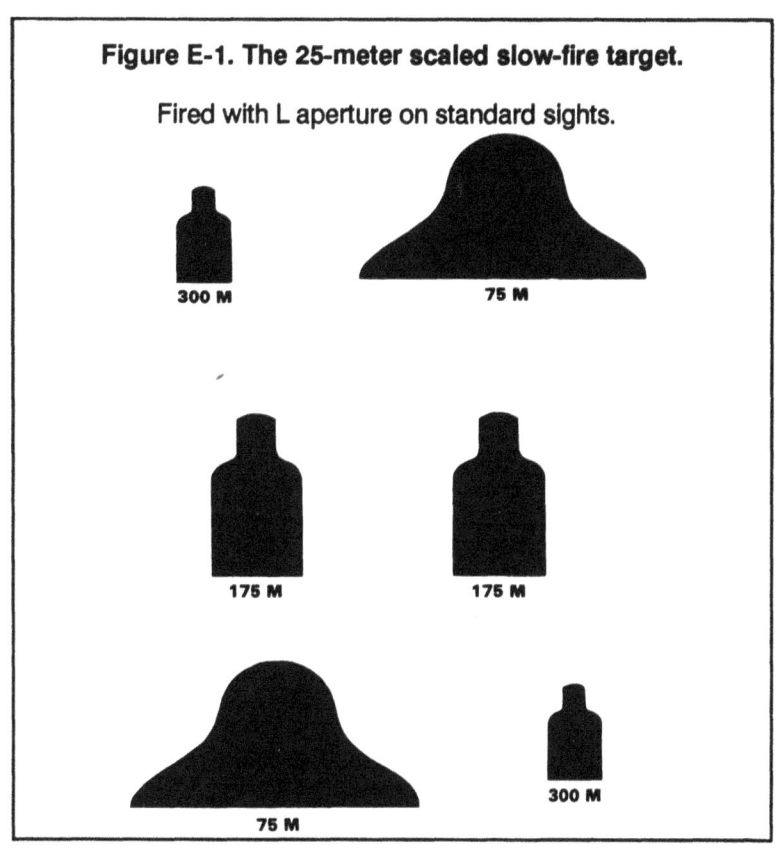

Figure E-1. The 25-meter scaled slow-fire target.

Fired with L aperture on standard sights.

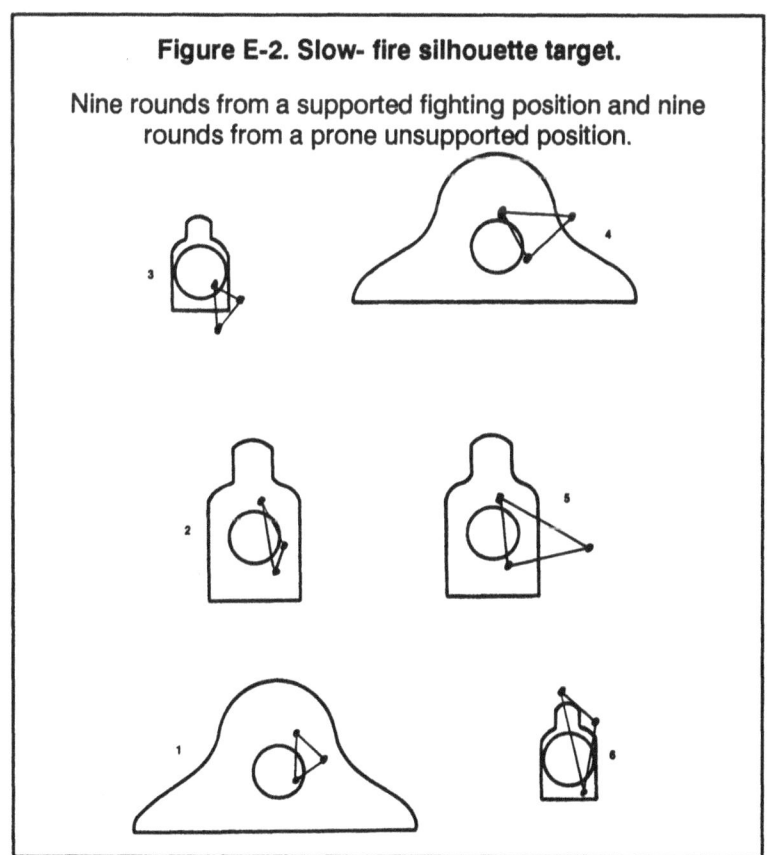

Figure E-2. Slow-fire silhouette target.

Nine rounds from a supported fighting position and nine rounds from a prone unsupported position.

can be expected to fire well on record fire. Following the critique of the second firing, these results are recorded and the targets are retained for future reference. No shot groups are fired on this target, but the concept of shot-group analysis is still valid. Considerable information can be obtained by checking the direction and distance of each 10 shots from target center as though they were a 10-shot group.

Targets in Figure E-1 have the 4-cm circle located at target center while the targets in Figure E-3 have the circle located either above or below target center, except for the 250-meter target. During the training exercise, these targets are scored as hits if the bullets hit within the black. However, the objective is to put all bullets within the circle. To progress at a pace that is not too fast for the soldier to understand, the initial targets use center-of-mass aiming so the soldier understands center of mass while aiming on various size targets. On the timed-fire targets, the option is provided for making adjusted aiming allowances for gravity. That is, the circles are placed on the timed-fire target at the aiming point that will place bullets at target center when firing a target at the range indicated. For example, the soldier must aim low on a 150-meter target to hit target center, and he must aim high on a 300-meter target to hit target center. Therefore, the option to use this adjusted aiming point is incorporated into the timed-fire target.

TRAINING

Units are flexible in conducting 25-meter firing exercises with scaled silhouette targets. The proper use of scaled silhouette targets can make a valuable contribution to a unit's marksmanship program by accomplishing the following:

Soldiers learn how best to aim at targets located at various ranges while receiving precise feedback concerning bullet strike—whether it is a target hit or miss. Soldiers can acquire knowledge of range estimation. What they see through the sights is similar to what they would see if they were firing a target at the actual distance.

- Soldiers learn that close targets can be hit with a quick shot while more distant targets take a more deliberate application of the four marksmanship fundamentals.

- The use of an adjusted aiming point to allow for gravity, target movement, or wind may be effectively practiced.

- The silhouette target provides a permanent record for analysis by the soldier and his leader to assist in identifying and correcting firing problems.

- Soldiers develop confidence in their ability to successfully engage pop-up targets located at actual range.

- This target serves as an important diagnostic checkpoint. If soldiers cannot hit the scaled silhouettes, they cannot hit targets at actual range. Therefore, remedial training is probably needed before field firing is allowed.

- Having developed good firing skills and knowing what happens to bullets while firing at silhouette targets from supported and unsupported positions, the soldier's pop-up field fire practice results in a worthwhile training experience.

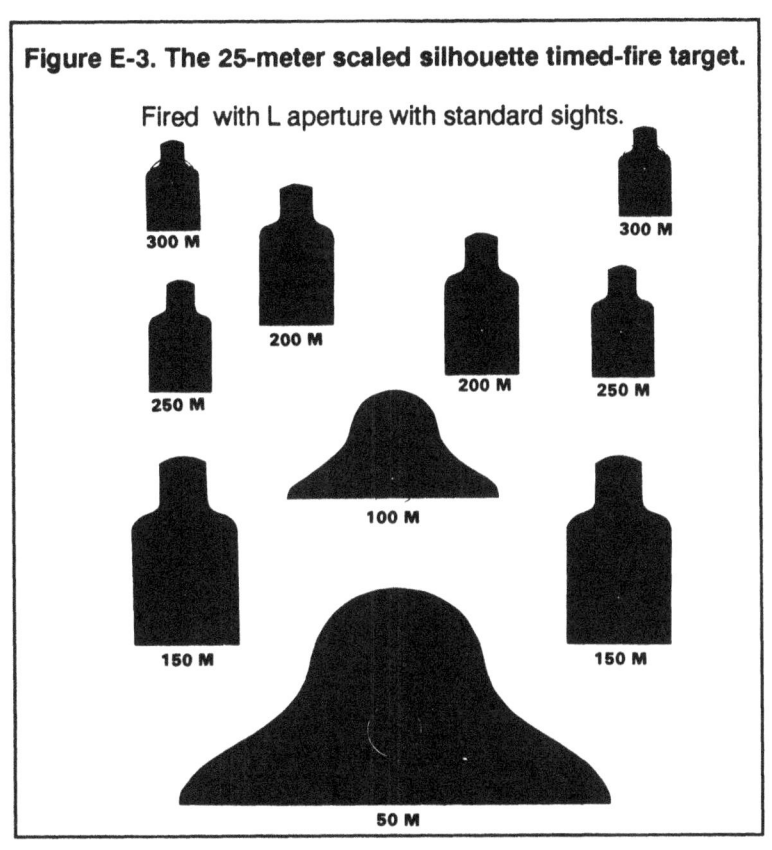

Figure E-3. The 25-meter scaled silhouette timed-fire target.

Fired with L aperture with standard sights.

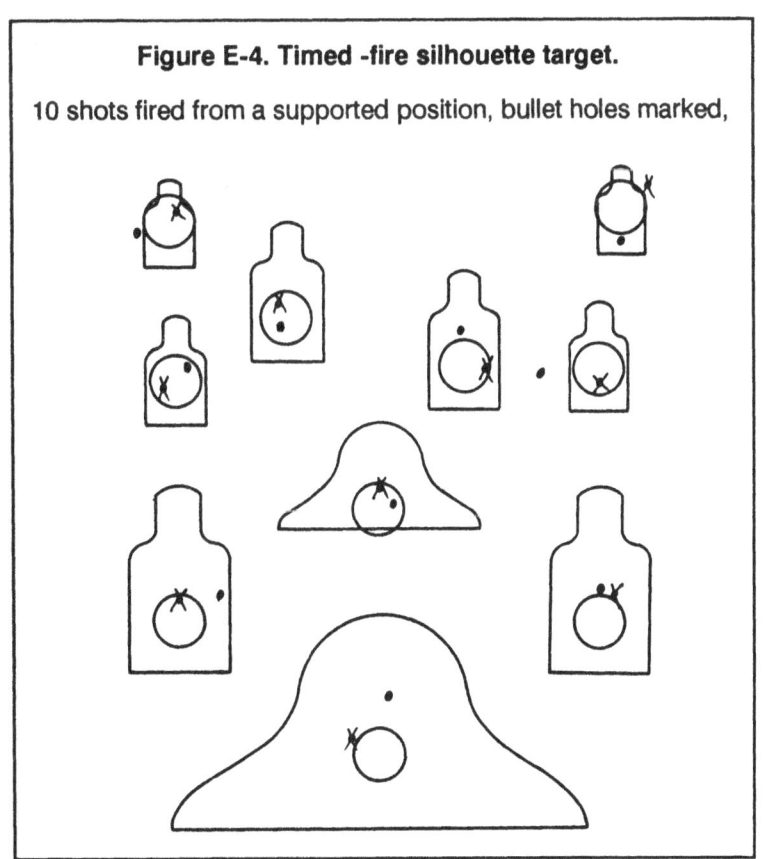

Figure E-4. Timed-fire silhouette target.

10 shots fired from a supported position, bullet holes marked,

The mode of fire may be either slow or timed, any firing position can be used, and any number of shots may be fired before checking the target. (Checking the target after only a few bullets have been fired is more beneficial than checking the target after several bullets have been fired.) When spotting scopes are available, the target can be checked from the firing line without clearing the firing line, which would allow a target check after each bullet fired. The scaled silhouette targets are also excellent for dry-fire training. They incorporate an adjusted aiming point to help the soldier learn the effects of gravity. They could be used, with instructor assistance, to learn about wind—for example, where should point of aim be for each target on the 10-target silhouette if there were a 10-mph full-value wind from the right.

All firing at the Army training centers is conducted with rifles that are equipped with standard sights. The long-range sight is used on the 250-meter range so the point of aim is equal to point of impact. The targets shown in Figures E-1 and E-3 are available for use by units that have rifles equipped with standard sights. For units that have some rifles equipped with the LLLSS, the slow-fire target shown in Figure E-5 and the timed-fire target shown in Figure E-6 are available. These targets are used the same as the previous targets. From the firing line, the soldier sees exactly the same thing and aims at the black silhouette exactly as he does the previous targets. The only difference is that bullet strike is evaluated based on the dotted circle and the dotted silhouette.

The target shown in Figure E-7, page E-8, is a silhouette target designed to be used as alternate course C at 25 meters. The target shown in Figure E-8, page E-8, is also an alternate course C target, but it has been scaled for 15 meters (50 feet or 600 inches) for use on indoor ranges. (A validated qualification course of fire for these targets is contained in Chapter 4.) Units may use any of the scaled silhouette targets to develop their own unique competitive program. Any action that encourages competition among soldiers can generate interest in developing good firing habits, and can motivate soldiers and sub-units to conduct practice required to develop good marksmanship skills.

While the targets scaled for live fire at 25 meters can serve several useful purposes on the 25-meter live-fire range, the perceived range to the target can be changed for dry-fire training by changing the distance to the target. Then, the targets are scaled based on a visual angle from the firing position, which means the 50-meter target is one-half actual size, the 100-meter target is one-fourth actual size, and so on. If the firer views the target from one-half the intended distance (12.5 meters), the perceived range to the target would be one-half—for instance, the 50-meter target would become a 25-meter target, and the 300-meter target would become a 150-meter target. Of course, the opposite would occur if the range to the target were doubled. When viewed from a range of 50 meters, the 50-meter target would appear as a 100-meter target, and the 300-meter target would appear as a 600-meter target. (The reduced targets in this manual could be used for indoor dry-fire training.) A simple procedure for finding the correct range is to adjust the distance while looking through the rifle sights until the 175-meter target appears to be the same size as the standard front sight post.

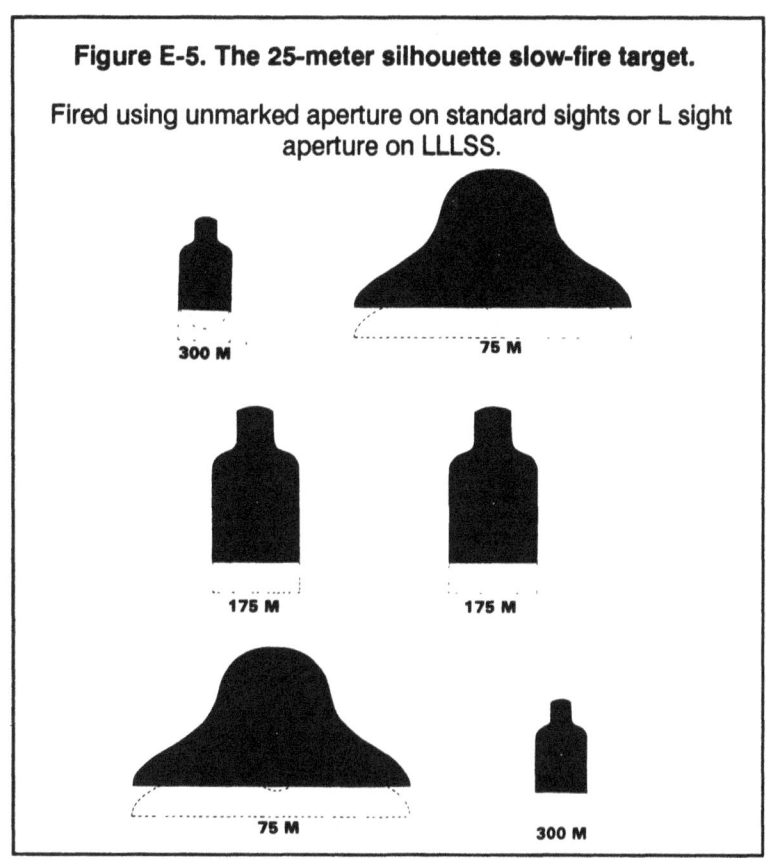

Figure E-5. The 25-meter silhouette slow-fire target.

Fired using unmarked aperture on standard sights or L sight aperture on LLLSS.

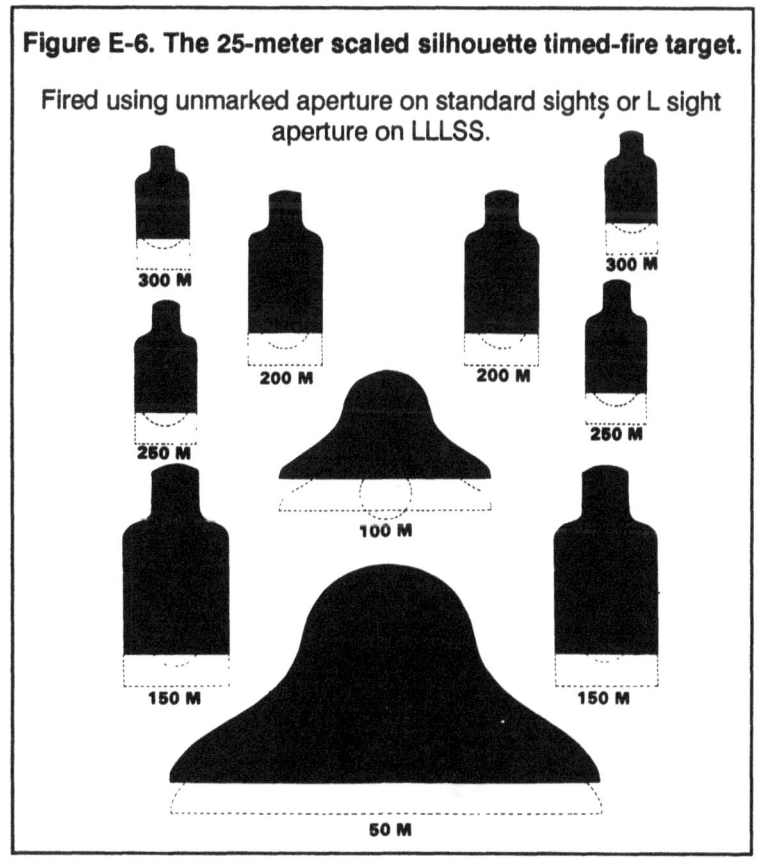

Figure E-6. The 25-meter scaled silhouette timed-fire target.

Fired using unmarked aperture on standard sights or L sight aperture on LLLSS.

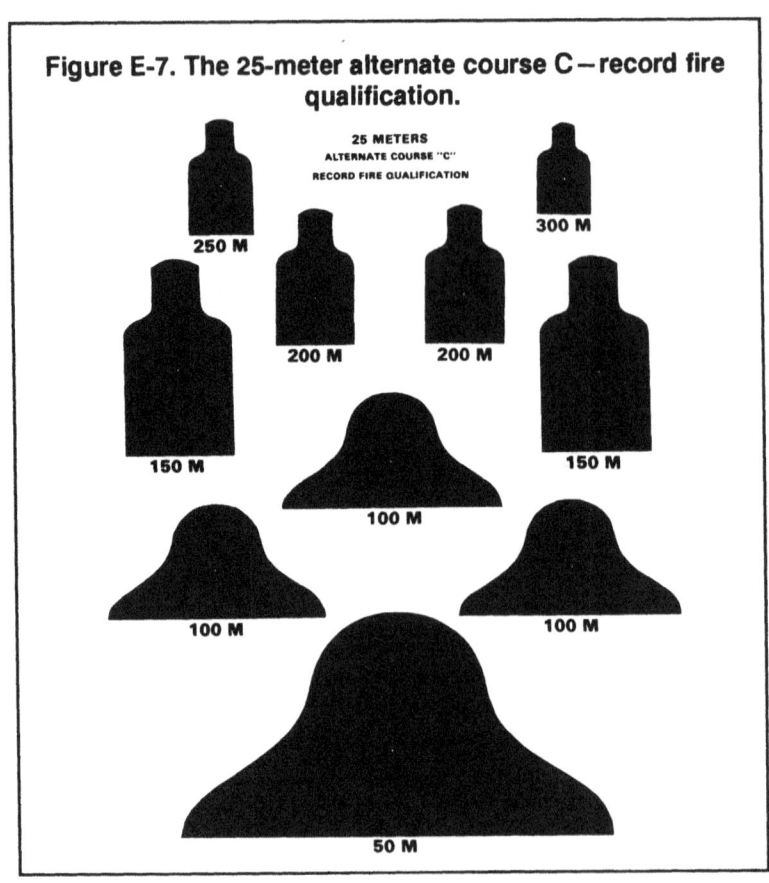

Figure E-7. The 25-meter alternate course C — record fire qualification.

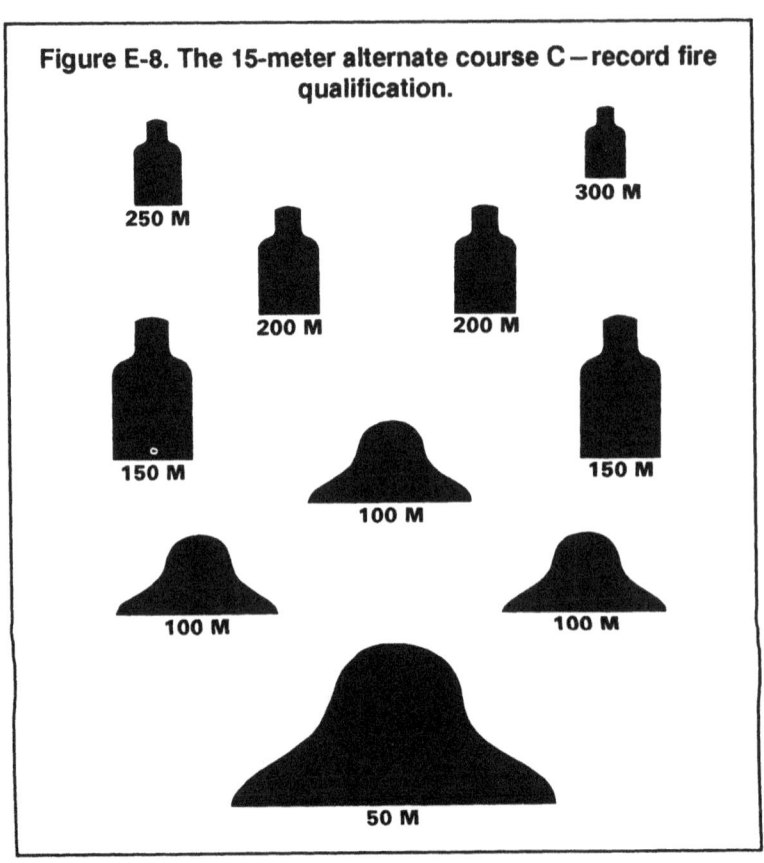

Figure E-8. The 15-meter alternate course C — record fire qualification.

FM 23-9

APPENDIX F

Precision Firing Information

This appendix provides information to assist instructors/trainers in effectively training M16A1 and M16A2 firers. Instructors/trainers must be knowledgeable on the effects of wind and gravity, ballistics, the elevation and windage rule, and bullet dispersion as they apply to firing proficiency.

Section I. EFFECTS OF GRAVITY

All soldiers should have basic knowledge of how the bullet is affected by gravity. Instructors/trainers must know the information contained herein.

EFFECTS ON AMMUNITION

The bullet begins to fall as soon as it leaves the muzzle of the rifle. The maximum speed or velocity of the bullet is at the muzzle, so it also begins to slow down as soon as it is fired. Figure F-1, page F-2, shows that the M193 ammunition drops 24 inches in slightly over one-third of a second. The chart shows the amount of drop relative to the departure line or bore line at 25-meter intervals and the time required to reach each range distance.

At first, the bullet travels fast, covering the first 25 meters at an average speed of almost 2,200 mph. The last 25-meter segment shown in Figure F-1 (275 to 300 meters) is covered at an average speed of about 1,450 mph. Gravity causes the rate of drop to increase as flight time increases. Since it takes the bullet more time to travel as the speed slows, the effects of gravity and wind increase as the range increases. Each band on the chart represents 25 meters — the width of the band indicates the length of time it takes to travel that 25-meter segment.

The drop in inches from the rifle bore line is shown on the left side in Figure 1. In this example, finding the number 5 on the left side of the chart and going across to where the bullet is indicated shows the bullet has dropped 5 inches when it reaches 150 meters and has taken .17 second (column at right) to arrive. The bullet drops 2 inches more between 150 meters and 175 meters, and it takes about .03 second to travel that extra 25-meter distance.

M16A1 and M193 ammunition.

Bullet drop. The bullet is affected by gravity just like any other falling object. Even though the bullet is traveling fast, once it has been fired from the rifle, it falls to the ground as though it were dropped by hand. Figure F-2 shows how much the bullet drops from the bore line of the rifle. If the bore of the rifle were to be lined up on a 450-meter target (the same as looking through the bore and aligning the bore with the target), the bullet would hit 64 inches below the spot where the bore was pointing. This much drop is important if soldiers are to be effective marksmen.

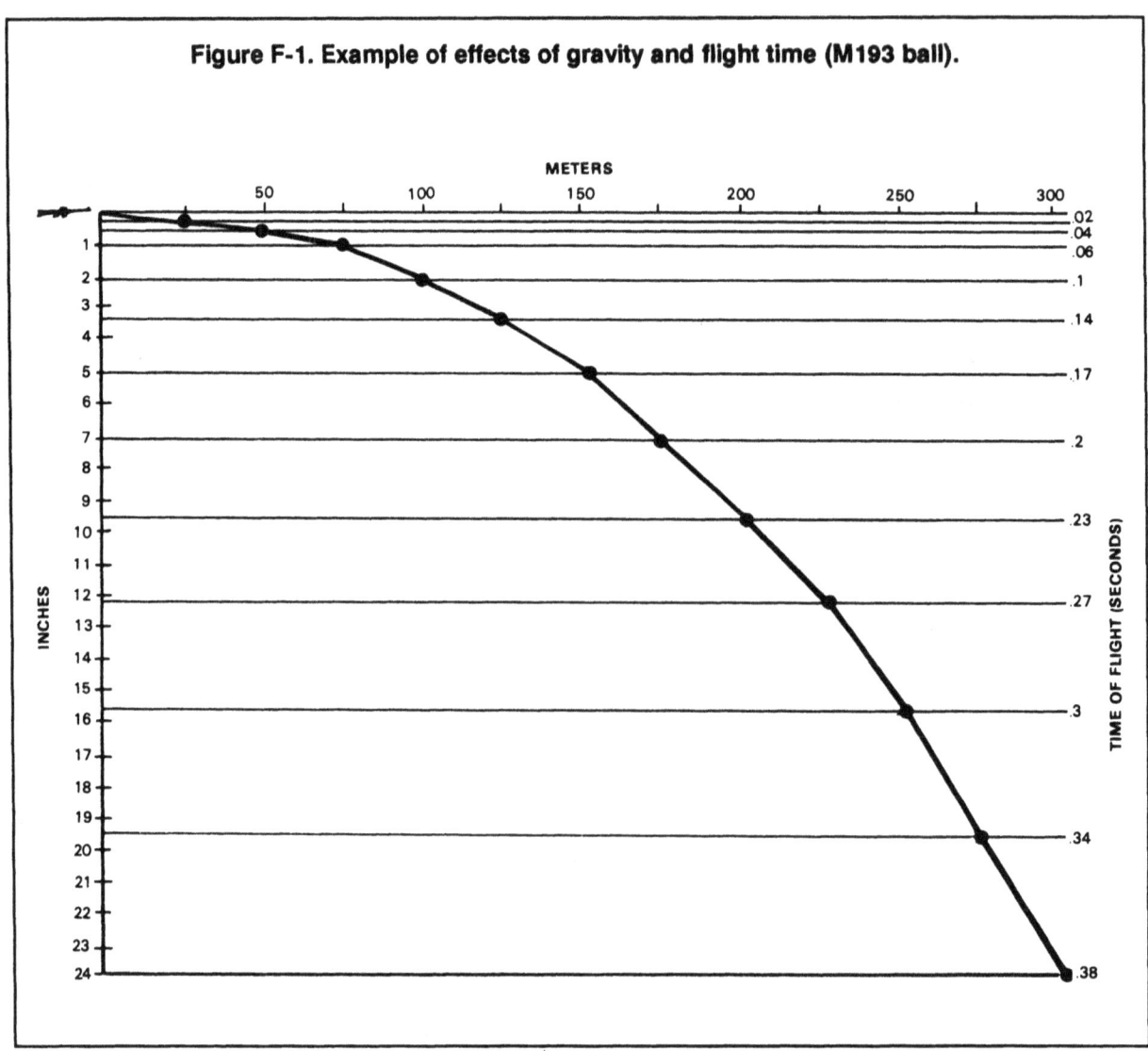

Figure F-1. Example of effects of gravity and flight time (M193 ball).

Compensation for gravity. The firer must compensate for the effects of gravity to engage high-priority targets. The objective of the battlesight zero is to find a zero range that allows for target engagement out to the maximum possible range while requiring minimum adjustments to the aiming point. A study of M16 trajectory data reveals that setting the sights to hit at 250 meters is the best compromise for hitting all targets from 0 to 325 meters without major adjustments to the aiming point. As shown in Figure F-2B, the bullet stays close to the line of sight, but this has been accomplished by pointing the bore well above the target. The bullet drop from the bore line applies equally in this figure. The bore is actually pointing 16 inches above the aiming point at 250 meters and 24 inches above the point where the bullet will strike at 300 meters. With graphics showing only line of sight and trajectory, it appears that the bullet rises and then falls, but this is not true. The bullet's line of departure is the bore line, and it always drops from that line as shown in Figure F-2A. The line of sight and bullet trajectory actually coincide at 42 meters (the bullet starts out about 2.6 inches below the sights) and again at 250 meters, giving the illusion that the bullet rises and falls.

Figure F-2. M16A1 short-range trajectory (250-meter zero).

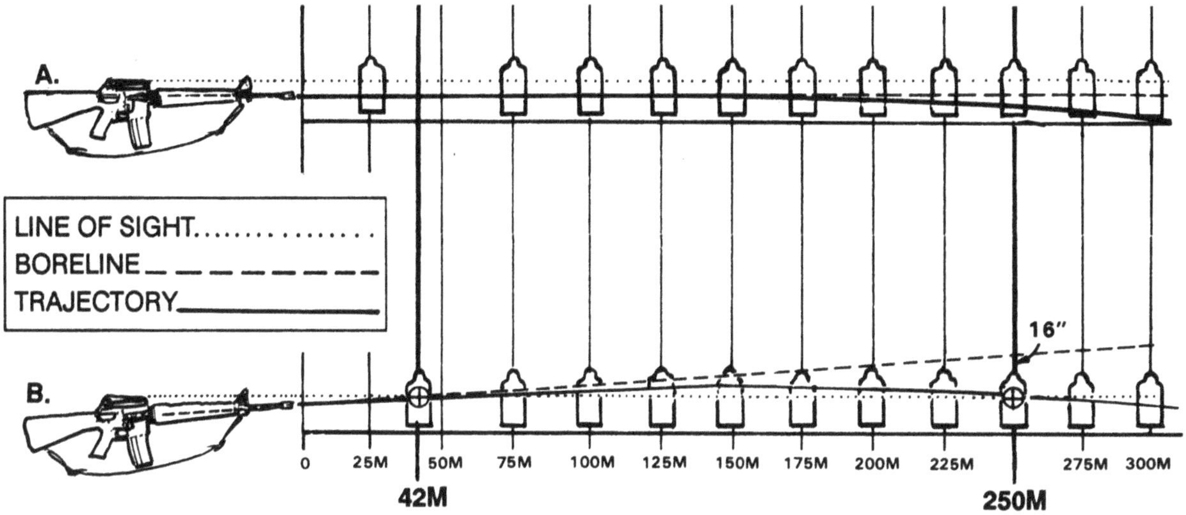

A. Illustrates that the bore line and line of sight are parallel, gravity causes the bullet to drop.
B. Illustrates that by raising the bore line above the line of sight, the effects of gravity are compensated for and the bullet hits all targets near center mass. Notice how the line of sight, bore line, and trajectory coinside at 42 meters, and how the line of sight and trajectory again coincide at 250 meters but the bore line is 16 inches above the target.

While the trajectory shown in Figure F-2B provides for target engagement out to 325 meters, targets beyond this range would require major adjustments in the aiming point. Therefore, after the 250-meter zero has been obtained, flipping to the long-range sight provides a zero of 375 meters as shown in Figure F-3. This provides for effective target engagement from 350 meters out to the maximum effective range of 460 meters.

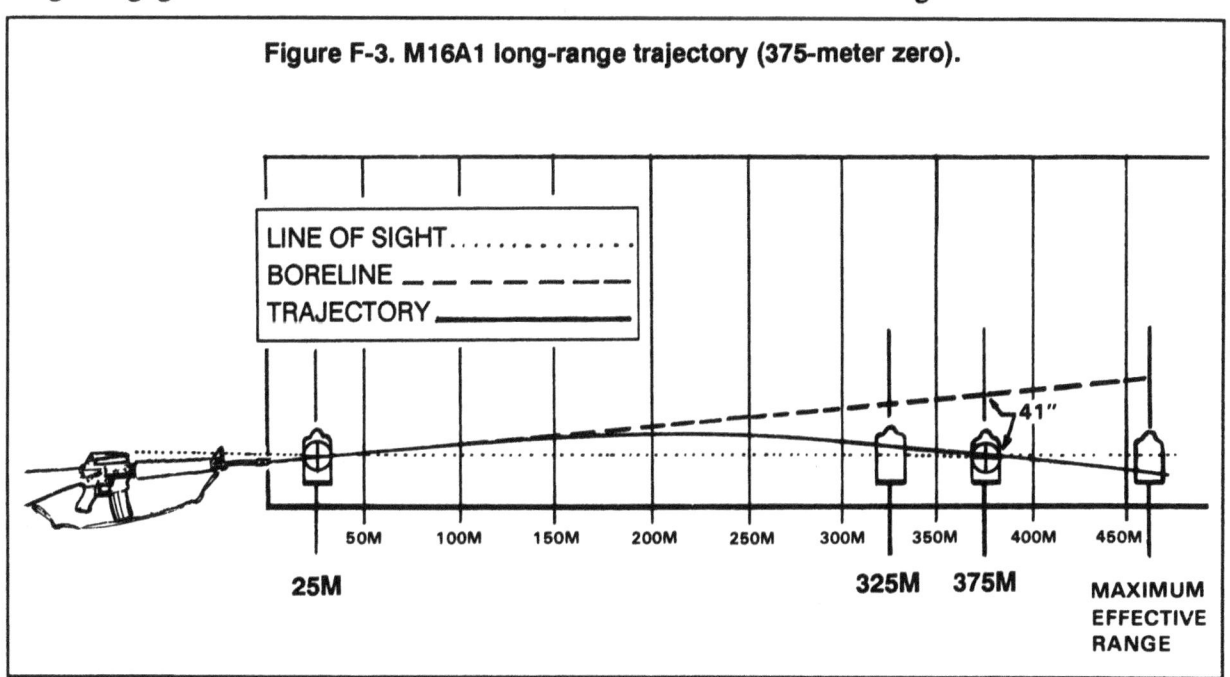

Figure F-3. M16A1 long-range trajectory (375-meter zero).

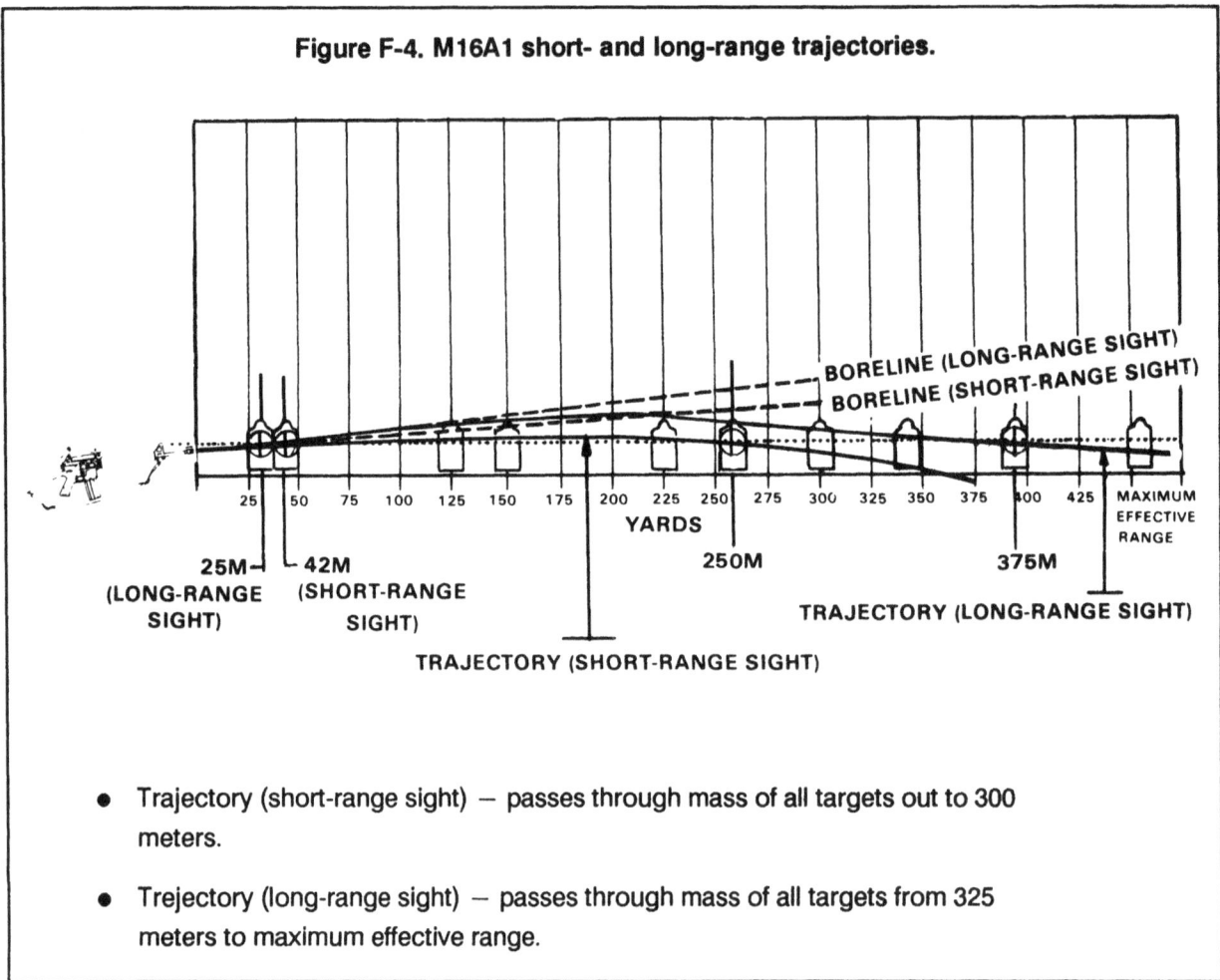

Figure F-4. M16A1 short- and long-range trajectories.

- Trajectory (short-range sight) — passes through mass of all targets out to 300 meters.

- Trejectory (long-range sight) — passes through mass of all targets from 325 meters to maximum effective range.

It should be noted that the same scale was used in Figures F-2, F-3, and F-4. Therefore, the relationship between the bore line and trajectory is similar in all figures. The bullet drops from the bore line the same way regardless of the zero range.

M16A2 and M855 ammunition.

Bullet drop. The bullet drop for the M16A2 is similar to that discussed for the M16A1. The M16A2 rifle is zeroed for 300 meters. Therefore, point of aim and point of impact are center of mass at 300 meters with subsequent rear sight elevation settings out to 800 meters. The adjusted aiming points for all targets less than 300 meters are shown in Figure F-5.

Trajectory. For engagement at ranges greater than 300-meters, setting the correct range on the rear sight elevation knob results in the point of impact being the same.

ADJUSTED AIMING POINT

Teaching an adjusted aiming point is intended to increase hit probability when properly presented. However, soldiers can be easily confused, which could result in degraded performance. Therefore, all soldiers should be taught to aim at target center unless they are confident they know the range to the target, or they have just engaged a close target and missed (presumably high), or they have just engaged a distant target and missed (presumably low), or the strike of the bullet had been observed. Soldiers must be given

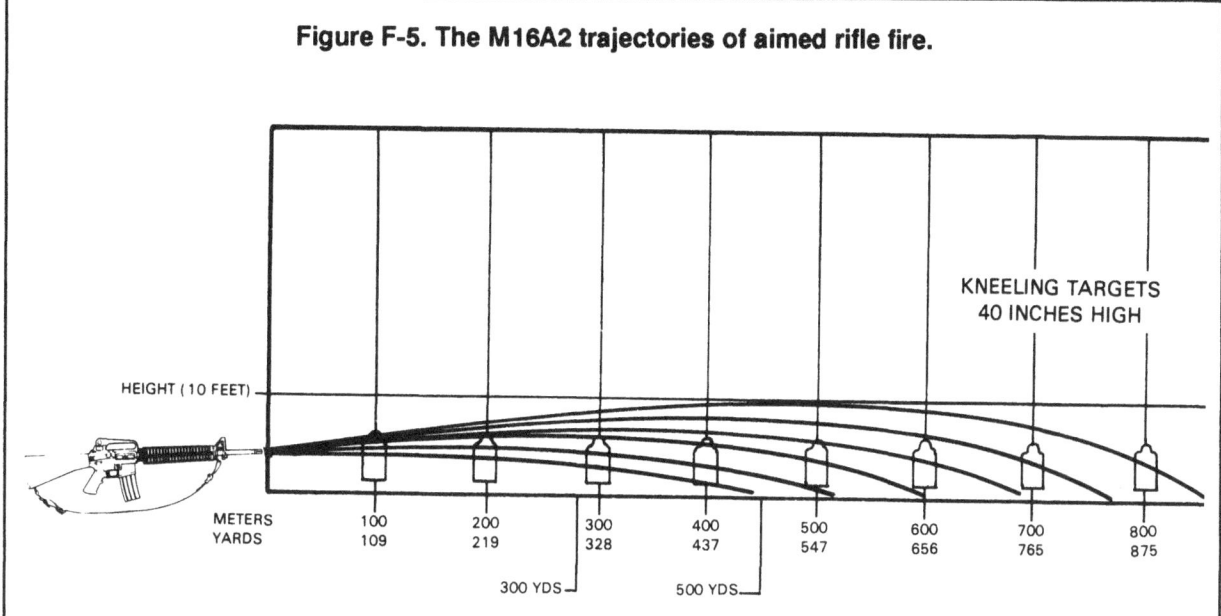

Figure F-5. The M16A2 trajectories of aimed rifle fire.

Point of aim at each target, at all ranges, is "center of mass" or in the center of the chest. Point of impact is also center of mass. Note that fire on target at 300 meters is also effective on 400 meter target. At 600 meters, fire does not exceed the height of a man at any point along its trajectory.

the correct information so they can improve their firing performance by adjusting their aiming point to allow for the effects of gravity.

The use of an adjusted aiming point could be more important for combat than for qualification since the combat environment is unstructured. For example, the soldier who must fire into a small bunker aperture located at a range of 150 meters could be more effective if he understands he must aim low. Also, a soldier who is trying to hit a head-and-shoulders type target at a range of 325 meters could increase his hit probability if he aimed high using a regular sight or aims low using the long-range sight. Usually a soldier should aim low for targets closer than 250 meters and aim high for targets farther away than 250 meters. This increases combat effectiveness while adding minimum complications to the marksmanship training program.

M16A1. Theoretically, using these aiming points places the center of each shot group (assuming a perfect zero) at target center, allowing for a maximum number of other variables and soldier error. These adjustments are small and should only be applied by competent firers who wish to improve their firing performance. The aiming points in Figure F-6, page F-5, are for use with the long-range sight. Since the sight is zeroed for 375 meters, the 350- and 400-meter targets can be engaged with center-of-mass aiming. As ranges exceed 400 meters, some adjustment of the aiming point is required for a better chance of hitting the target.

Figure F-6. M16A1 aiming points.

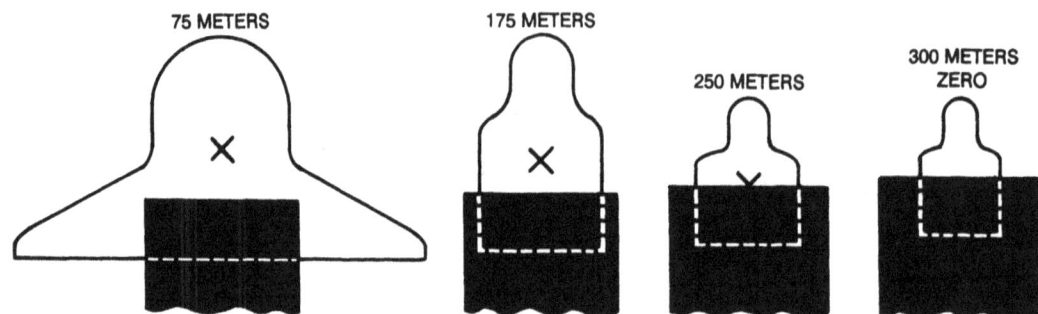

Use of the M16A1 short-range sight—the aiming points shown result in the highest probability of bullets hitting target center.

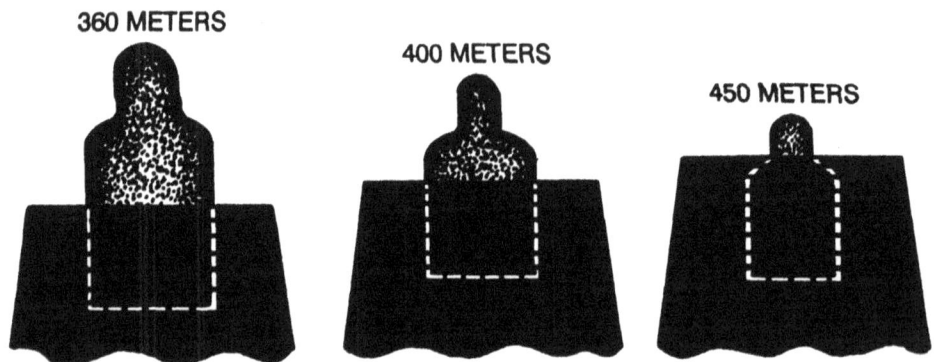

" L" SIGHT

Use of the M16A1 long-range sight—the aiming points shown result in the highest probability of bullets hitting target center.

M16A2. With the rifle zeroed for 300 meters, the correct adjusted aiming point for targets less than 300 meters is shown in Figure F-7.

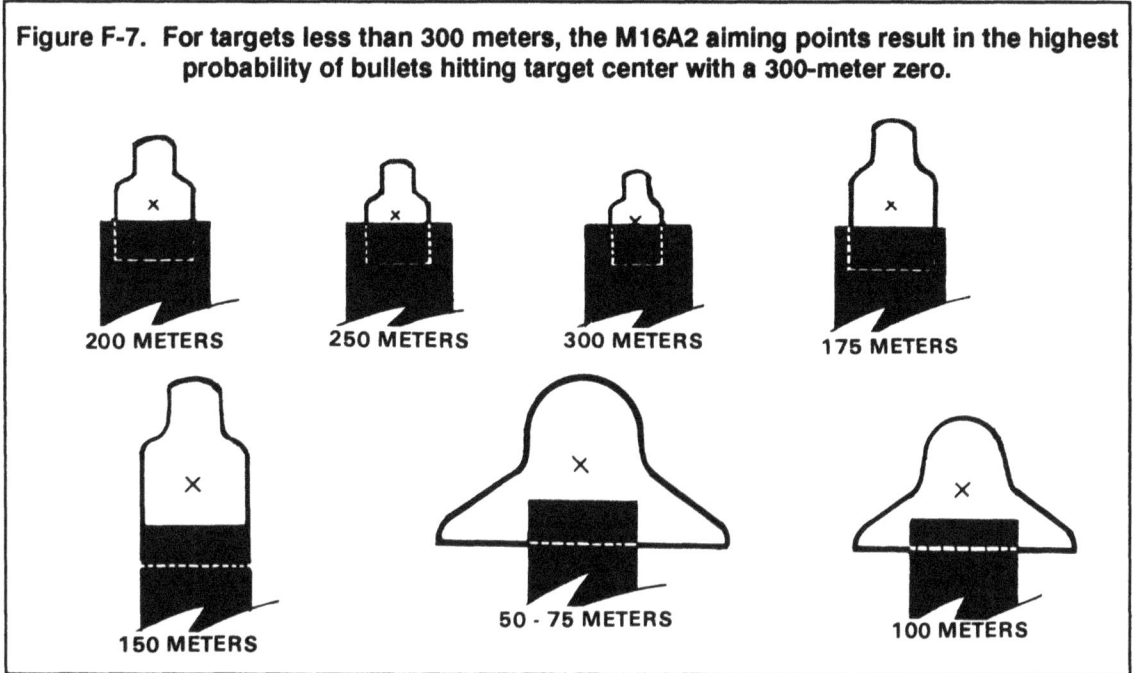

Figure F-7. For targets less than 300 meters, the M16A2 aiming points result in the highest probability of bullets hitting target center with a 300-meter zero.

Section II. EFFECTS OF WIND

Marksmanship instructors/trainers should know how the effects of wind influence the flight of the bullet, and soldiers should know how to compensate for such bullet displacement. This instruction is appropriate for all marksmanship training and concurrent training. Wind affects the bullet similar to gravity — as range increases, the effects of wind increase.

WIND SPEED AND DIRECTION

The effects of wind vary depending on changes in wind speed and direction. Wind is classified by the direction it is blowing in relationship to the firer/target line. The clock system is used to indicate wind direction and value (Figure F-8, page F-8). Winds that blow from the left (9 o'clock) or right (3 o'clock) are called full-value winds, because they have the most effect on the bullet. Winds that blow at an angle from the front or rear area are called half-value winds, because they have about one-half the effect on the bullet as full-value winds. Winds that blow straight into the firer's face or winds that blow straight into the target are termed no-value winds, because their effect on the bullet is too small to be concerned with.

Figure F-9, page F-8, illustrates how the effects of wind on the bullet are similar to the effects of gravity — as range increases, the effects of wind increase.

A 10-mph full-value wind moves an M16A1 (M193) bullet about 1/2 inch at 25 meters to about 46 inches at 475 meters. (Using the data presented, wind effects for all

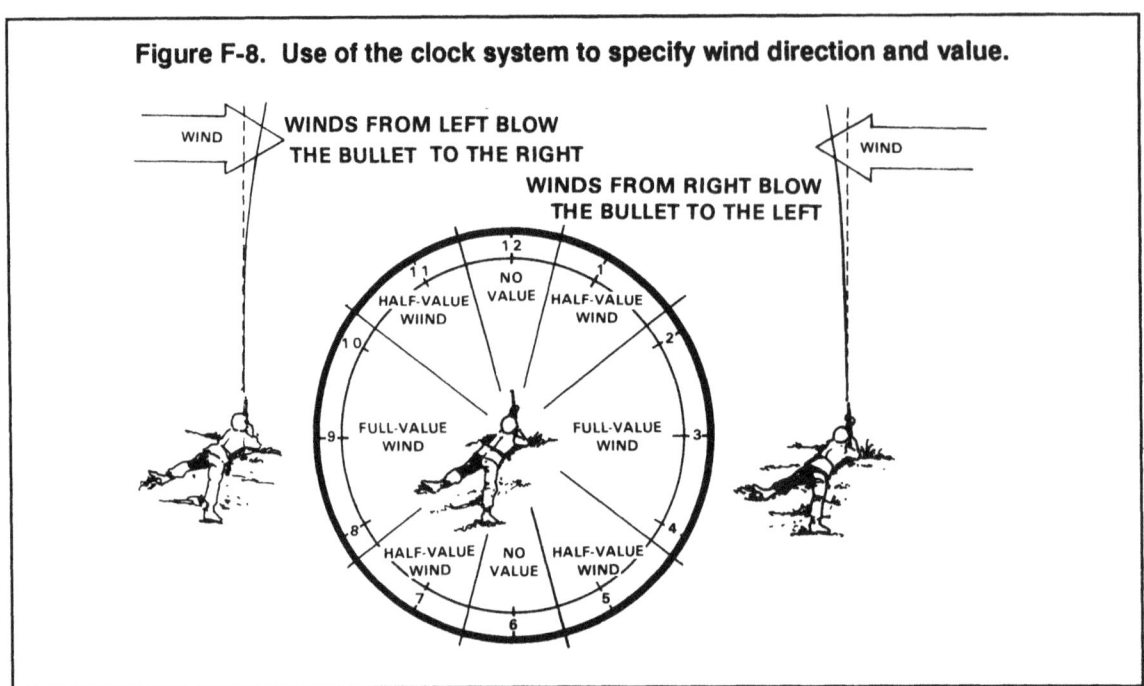

Figure F-8. Use of the clock system to specify wind direction and value.

conditions can be determined.) A wind of greater speed increases bullet movement by a uniform amount — a 15-mph wind moves the bullet at all ranges 1 1/2 times more than a 10-mph wind. Using these data, firers can determine wind effects for all conditions.

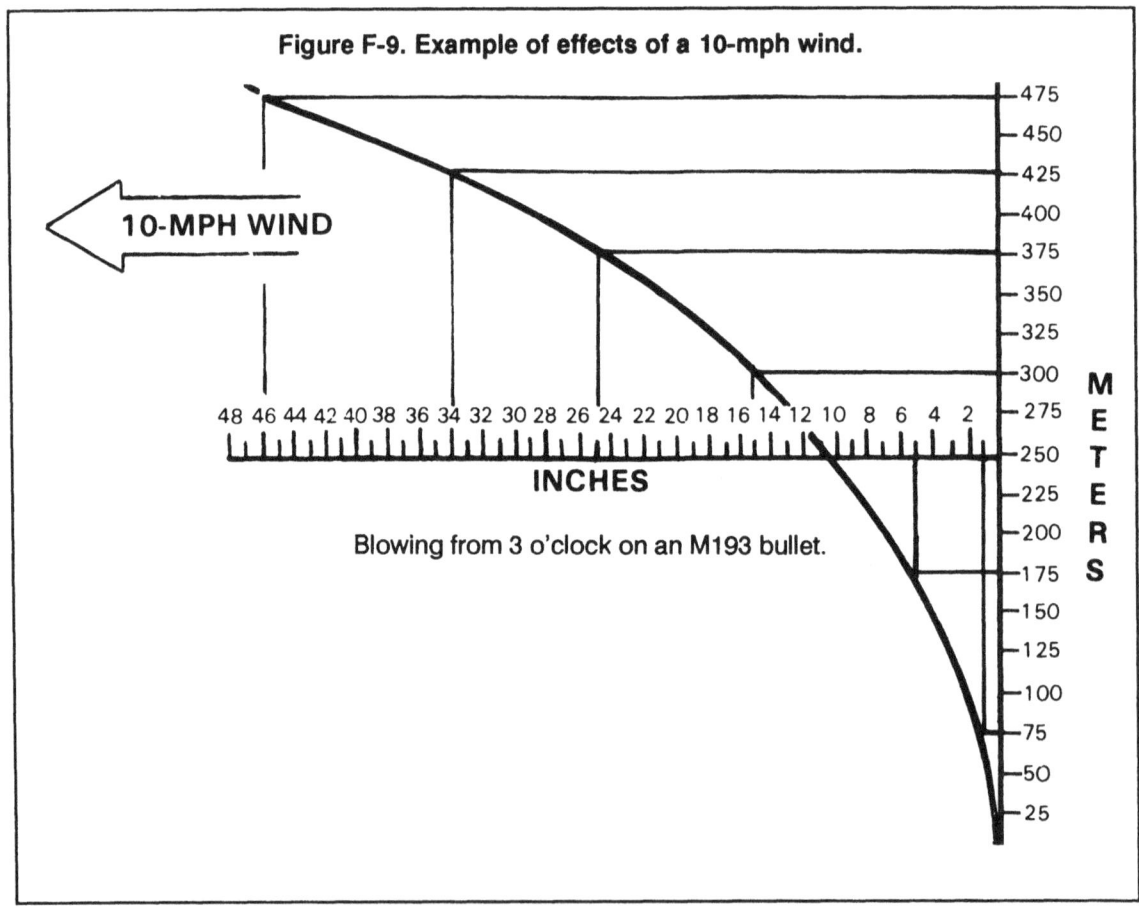

Figure F-9. Example of effects of a 10-mph wind.

Blowing from 3 o'clock on an M193 bullet.

Effects of wind are much greater at longer ranges; however, they are uniform in relation to speed—for example, a 5-mph full-value wind would have exactly one-half the effect shown in the figure, moving the bullet 5 inches at 250 meters, then 10 inches, and so on. A wind of greater speed would move bullets at all ranges 1 1/2 times more than a 10-mph wind, or 7 1/2 inches at 175 meters. The same rule also applies to a half-value wind. A 5-mph half-value wind would move bullets one-fourth the amount shown in Figure F-9—3 3/4 inches at 300 meters.

An easy way to remember the effects of wind is that a 10-mph wind moves the bullet 10 inches at the battlesight zero range of 250 meters. If this information is taken to the standard field fire range with targets at 75, 175, and 300 meters, it is easy to remember that a 10-mph wind moves the bullet 1, 5, 10, and 15 inches at the ranges of 75, 175, 250, and 300 meters, respectively. These numbers can be converted to a 1-mph wind—1/10 inch at 75 meters, 1/2 inch at 175 meters, 1 inch at 250 meters, and 1-1/2 inches at 300 meters—so that when the wind speed has been determined, it can be multiplied with the mph figure to determine bullet displacement.

WIND MEASUREMENT

A wind gage can be used for precise measurement of wind velocity. When a gage is not available, velocity is estimated by one of the following methods:

Flag method. If the firer can observe a flag or any clothlike material hanging from a pole, he should be able to estimate the angle formed at the juncture of the flag and pole. As shown in Figure F-10, dividing this angle by the constant number 4 equals the wind velocity in mph.

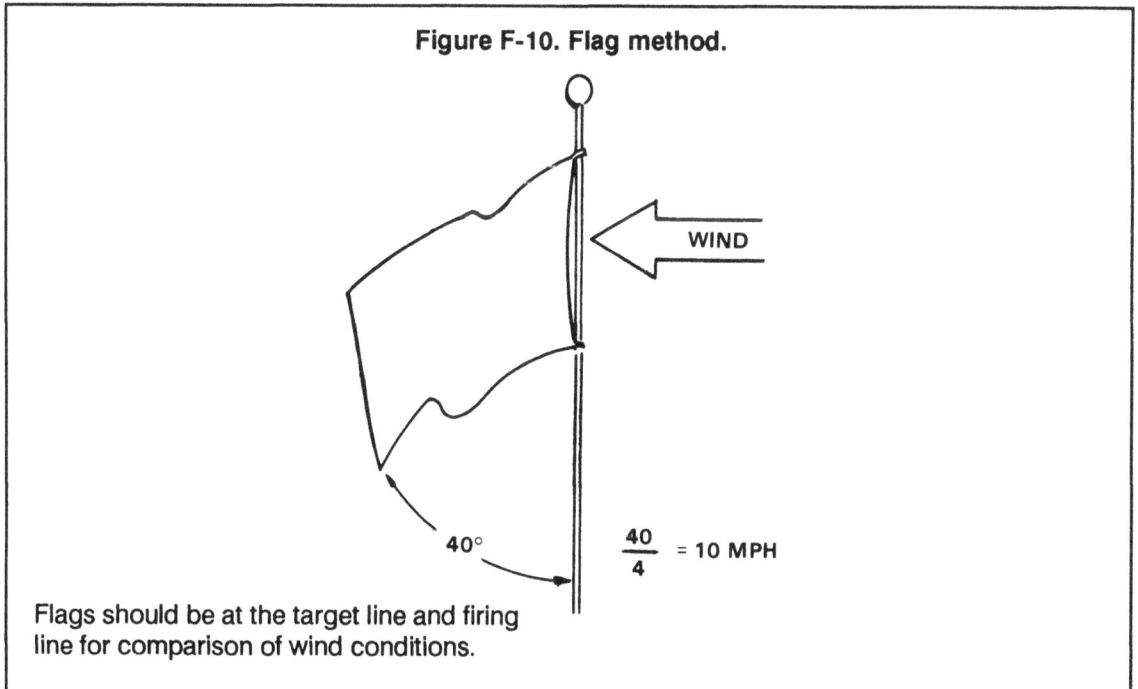

Figure F-10. Flag method.

Flags should be at the target line and firing line for comparison of wind conditions.

Pointing method. If a flag is not visible, a piece of paper or other light material can be dropped from the shoulder. By pointing directly at the spot where it lands, the angle can be estimated. As shown in Figure F-11, this angle is also divided by the constant number 4 to determine the approximate wind speed in mph. This indicates conditions at the firing position, which could be different at the target position.

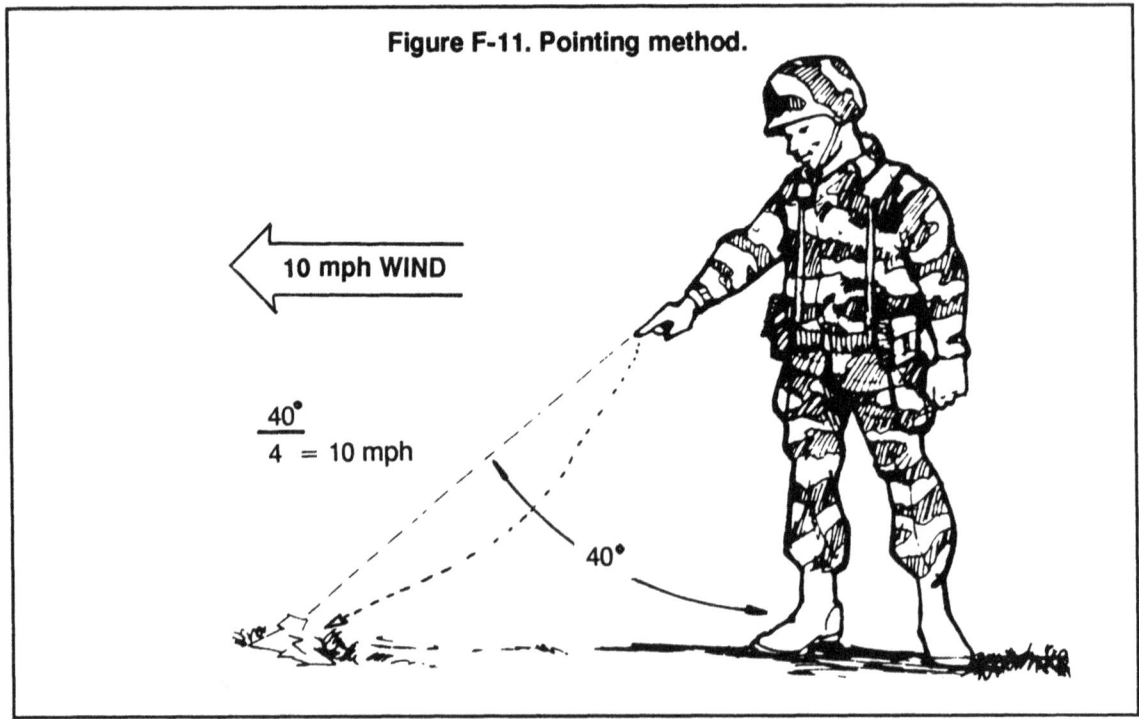

Figure F-11. Pointing method.

Observation method. If the two methods already described cannot be used, the following information can assist in determining wind velocities:

- Winds under 3 mph can barely be felt by the firer, but the presence of slight wind can be determined by drifting smoke.
- Winds of 3 to 5 mph can be felt lightly over the firer's face.
- Winds of 5 to 8 mph constantly move the leaves of trees.
- Winds of 8 to 12 mph raise dust and loose paper.
- Winds of 12 to 15 mph cause small trees to sway.

Wind is highly variable and sometimes quite different at the firing position than at the target position. Even though the wind is blowing hard at the firing line, the bullet path could be protected by trees, brush, or terrain. The wind can vary by several mph between the time a measurement is taken and when the bullet is fired. Therefore, trianing time should not be wasted trying to teach soldiers an exact way to measure wind speed. Soldiers should understand that the wind can blow the bullet off course but not to overcompensate and miss targets because of applying too much holdoff.

ADJUSTED AIMING POINT

Given the nature of the record fire course and combat, it is not appropriate to make sight adjustments for wind; therefore, the holdoff technique must compensate for the effects of wind. Placement of the aiming point causes bullets to hit target center when firing the M16A1 rifle in a full-value 10-mph wind (Figure F-11). The firer displaces the center of the front sight post the number of inches shown in Figure F-12 for that particular range.

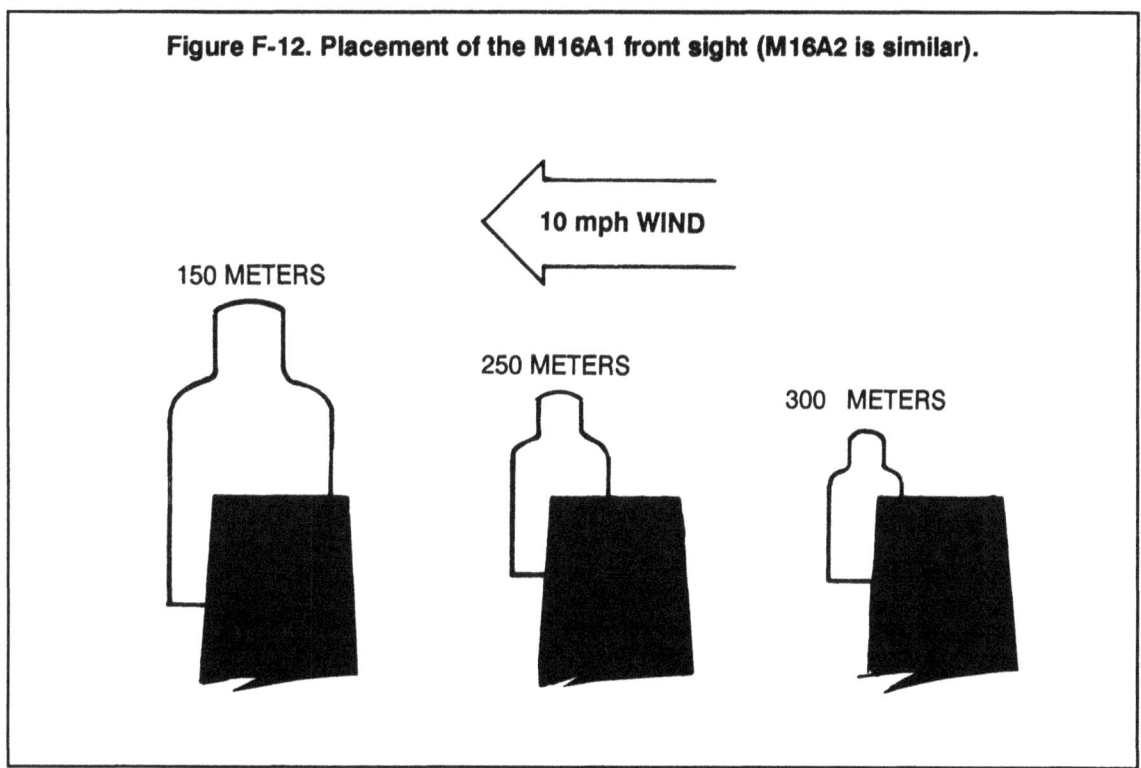

Figure F-12. Placement of the M16A1 front sight (M16A2 is similar).

Wind has a minor effect on the M16 bullet relative to the size of the target at ranges out to 100 meters. When engaging targets in excess of 150 meters in heavy winds, adjusting the aiming point into the wind increases the probability of a hit. Wind effects are uniform in relation to speed, that is, a 5-mph wind has one-half the effect of a 10-mph wind, and a 20-mph wind has twice the effect of a 10-mph wind.

Firers must adjust their aiming point into the wind to compensate for the effects of wind. If they miss a distant target and the wind is blowing from the right, they should aim to the right for the next shot. A guide for the initial adjustment is to split the front sight post on the edge of the target facing the wind.

The newly assigned soldier should aim at target center for the first shot and then adjust for wind when he is confident that wind caused the target miss. Experienced firers should be able to apply the appropriate holdoff for the first shot, but the basic rule must be followed—when in doubt, aim at target center.

Section III. ELEVATION AND WINDAGE RULE

The elevation and windage rule states that one click of elevation or windage moves the strike of the bullet a specific distance at a specific range. At a range of 25 meters, one click of windage moves the strike of the bullet .33 cm, and one click of elevation on the front sight moves the strike of the bullet .83 cm. To compute the distance (D) that one click of elevation (front sight) or windage moves the strike of a bullet at a given range

FM 23-9

(R), divide the range (expressed in meters) by 25, and multiply by either .33 cm for windage or .83 cm for elevation.

Windage: $D = \dfrac{R}{25\,m} \times .33$ D = Distance in centimeters

Elevation: $D = \dfrac{R}{25\,m} \times .83$ R = Range in meters

WINDAGE

To compute the distance that one click of windage moves the strike of the bullet at a range of 300 meters, divide 300 meters by 25 meters and multiply by .33 cm.

$D = \dfrac{300\,m}{25\,m} \times .33 = 12 \times .33 = 4\,cm\ (4.0\ or\ 3.96)$

Therefore, one click of windage moves the strike of the bullet 4 cm at a range of 300 meters. Table F-1 illustrates the amount of change in both centimeters and inches of one click of windage at various ranges.

| Table F-1. Windage measurements. ||
Distance (in meters)	Impact*
25	0.33 cm (1/8 inch)
100	1.45 cm (5/8 inch)
200	2.9 cm (7/8 inch)
300	4.4 cm (1 3/4 inches)
400	5.8 cm (2 1/4 inches)
500	7.3 cm (2 7/8 inches)
600	8.7 cm (3 1/2 inches)
700	10.2 cm (4 inches)
800	11.6 cm (4 5/8 inches)

*All values rounded off.

ELEVATION

Front sight. To compute the distance that one click of elevation (front sight) moves the strike of the bullet at a range of 300 meters, divide 300 metes by 25 meters and multiply by .83 cm.

$D = \dfrac{300\,m}{25\,m} \times .83 = 12 \times .83 = 10\,cm$

Therefore, one click of elevation on the front sight moves the strike of the bullet 10.9 cm at a range of 300 meters. Table F-2 shows the amount of change in both centimeters and inches of one click of elevation on the front sight at various ranges.

Table F-2. Change in elevation for one click on front sight.

Distance (in meters)	Impact*
25	0.83 cm (3/8 inch)
100	3.6 cm (1 3/8 inches)
200	7.3 cm (2 7/8 inches)
300	10.9 cm (4 1/8 inches)

*All values rounded off.

Rear Sight. The elevation knob adjusts elevation 1.1 inch for each click at 100 meters. If the scale on the elevation knob were cut in half and flattened out it would look like Figure F-13, with each dot and each number representing one click of elevation.

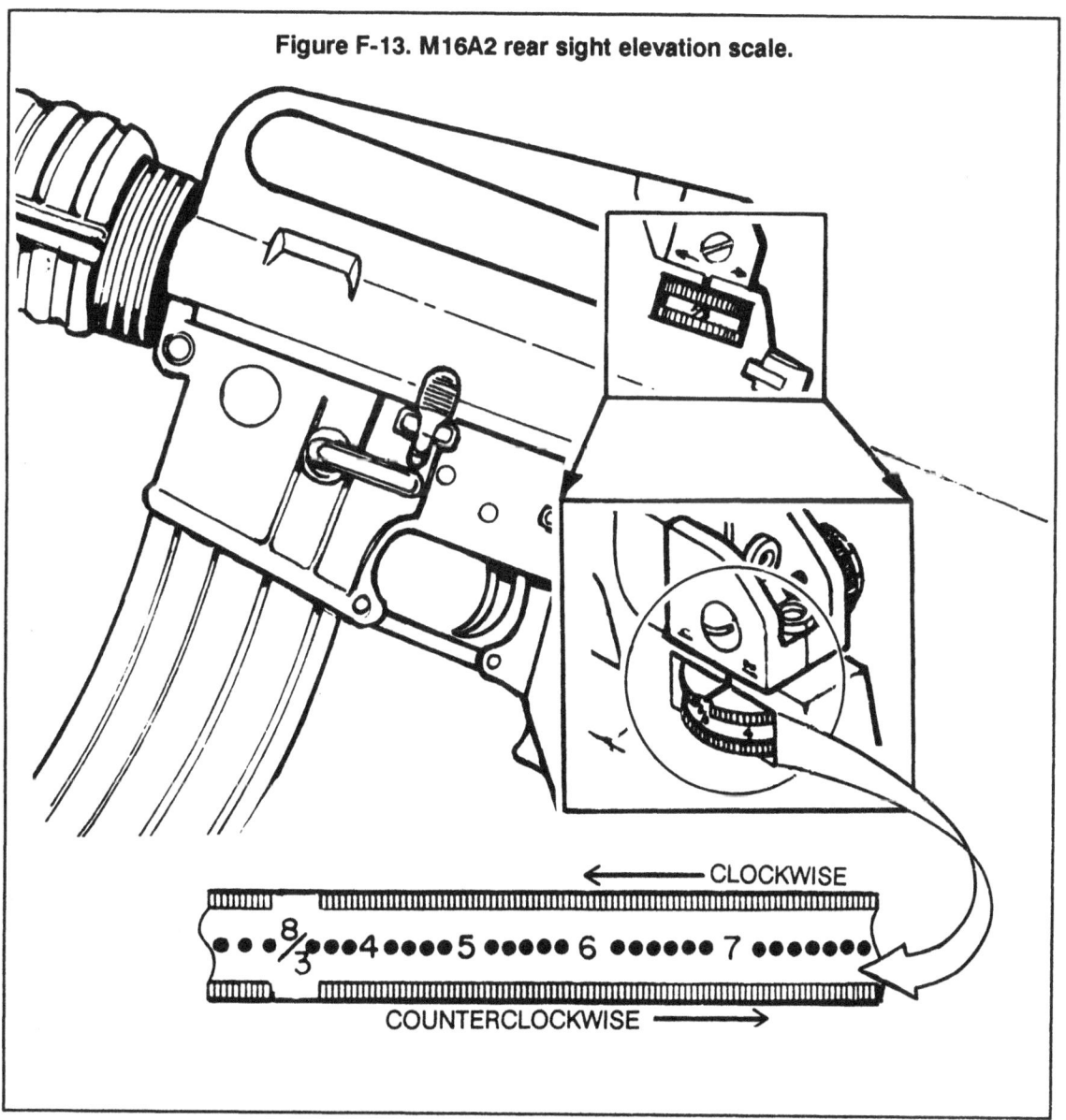

Figure F-13. M16A2 rear sight elevation scale.

Table F-3 shows the amount of change in elevation of the strike of the bullet for each click at various ranges — one click equals one minute of angle.

Table F-3. Change in elevation for one click at various ranges on rear sight.	
Distance	Change in Elevation for Each Click
When aiming at —	
300 meters, 1 click =	3.3 inches (7.9 cm)
400 meters, 1 click =	4.4 inches (10.5 cm)
500 meters, 1 click =	5.5 inches (13.1 cm)
600 meters, 1 click =	6.6 inches (15.7 cm)
700 meters, 1 click =	7.7 inches (18.3 cm)

APPLICATION

The squares on the M16A2 25-meter zero target are designed so that one click of elevation moves the strike of the bullet one square on the target. Three clicks of windage move the strike of the bullet one square on the target.

Section IV. BALLISTICS

Commanders and marksmanship trainers must understand some aspects of ballistics to teach the principles of zeroing and engagement of long-range targets. Ballistics is a science dealing with the motion and flight characteristics of projectiles. The study of ballistics in rifles is divided into three categories: **internal, external, and terminal.**

- **Internal ballistics** concerns what happens to the bullet before it leaves the muzzle of the rifle.

- **External ballistics** deals with factors affecting the flight path of the bullet between the muzzle of the rifle and the target.

- **Terminal ballistics** deals with what happens to the bullet when it comes in contact with the target.

INTERNAL BALLISTICS FACTORS

The overall dimensions of the combat service 5.56-mm cartridges are the same, which allow cartridges to be fired safely in M16A1 or M16A2 rifles. There are internal differences that affect firing accuracy. An ammunition comparison is provided in Figure F-14.

This increase in projectile length, weight, and configuration requires different twists in the barrels, lands, and grooves to stabilize the M855 bullet in flight (Figure F-15). The M16A1 has a 1:12 barrel twist. (The bullet rotates by the lands once for

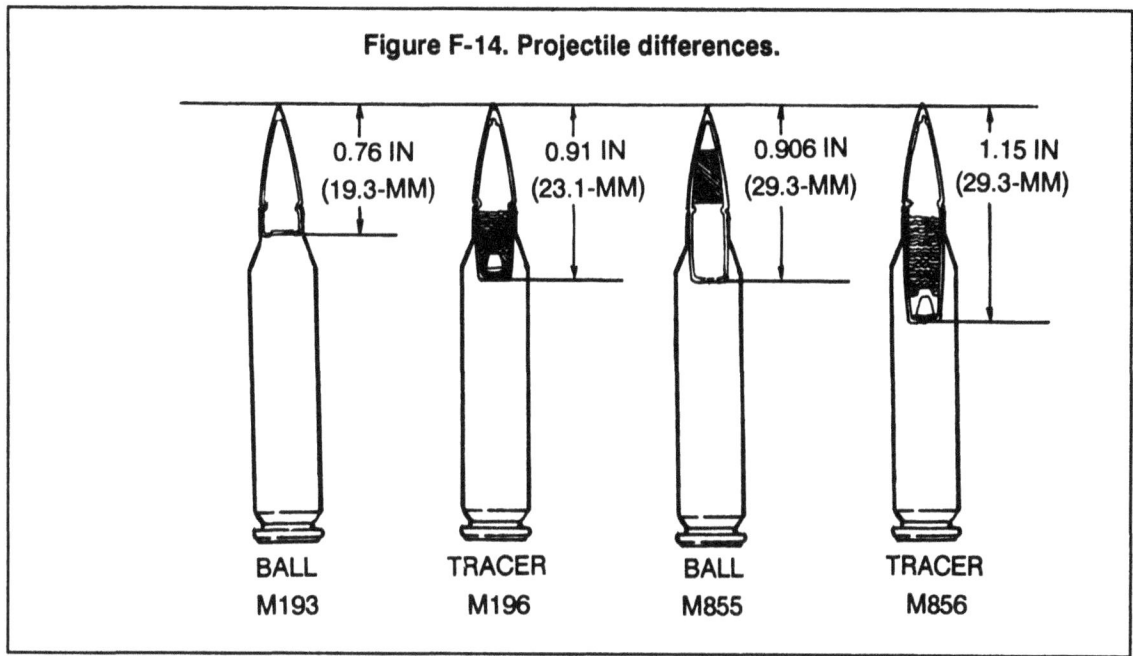

Figure F-14. Projectile differences.

every 12 inches of travel down the barrel.) The M16A2 has a 1:7 barrel twist. (The bullet rotates once for every 7 inches of travel down the barrel.)

The M16A1, with its 1:12 twist, does not put enough spin on the heavier M855 bullet to stabilize it in flight, causing erratic performance and inaccuracy for training or full combat usage (30.48- to 35.56-cm shot group at 91.4 meters and 72-inch shot group at 274.2 meters)(Figure F-15A). Therefore, while it is safe to fire the M855

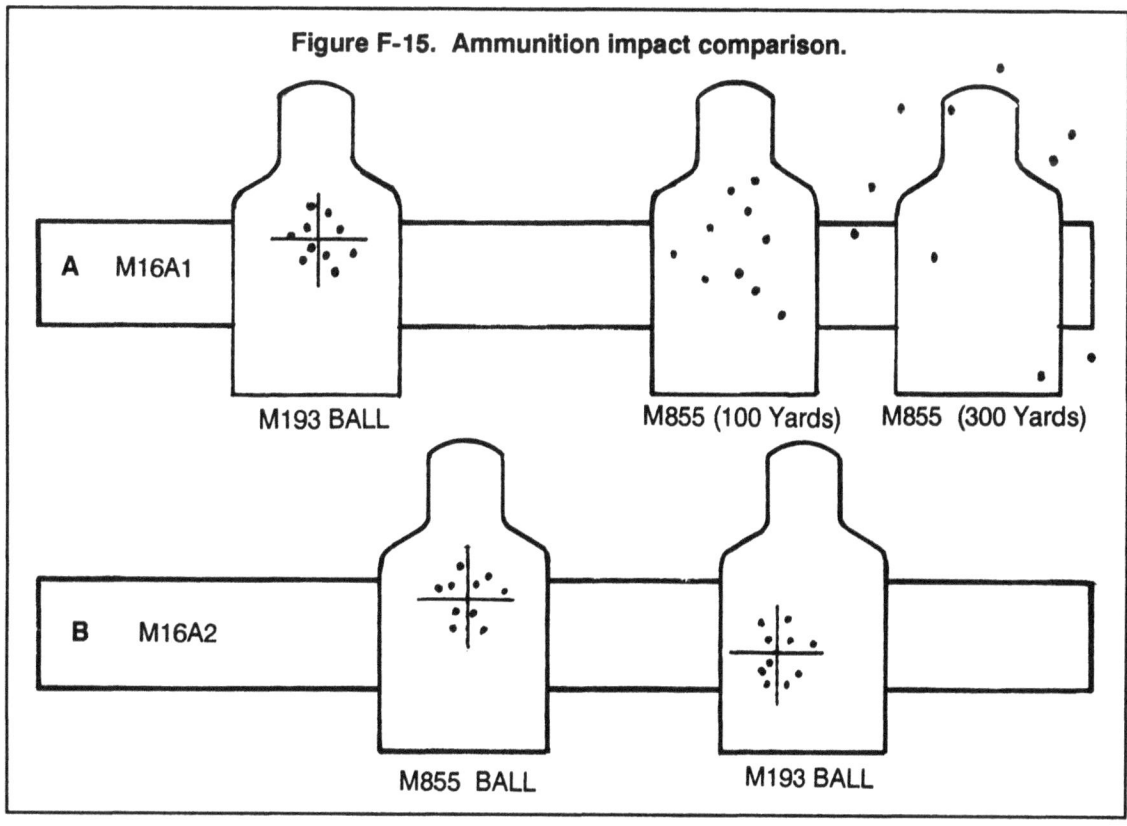

Figure F-15. Ammunition impact comparison.

cartridge in the M16A1 rifles, it should only be used in a combat emergency, and then only for close ranges of 91.4 meters or less.

The M16A2 rifle with its 1:7 twist fires both types of ammunition with little difference in accuracy out to a range of 500 meters. The M16A2 and its ammunition are more effective at ranges out to and beyond 500 meters due to a better stabilization of the round.

The two 10-round shot groups in Figure F-15A were fired by a skilled marksman at a distance of 274.2 meters, using the same M16A1 rifle. The 25.4-cm shot group on the left was fired (and zeroed) with M193 ammunition. The 6-foot shot group on the right was fired after substituting M855 ammunition.

Figure F-15B shows two 25.4-cm shot groups fired by the same skilled marksman at a distance of 274.2 meters, using an M16A2 rifle. The shot group on the left was fired (and zeroed) with M855 ammunition. The shot group on the right was fired after substituting M193 ammunition.

NOTE: Both the M193 and M855 ball ammunition can be used in training and accurately function in the M16A2; however, do not substitute between the types during firing. Do not zero with one type and then fire the other for any type of training. Due to the different characteristics of each round, be sure to zero with the same type ammunition that will be used for training. Figure F-15B illustrates how the group changes when an M16A2 is zeroed with M855 ammunition but fires M193 ammunition. The reverse occurs if zeroed with M193 and fires M855.

A simple rule of thumb to use to preclude any problems is:

- **When firing the M16A1 rifle, use only the ammunition specifically designed for the rifle (M193 ball ammunition).**

- **When firing the M16A2 rifle, use the ammunition that was designed for the weapon (M855 ball ammunition).**

EXTERNAL BALLISTICS FACTORS

When the bullet is launched into the earth's atmosphere at some 2,200 mph, its path is influenced by various forces and elements. As the temperature rises, the bullet hits higher on the target. As the atmospheric pressure rises, the bullet hits lower—the higher the humidity, the lower the bullet strikes. A strong wind from the rear causes the bullet to hit high while a strong head wind causes the bullet to hit low. Firing uphill or downhill normally causes the bullet to hit high. Changing light conditions (bright to cloudy, different sun angles) can affect aiming and cause the bullet to hit in different locations.

These factors combined with slight differences in bullet shape and weight, powder charge, chamber pressure, muzzle velocity, barrel erosion influence the flight of the bullet. For these reasons, the firer will probably never see three bullets in the same target hole.

Some factors, such as temperature, produce only small effects at range—for example, a bullet that hits the center of a target at 250 meters when the temperature is 0 degrees Fahrenheit would strike the target 1.905 cm higher when the temperature is increased to 120 degrees Fahrenheit. One click of elevation at this range is almost 7.62 cm. Soldiers should not try to adjust their point of aim by 1.905 cm on a 250-meter target. However, there are a few factors that can influence shooting performance to such an extent that they should be discussed in detail.

The instructor/trainer must know that several variables exist. Some of these variables have a small influence on the bullet, and can complicate and confuse marksmanship training. Time should be spent only on mastering the most significant factors such as effects of gravity and flight time.

TERMINAL BALLISTICS FACTORS

Bullet penetration depends on the range, velocity, bullet characteristics, and target material. Greater penetration does not always occur at close range with certain materials since the high velocity of the 5.56-mm bullet causes it to disintegrate soon after impact (see Table F-4).

Table F-4. Example penetration comparisons.

	50 meters	300 meters
Pine boards:		
M193	7.5 inches (190.50 mm)	10.5 inches (265.00 mm)
M855	9.0 inches (231.30 mm)	12.0 inches (304.80 mm)
Mild steel plate:		
M193	0.413 inch (10.5 mm)	0.137 inch (3.5 mm)
M855	0.551 inch (14.0 mm)	0.413 inch (10.5 mm)

BULLET DISPERSION AT RANGE

The effects of bullet dispersion and accuracy at various ranges are discussed in this paragraph. Instructors/trainer must have a working knowledge of normal shot dispersion.

Minute of Angle. Minute of angle is a term used to discuss shot dispersion. It is the standard unit of measurement used in adjusting rifle sights and other ballistic-related measurements. It is also used to indicate the accuracy of a rifle. A circle is divided into 360 degrees. Each degree is further divided into 60 minutes, so that a circle contains 21,600 minutes. A minute of angle is an angle beginning at the muzzle that would cover 2.540 cm at a distance of 91.4 meters (Figure F-16, page F-18). When the range is increased to 182.8 meters, the angle covers twice the distance, or 5.08 cm. The rule applies as range increases: 7.62 cm at 274.2 meters, 10.16 cm at 365.6 meters, and so on.

Increase of Shot-Group Size. Just as the distance covered by a minute of angle increases each time the range changes, a shot group can be expected to do the same. If there is 2.540 cm between bullets on a 25-meter target, then there will be an additional 2.540 cm of dispersion for each 25 meters of range—a 2.540-cm group at 25 meters (about 3.5 minute of angle) is equal to a 25.4-cm group at 250 meters (Figure F-17).

The 25-Meter Zero Standard (Figure F-18, page F-18). A standard E-type silhouette is about 48.26 cm wide; a circle (angle) that is 48.26 cm at 300 meters is 4 cm at 25 meters. Therefore, a soldier that can fire all bullets well within a 4-cm circle at 25 meters and adjusts the sights for zero will hit the target at all ranges out to 300 meters.

FM 23-9

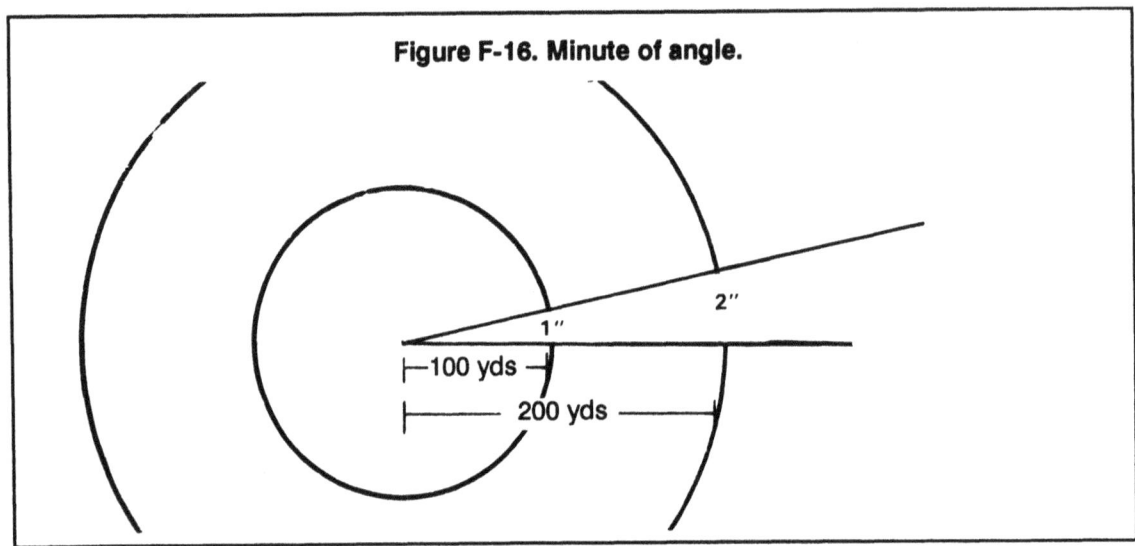
Figure F-16. Minute of angle.

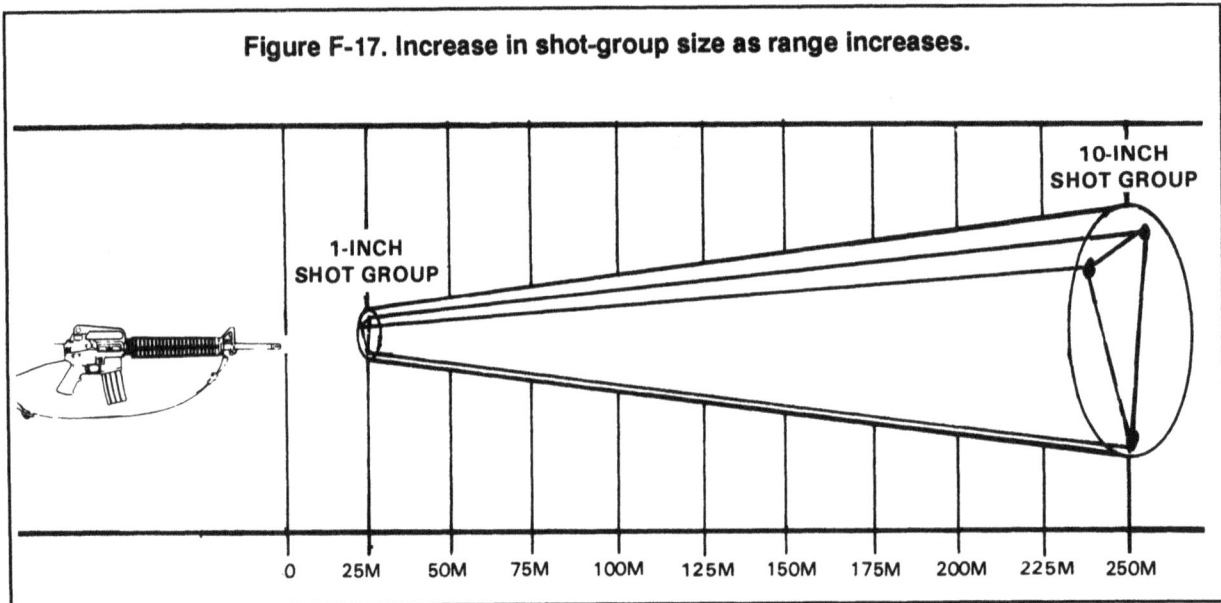
Figure F-17. Increase in shot-group size as range increases.

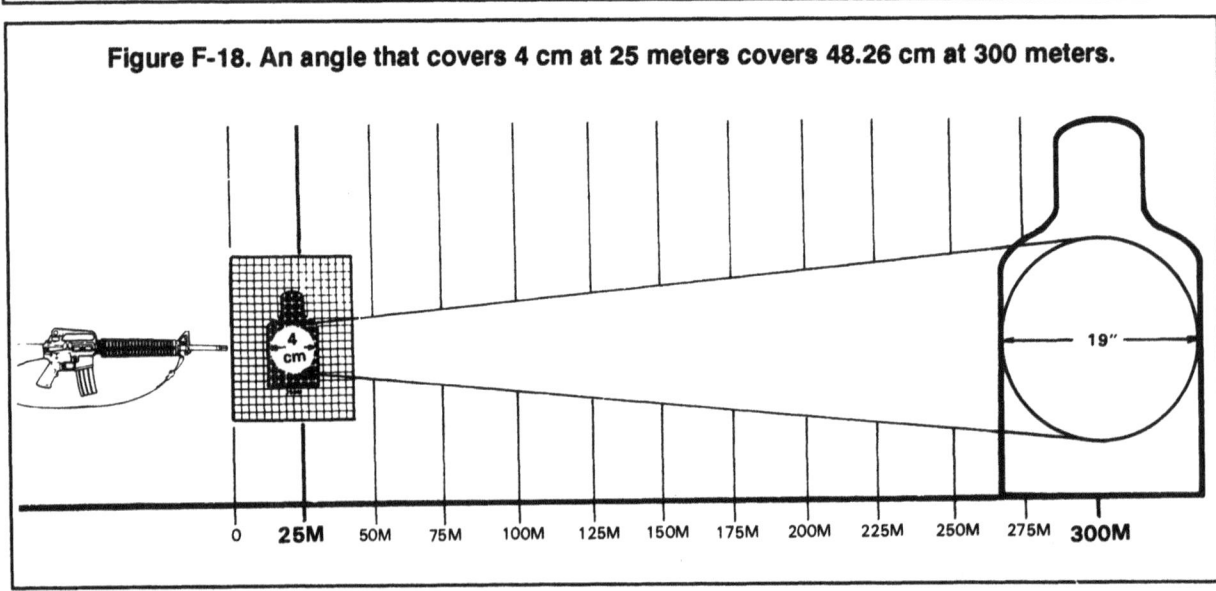
Figure F-18. An angle that covers 4 cm at 25 meters covers 48.26 cm at 300 meters.

FM 23-9

APPENDIX G

LIVE-FIRE EXERCISES

This appendix contains guidelines for the instructor/trainer for organizing ranges, conducting training, and selecting personnel and equipment to establish a live-fire range during rifle marksmanship training. It also contains the intent, requirements, and brief description of each range.

NOTE: Any change to the authorized qualification courses must be approved by the TRADOC commander. All questions concerning authorized qualification courses should be forwarded to: Commandant, US Army Infantry School, ATTN: ATSH-I-V, Fort Benning, GA 31905.

Section I. GROUPING RANGE

This section provides guidelines for the instructor/trainer for the conduct of a grouping range. It is designed to include organization, conduct of fire, analysis of shot groups, and shot group sizes.

CONDUCT OF GROUPING RANGE

Fundamental exercises are conducted in two situations: during IET at the Army training centers, and as part of the soldier's individual and collective sustainment training programs within his unit. The organization and conduct of a grouping range are based on the availability of ammunition and the firing ability of personnel in training.

Initial training consists of an explanation of the purpose of a grouping exercise. It highlights individual actions needed to receive maximum benefit from the expended ammunition.

Each shot is fired using exactly the same aiming point (target center of mass). The objective is to fire tight shot groups and to place those shot groups in a consistent location (the actual location of groups on the target is not important). Each three-round shot group must be connected with lines and labeled in sequence—for example, 1, 2, 3 and so on.

Initial firing is conducted from an individual fighting position or from a prone supported position. Once firing proficiency has been demonstrated from the supported position, grouping exercises can be conducted from the prone unsupported position. For example, if 27 rounds are available for the grouping exercise, 18 rounds can be fired from a supported position and the remaining 9 rounds from an unsupported position. If proficiency is not demonstrated for the first 18 rounds, the soldier continues in the supported position for the remaining 9 rounds.

RANGE ORGANIZATION

Organization for training is based on a 200-man unit and a standard 25-meter range of 110 firing points. The unit is divided into two orders—the first order fires while the second order performs coaching duties. The extra 10 firing points are used to conduct corrective instruction. (The organization described herein refers to Army training centers with proper facilities. When using smaller ranges, the unit is divided into three or more orders.)

On 25-meter ranges, a firing position and sandbags should be provided at each firing point so instruction in firing from supported positions can be conducted (see Figure G-1).

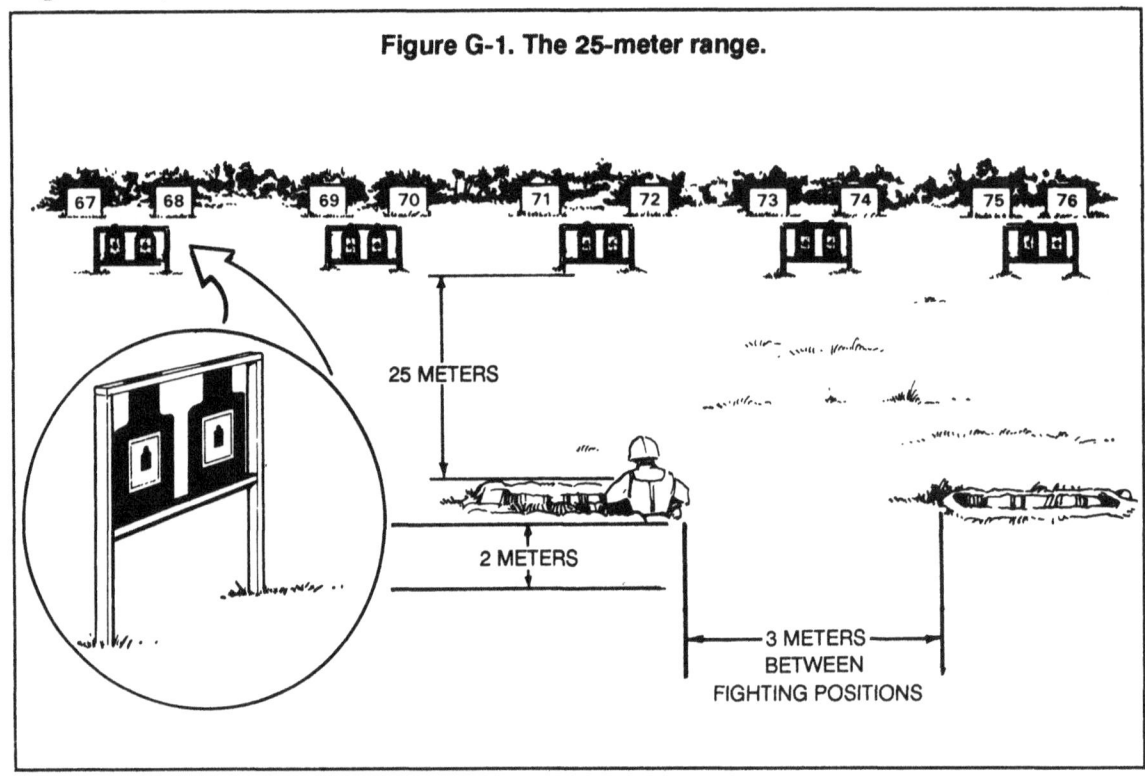

The control tower manages live-firing operations and monitors the progress of training. It is not intended that instruction and training be performed from the tower — training is conducted on the firing line by the instructor/trainer. Before each live-fire exercise, personnel must be briefed on range safety regulations and have hearing protection.

A control or point tower should be centrally located to the rear of the firing line. It should be elevated to permit unrestricted observation of the range, both to the rear of the firing line and to a safe distance beyond the line of targets to the front. All firing commands are issued from the control tower and must be obeyed immediately. The only exception is if an unsafe act occur — the first soldier to see such an act should command CEASE FIRE.

CONDUCT OF TRAINING

The instructor/trainer must understand how to correctly analyze shot groups.

Shot-Group Marking. If the soldier is to benefit from this exercise and if the instructor/trainer (or coach) is to provide useful guidance, the soldier must mark each shot group for a clear record of his firing practice (Figure G-2). He connects his three bullet holes on his target with a straight line and places a number by the shot group. Then the soldier marks the approximate center of the shot group with an X. (Various shot groups are shown in Figure G-3.)

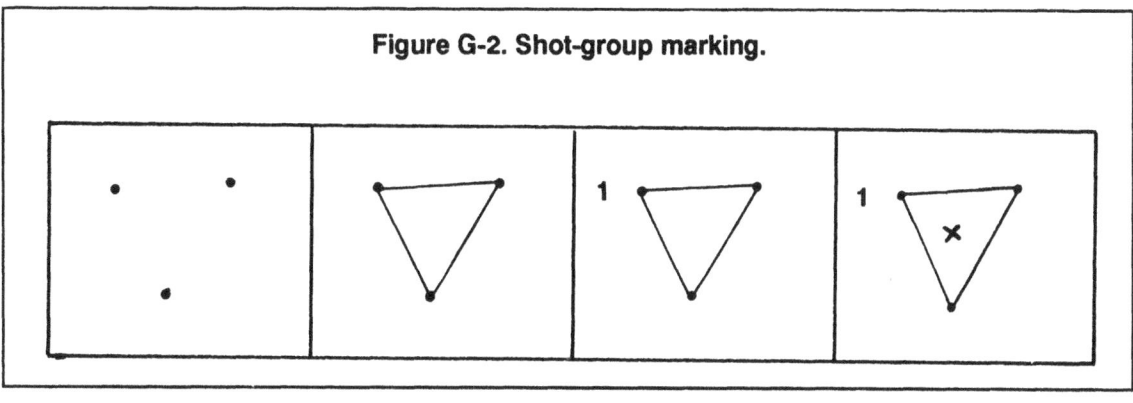

Figure G-2. Shot-group marking.

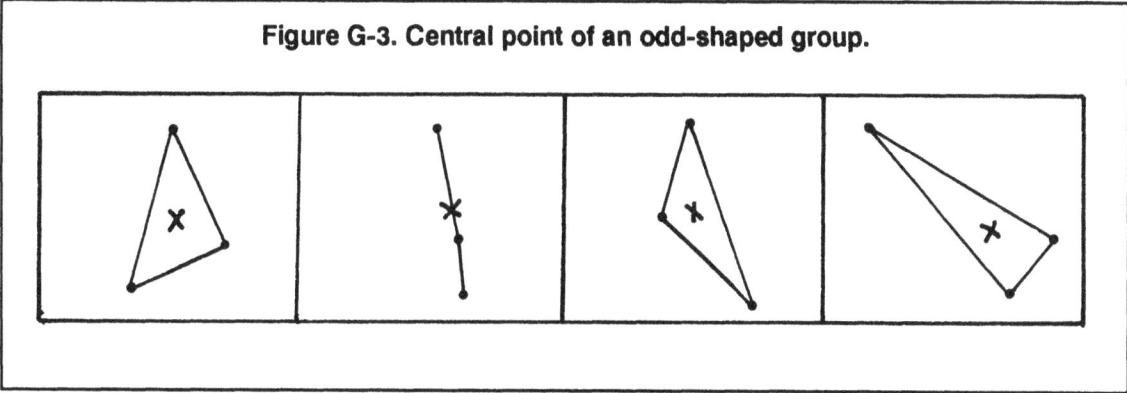

Figure G-3. Central point of an odd-shaped group.

Marking the X. The X represents the center of the three shots. When two shots are near one end of the group and the third shot is toward the other end, the X is placed closer to the two near shots. This is not a precise marking that requires a measurement but is a procedure to help in shot-group analysis. The three-round shot group confirms the variation inherent in the rifle, ammunition, and firer. While some of the variation within a single shot group is due to the rifle and ammunition, the firing of more shot groups indicates any firer errors. (Figure G-4 shows three marked shot groups.)

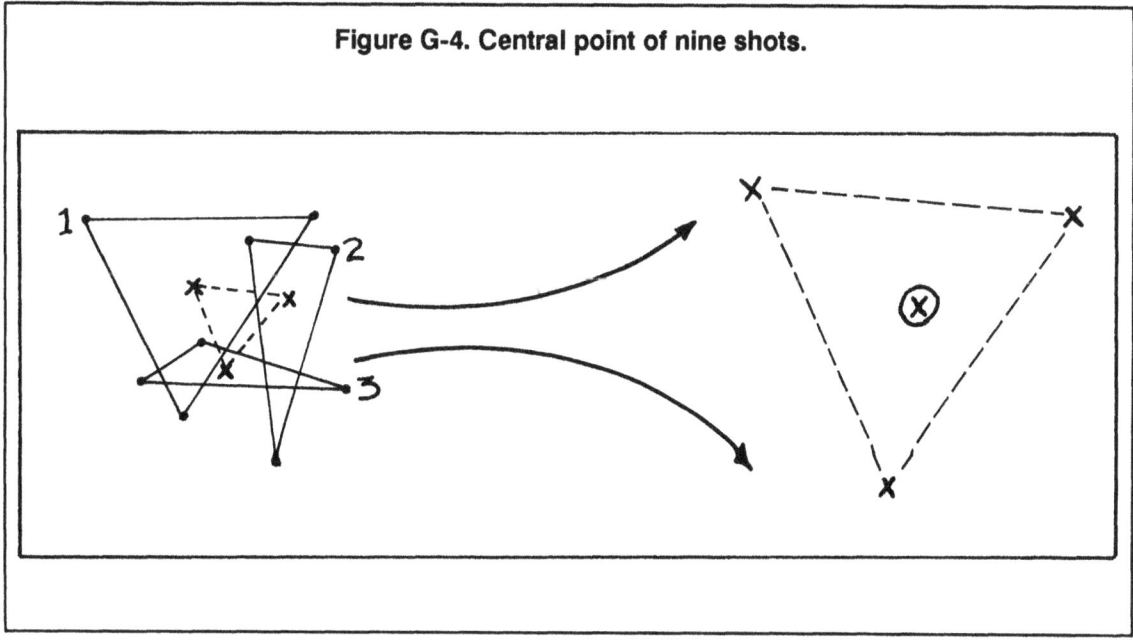

Figure G-4. Central point of nine shots.

Locating center of shot groups. The central point of all nine bullets fired can be found by treating the X as a single bullet. If the soldier maintains the same point of aim for each shot, this indicates firing errors or needed sight changes.

Shot-Group Analysis. Shot-group analysis begins by the instructor/trainer observing the soldier while he fires. The purpose of shot-group analysis is to determine firer errors so that the soldier can apply corrections for the next shot group. The target shown in Figure G-5 illustrates that a match-grade quality rifle-ammunition combination, which places all bullets in almost the same hole, helps detect the slightest error. When firing a standard service rifle-ammunition combination, the soldier displays a dispersion pattern that is discounted as firer error.

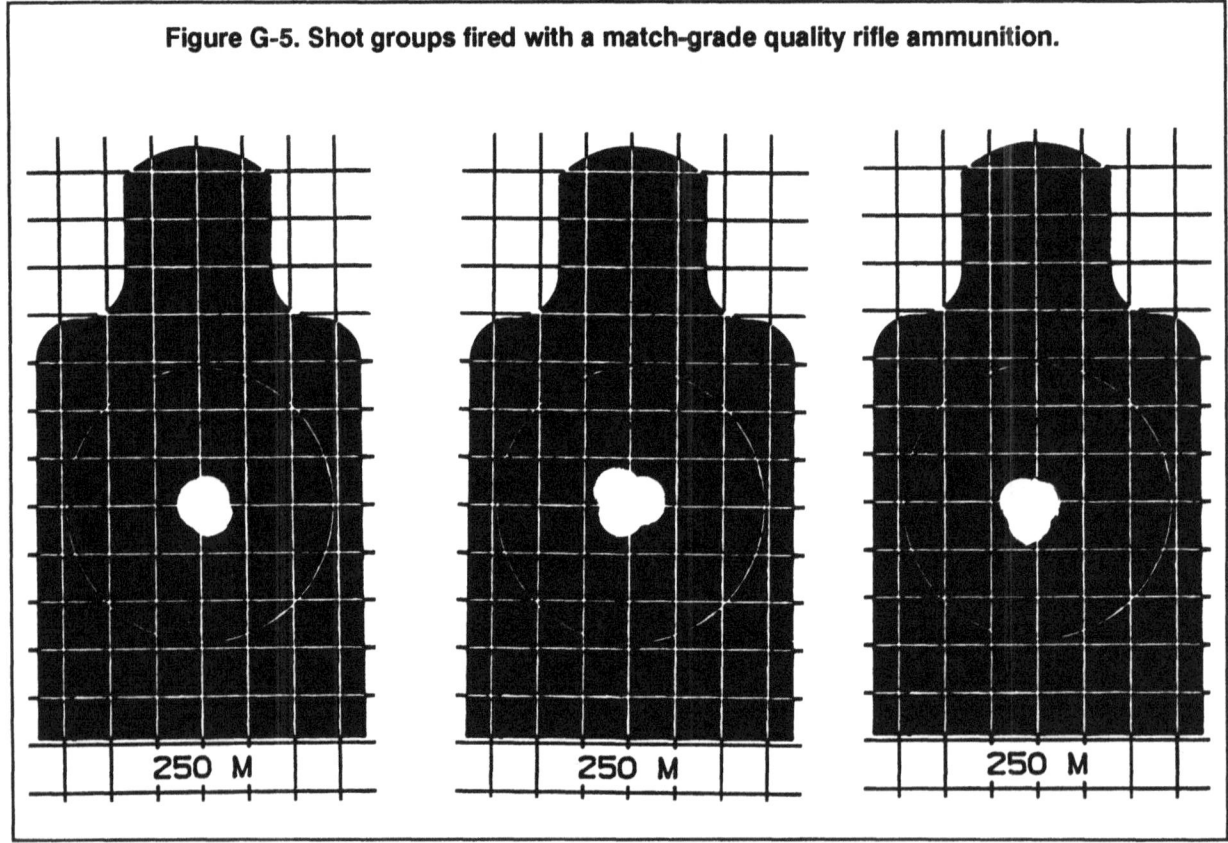

Figure G-5. Shot groups fired with a match-grade quality rifle ammunition.

The targets shown in Figure G-6 reflect possible 25-mm zero performances by standard rifle-ammunition combinations and proper soldier performance. The soldier can fire on all combat targets out to ranges beyond 300 meters. Because the variability of the standard rifle must be considered for accurate and useful shot-group analysis, the instructor/trainer must promote confidence in the soldier and his rifle.

Shot-Group Size. A key to analyzing shot groups is shot-group size. Figures G-7 and G-8 show (near) actual shot-group size. The circles on the reduced targets shown in Figures G-8 through G-11 are 4 cm in diameter, and each block on the target is .7 cm. As in Figure G-7, any shot group within 2 cm (about three squares on the target) indicates that no firer error is involved or that none can be detected. Regardless of the arrangement of the three bullets within the group, no useful information is provided to improve the soldier's firing performance.

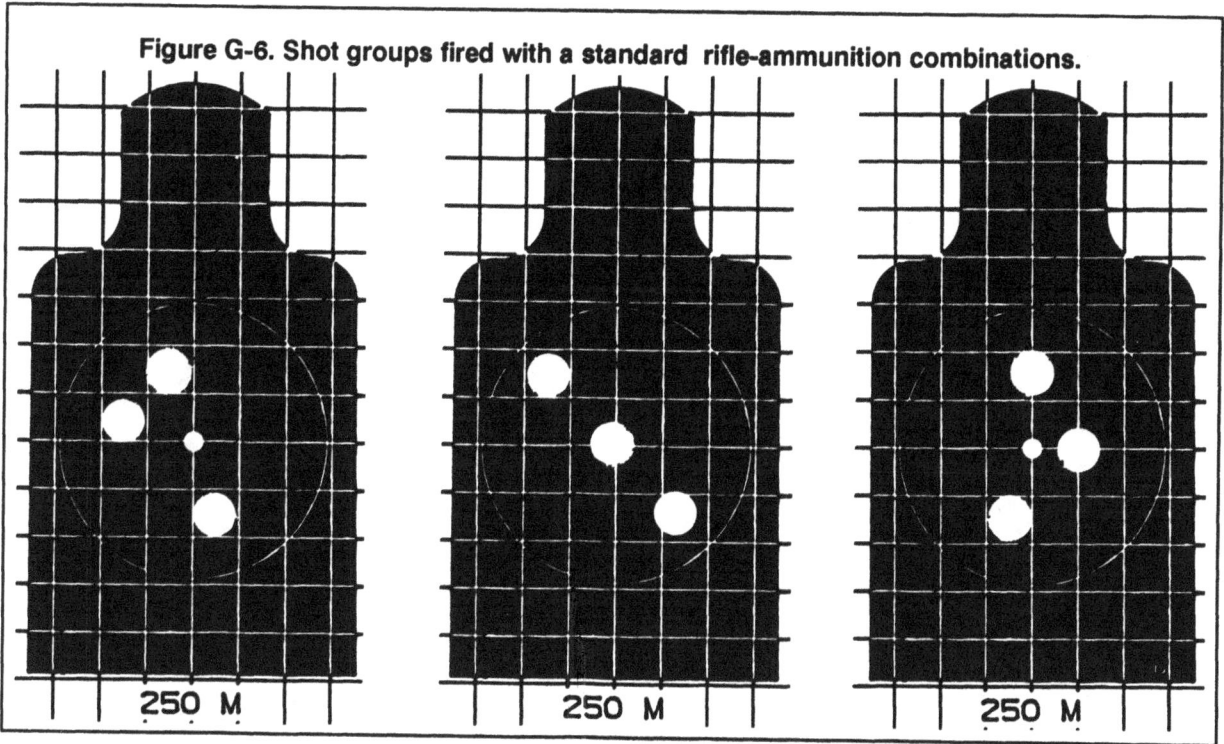

Figure G-6. Shot groups fired with a standard rifle-ammunition combinations.

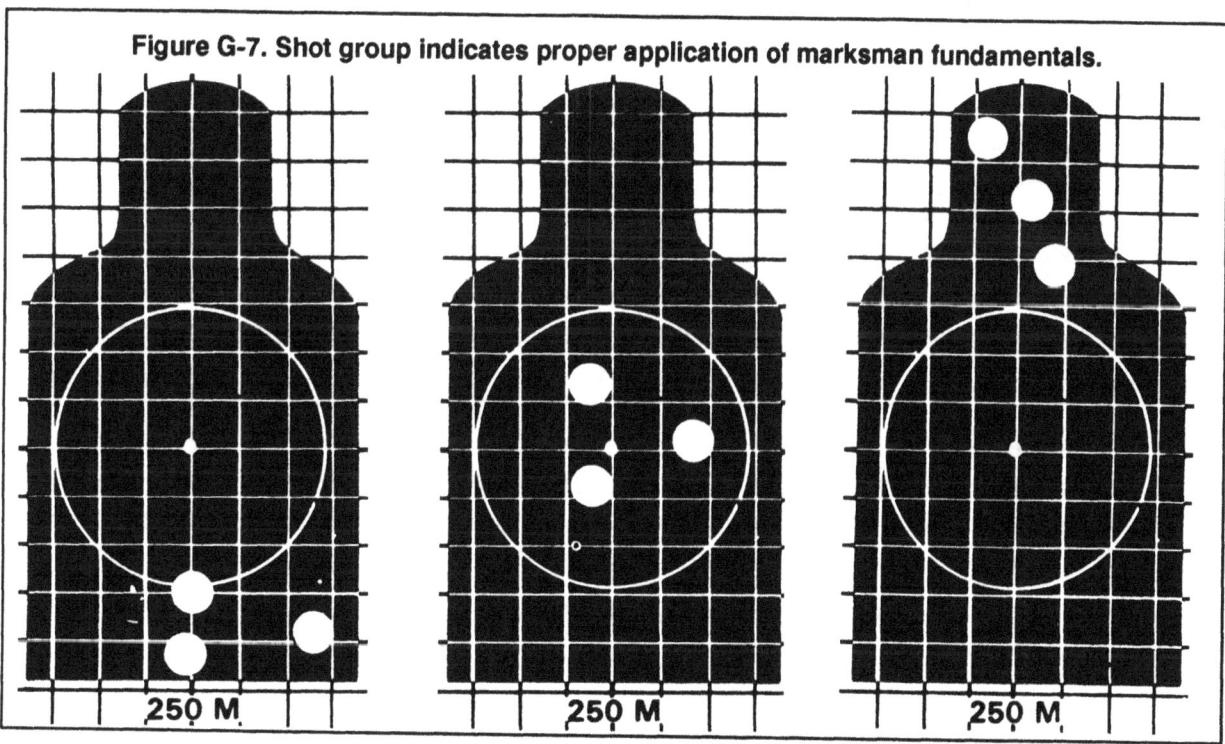

Figure G-7. Shot group indicates proper application of marksman fundamentals.

Shot groups — about 3 cm. The targets shown in Figure G-8 represent acceptable firing performances and possibly the ability of some rifles. Regardless, a better firing performance should be expected, and the instructor/trainer should ensure that the soldier is properly applying the four marksmanship fundamentals. Also, the he should explain that group size is not due to weapon performance (an experienced firer tests the rifle). The placement of shots within these groups (about 3 cm or four squares on the target) reflects no useful information. Firing another shot group (without a sight change) could provide more information to determine possible firer error.

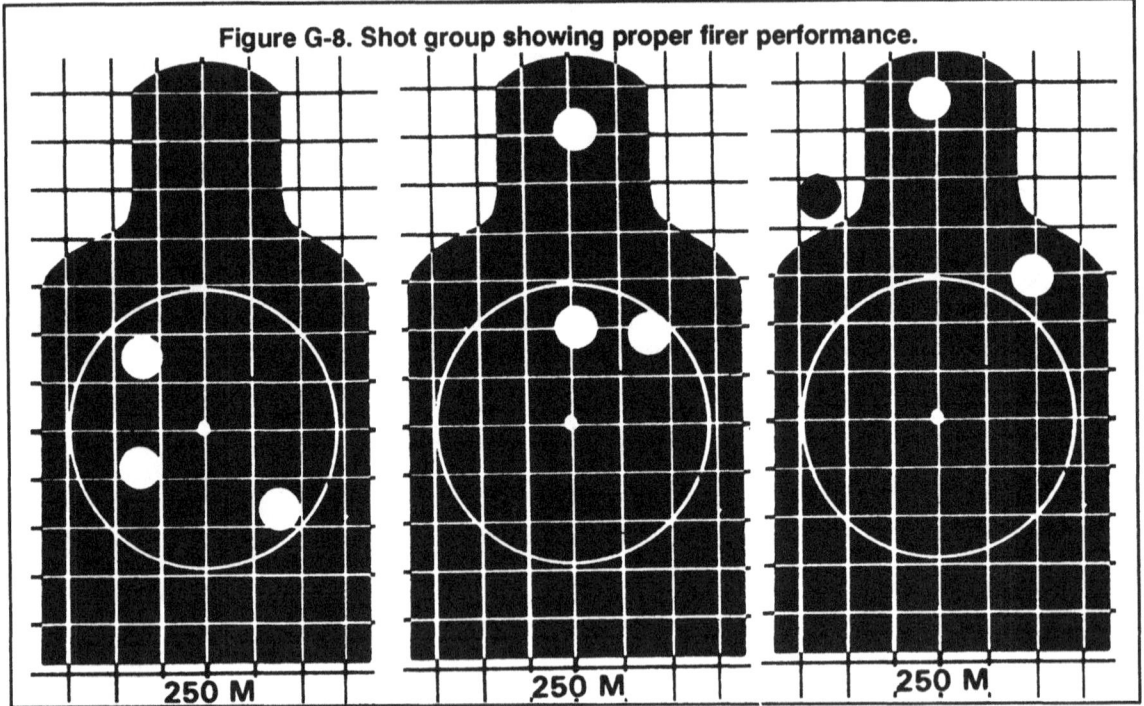

Figure G-8. Shot group showing proper firer performance.

Shot groups — about 4 cm. The three shot groups shown in Figure G-9 are about 4 cm (six squares on the target) and indicate firer error.

Firers are checked for a proper steady position, that the eye is focused on the front sight post tip (which is placed at target center for each firing), that the breath is locked during trigger squeeze, and that the trigger squeeze is correct.

More shot groups provide better information from which to direct remedial training. The vertical aiming point may not be the same for each shot since the vertical dispersion is greater than the horizontal dispersion. Any problem with finding target center of mass is probably in the vertical plane and not in the horizontal plane. If the soldier has his eye focused on the front sight post, his vertical aiming error should be minimal, which usually cannot be detected on the target. Any of these three shot groups could have one round that was pulled by the firer or was a flier (an erratic round over which the firer has no control). Therefore, another shot group must be fired.

NOTE: Location of the shot group is not important when conducting a grouping exercise. Size of the group and the ability to place two or more groups in the same location are important.

Shot groups — 6 cm or larger. The three shot groups shown in Figure G-10 are the easiest to analyze, indicating obvious firer error — improper trigger squeeze is part of the problem. Shot groups that are about 6 cm or larger (about nine squares on the target) are normally the result of the soldier knowing when the rifle is going to fire. Therefore, the instructor/trainer tries to improve trigger squeeze by using the ball-and-dummy technique so the soldier can tighten his shot group (Appendix C). Firers with these shot groups should receive extensive dry-fire training to help correct firing problems.

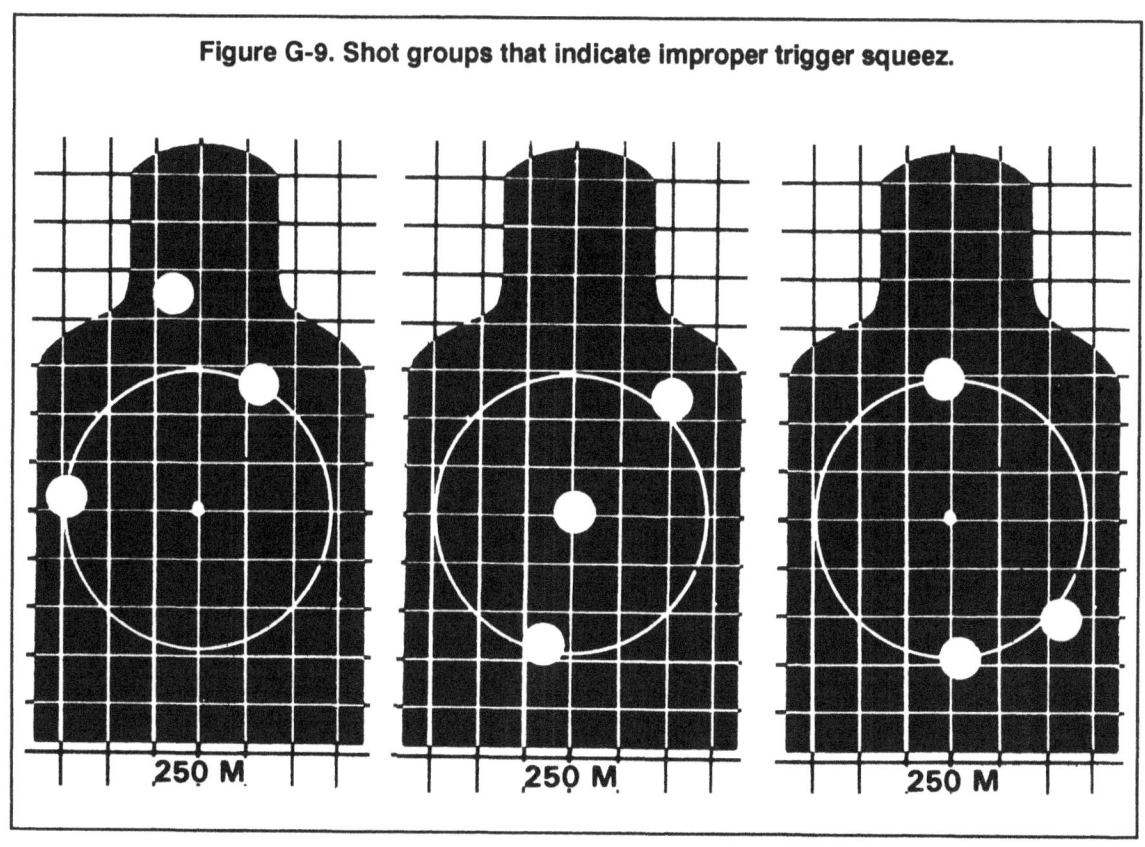

Figure G-9. Shot groups that indicate improper trigger squeez.

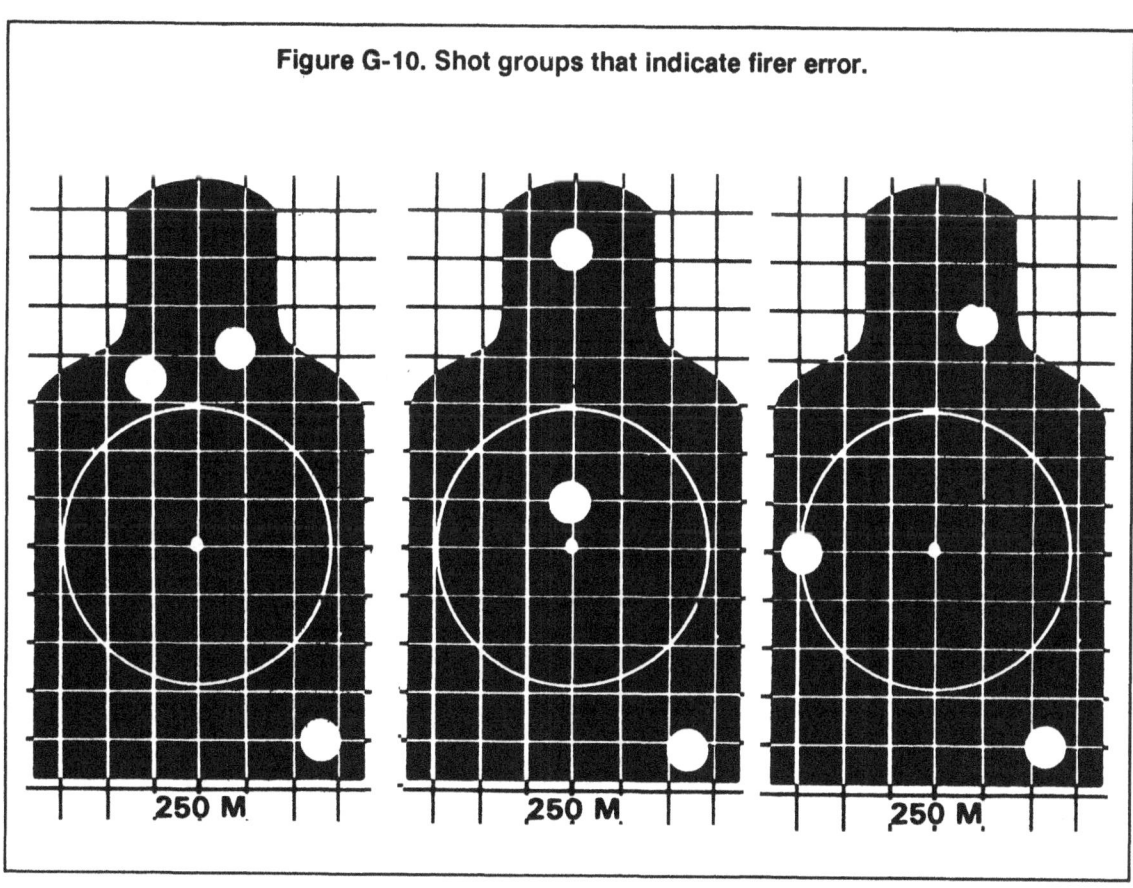

Figure G-10. Shot groups that indicate firer error.

Grouping performance – example 1. The shot groups in Figure G-11 represent acceptable shot groups (4 cm or less) in the same location. It is appropriate to make a sight change of left 4 and down 6. Any change should be clearly marked on the target and saved for reference.

Grouping performance – example 2. The groups in Figure G-12 indicate that proper firing fundamentals are being applied by the soldier for each shot group, but that the soldier could be using a different aiming point each time a shot group is fired. The soldier's understanding of the aiming process is questioned, and his position is checked for consistency. The instructor/trainer cannot determine which shot group best represents the firer's zero; therefore, a sight change should not be made.

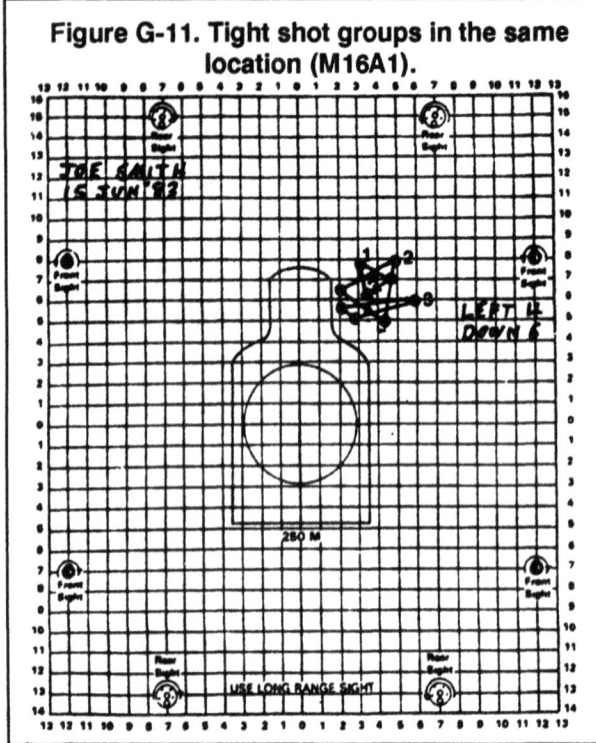

Figure G-11. Tight shot groups in the same location (M16A1).

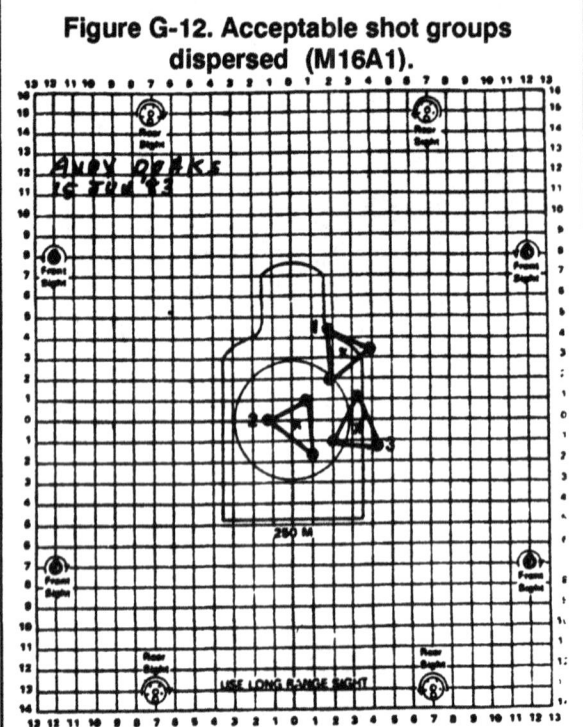

Figure G-12. Acceptable shot groups dispersed (M16A1).

Grouping performance – example 3. The groups in Figure G-13 indicate consistent aiming, but the soldier probably knows when the rifle is going to fire (improper trigger squeeze) or he is firing from an unsteady position.

Grouping performance – example 4. The groups shown in Figure G-14 indicate problems with shot-group size and with consistent placement of groups. The four marksmanship fundamentals should be checked.

Grouping performance – example 5. The shot groups shown in Figure G-15, when viewed as nine shots, reflect proper horizontal placement of shots but unsatisfactory vertical dispersion. This indicates a failure to aim at target center of mass for each shot. The soldier's aiming procedure is checked along with other marksmanship fundamentals.

Grouping performance – example 6. The shot groups shown in Figure G-16 are proper groups, but vertical dispersion indicates that a different aiming point is used for each group. The soldier's understanding of the target center of mass and aiming process should be questioned.

Figure G-13. Improper shot groups with consistent placement (M16A1).

Figure G-14. Improper shot groups dispersed (M16A1).

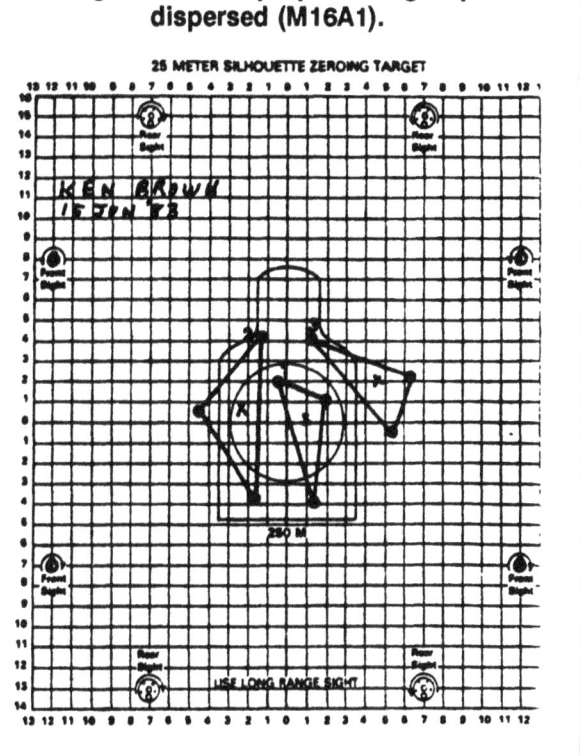

Figure G-15. Marginal shot groups; improper vertical placement and proper horizontal placement (M16A1).

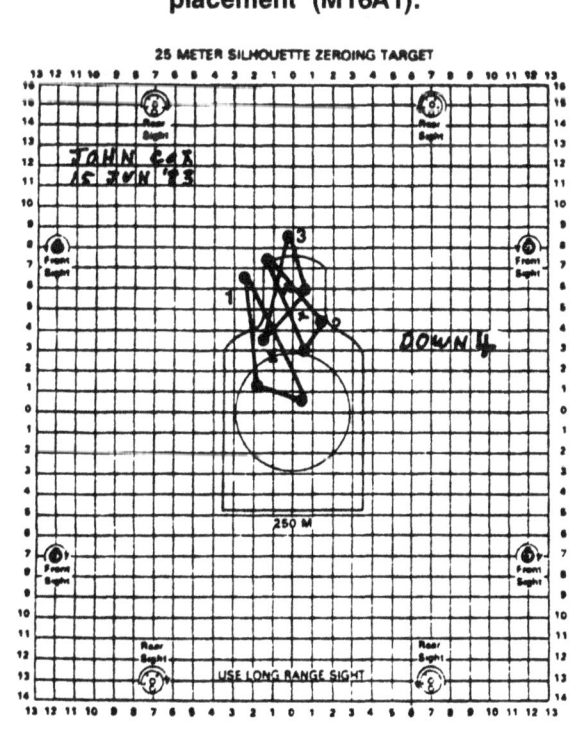

Figure G-16. Proper shot groups; improper vertical dispersion (M16A1).

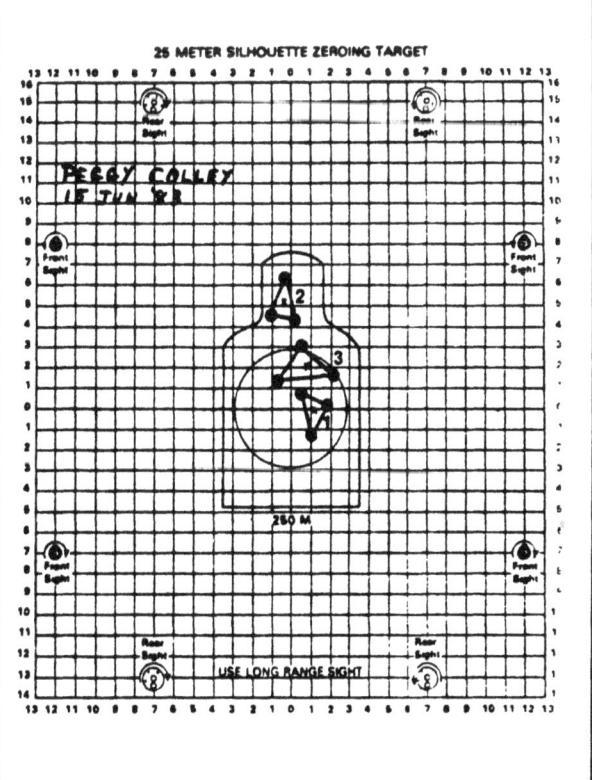

Grouping performance — example 7. The shot groups in Figure G-17 are improper groups. A sight change is made to bring the groups closer to target center. If the initial group is close to the paper's edge, the groups are still completed to provide the needed information for a sight change. A large sight change moves the groups close to target center of mass. Assuming that the last bullet in the third group is to the right of the target, a change of 10 clicks left and 10 clicks up is indicated.

Grouping performance — example 8. The groups shown in Figure G-18 indicate improper firing. Trigger jerk is indicated as a probable cause, but all fundamentals are checked.

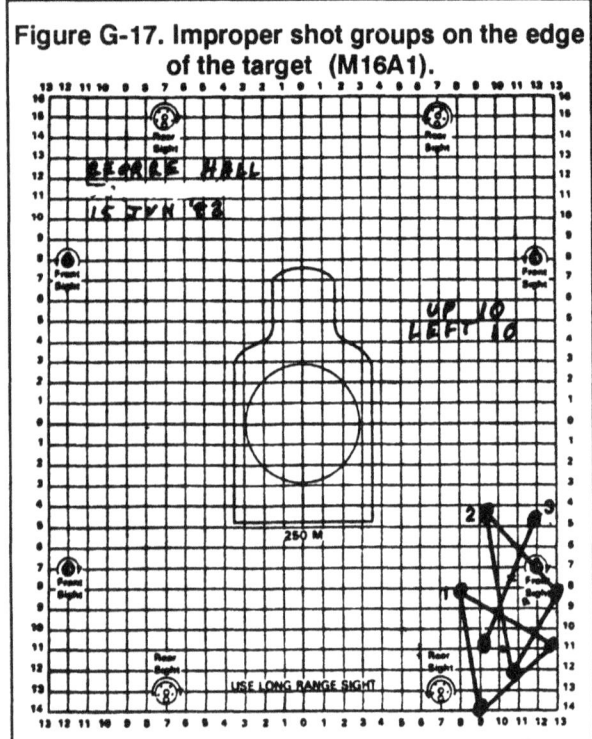

Figure G-17. Improper shot groups on the edge of the target (M16A1).

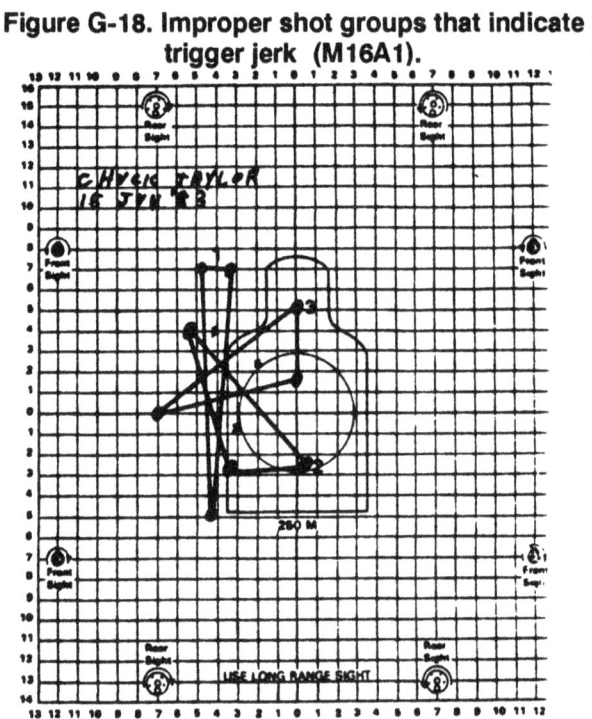

Figure G-18. Improper shot groups that indicate trigger jerk (M16A1).

NOTE: Targets are changed after firing the first nine rounds. These targets are saved since all bullets fired are important for self-analysis, coaching, or future sight changes.

Section II. ZERO RANGE

This section establishes guidelines for the instructor/trainer on the conduct of a zero range. It contains information on the setup, operation, and conduct of training.

CONDUCT OF ZEROING ON A 25-METER RANGE

When the soldier can consistently place two consecutive three-round shots within the same general area at 25 meters, he is ready to zero his rifle.

The soldier fires a three-round shot group at the 25-meter zero target. The firing line is cleared, and he moves downrange to examine the shot group. The soldier connects the bullet holes with a pencil line and marks the holes with the number 1. With the instructor/trainer, the soldier examines the shot group for size placement and fundamental errors. If the shot group is on the paper target, no sight changes should be made.

The soldier then returns to the firing line and fires a second three-round shot group. Again, he moves downrange and examines the second shot group. The soldier connects the second set of bullet holes and marks them with the number 2. He then determines, along with the instructor/trainer, if any sight changes are needed. To make a sight change, both shot groups should be in the same location and no larger than 4 cm. All sight changes should be recorded on the target. A review of previous firings (groupings) is helpful.

After the sight changes are made, two more three-round shot groups are fired, using the same procedure to confirm the zero. The shot groups must be centered within the zero circle. When the instructor/trainer is satisfied that the soldier has achieved the best possible zero, the soldier is removed from the firing line.

Using this information, the zeroing process would be conducted as outlined in the following example:

Using the L-sight and applying the fundamentals, the soldier consistently aims target center of mass as shown in Figure G-19. The soldier fires two separate three-round shot groups, as shown in Figure G-20, and numbers them. Based on the location of these two groups (Figure G-20), the soldier would make the sight adjustments shown in Figure G-21.

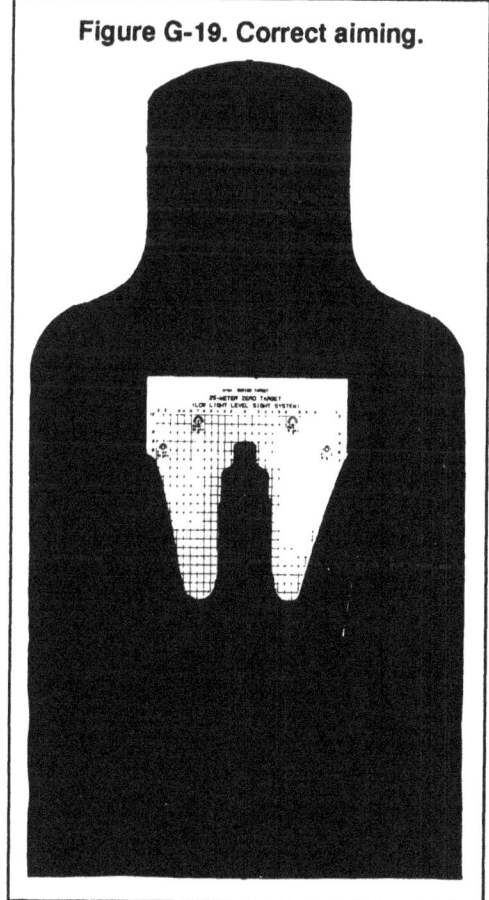

Figure G-19. Correct aiming.

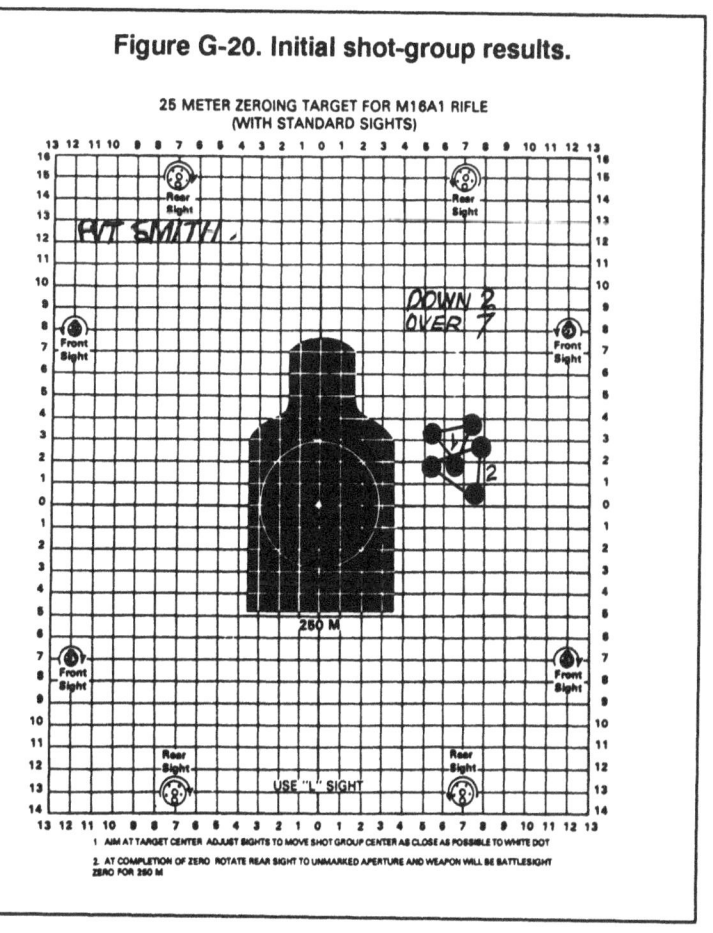

Figure G-20. Initial shot-group results.

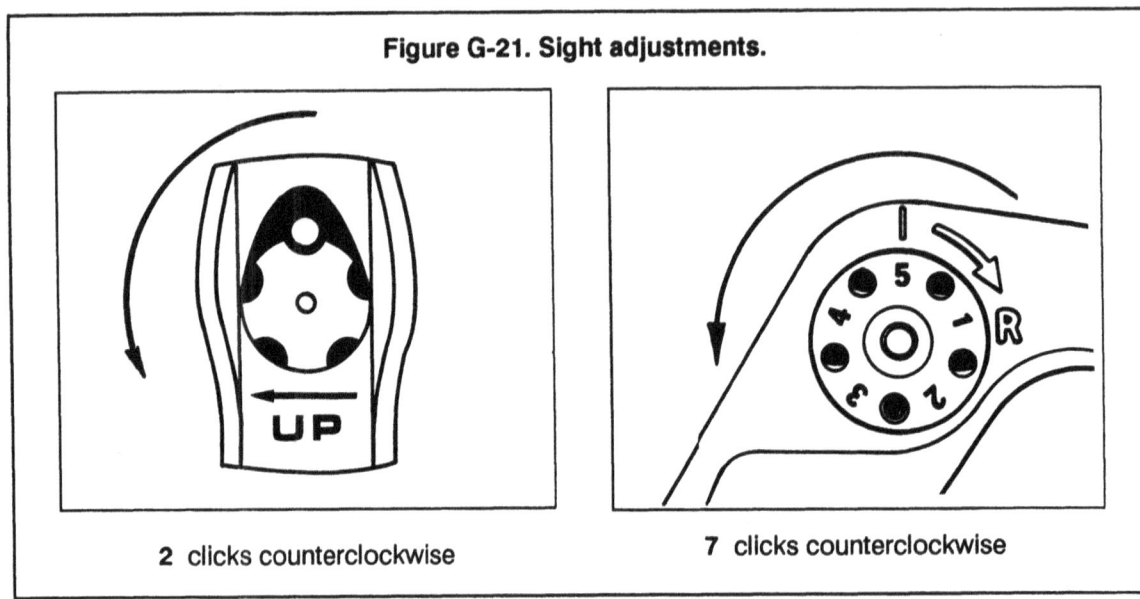

Figure G-21. Sight adjustments.

2 clicks counterclockwise   7 clicks counterclockwise

After making the correct sight changes, the soldier fires two more separate three-round shot groups to confirm that the adjustments have aligned the sights with the center of the target and that the bullets are well within the 4-cm circle (Figure G-22).

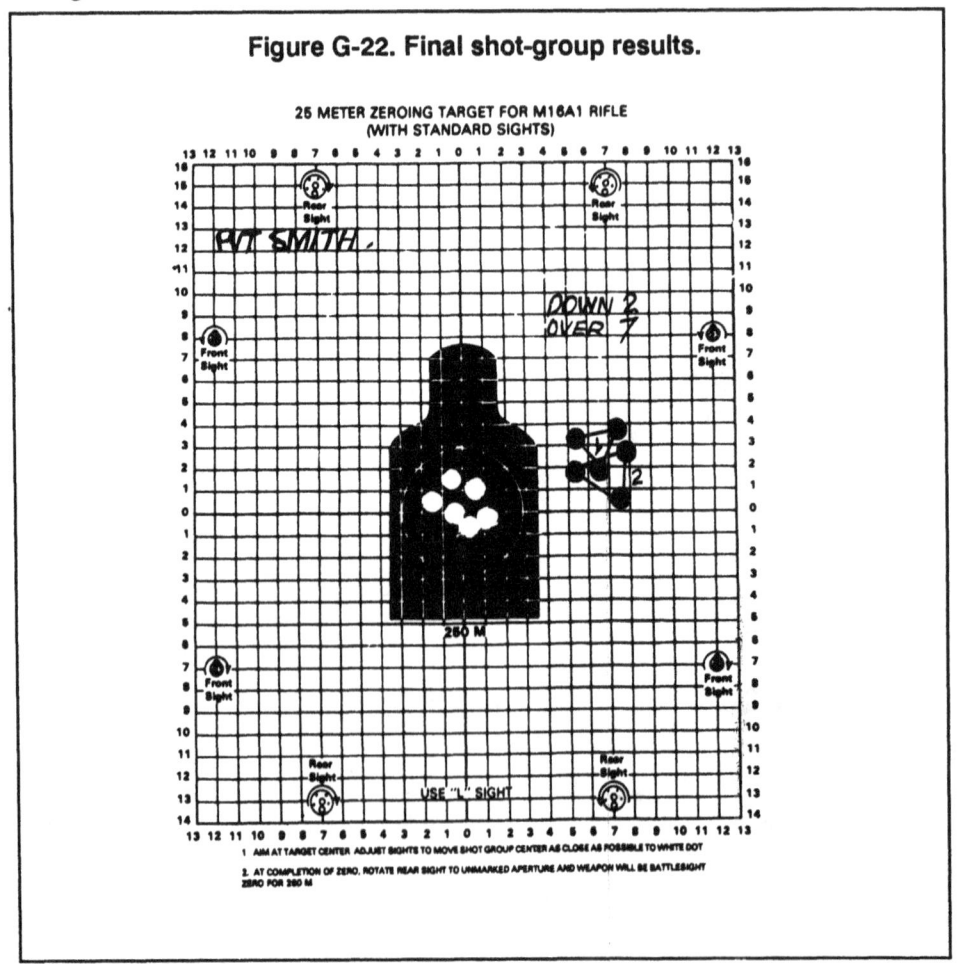

Figure G-22. Final shot-group results.

MECHANICAL ZERO

To mechanically zero the weapon, the firer adjusts the front sight post up or down until the base of the front sight post is flush with the well. Then he adjusts the front sight post 11 clicks in the direction of UP. This moves the post down into the well 11 clicks. The soldier turns the rear sight windage drum until it moves all the way to the left side and locks. Then, he turns the windage drum back (right) 17 clicks so the rear sight is approximately centered (see Figure G-23).

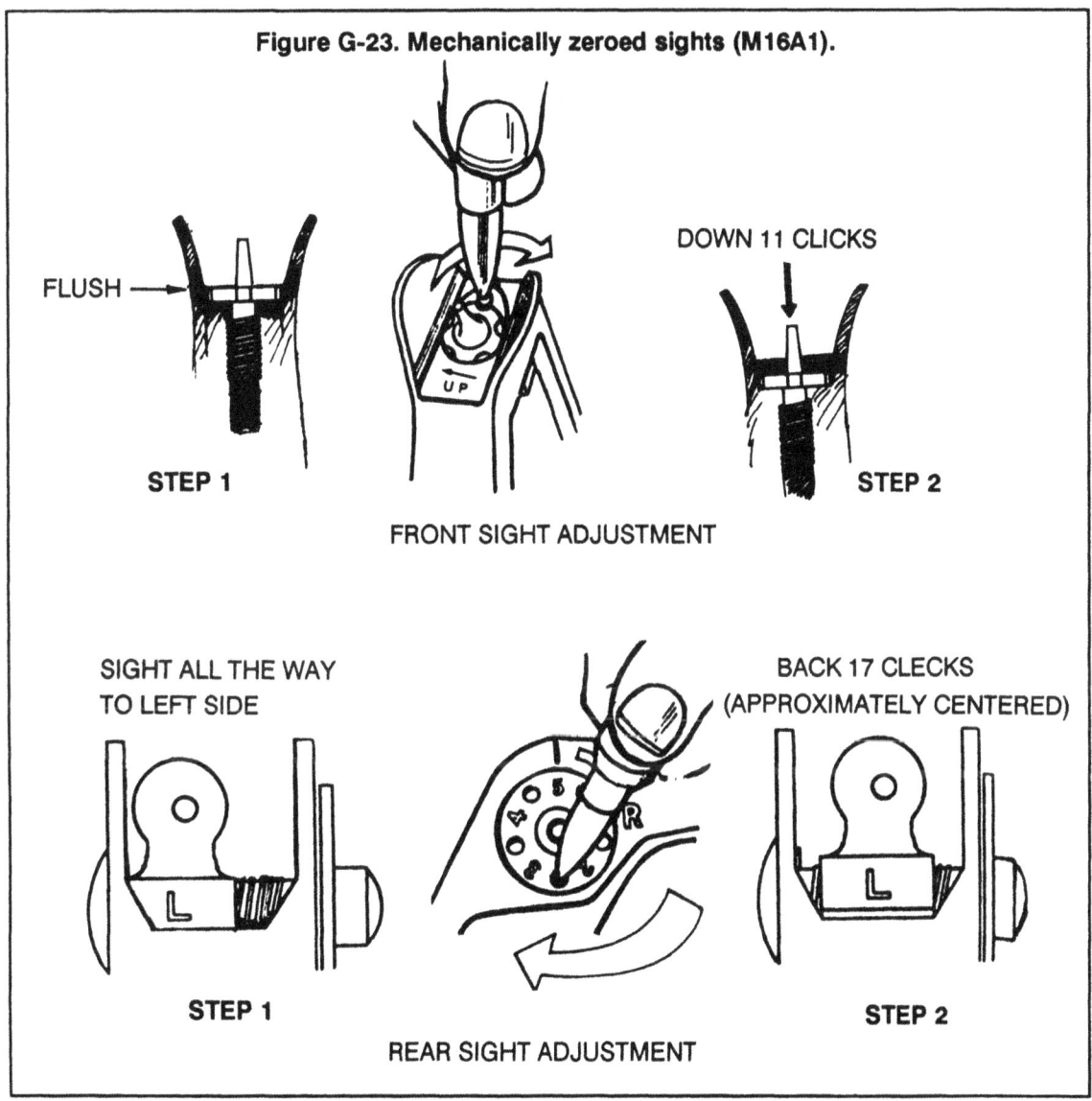

Figure G-23. Mechanically zeroed sights (M16A1).

RECORDING OF ZERO, M16A1

Using the example in Figure G-22 and assuming that the initial setting was correctly adjusted, the soldier would record his zero setting as follows:

Front sight.

Elevation zero. Since the center of the shot groups was determined to be two squares high, the correction would be two clicks down. With the initial setting of 11, this would be 11 minus 2, or 9 and should be recorded as 9 UP.

Rear sight.

Windage zero. Since the center of the shot groups was determined to be seven squares to the right, the correction would be seven clicks left. This would be 17 minus 7 or 10 and should be recorded as R 10.

The data recorded for the firer in this example would be as shown in Figure G-24.

Figure G-24. Recording of data.

```
SFC GARY ROANE
   21 APR 89
   V9 —— R10
WPN # 1234567
```

NOTE: These data are not transferable to another rifle that the soldier is assigned.

INITIAL SIGHT SETTING, 25METERS

Before firing, the soldier sets the initial sight setting for 25-meters on his weapon. (This is equivalent to using the L sight on the M16A1.) To do this, he adjusts the front sight up or down until the base of the sight post is flush with the sight post well. He, then, turns the rear windage knob until the index mark on the 0-2 sight is aligned with the rear sight base index (Figure G-25). The elevation knob is turned one click past the 8/3 mark (8/3 + 1). The elevation knob remains in this position until the battlesight zero is obtained.

During zeroing for elevation, the soldier makes all adjustments with the front sight post. To place the actual 300-meter zero on the rifle, he rotates the elevation knob down one click so that the 8/3 mark on the range scale is aligned with the alignment mark on the side of the rifle receiver. Once zero is accomplished, for engagements at ranges greater than 300 meters, he places the correct range setting on the elevation knob to obtain point of aim/point of impact.

INITIAL SIGHT SETTING FOR FIELD FIRE, M16A2

When zeroing on either a modified field fire range or a KD range using the M16A2 300-meter feedback targets, the initial sight setting field fire must be placed on the weapon. To do this, the soldier flushes the front sight post, centers the 0-2 aperture on the rear sight, and sets the elevation knob to the 8/3 mark. Point of aim should be on the 300-meter feedback target center of mass. During zeroing, the soldier makes all adjustments for elevation with the front sight post. If the rifle has been zeroed by another soldier, the weapon is closer to being zeroed than if returned to the initial sight setting. Therefore, the sight setting on the rifle is usually used as received by the soldier.

Figure G-25. Sight adjustments.

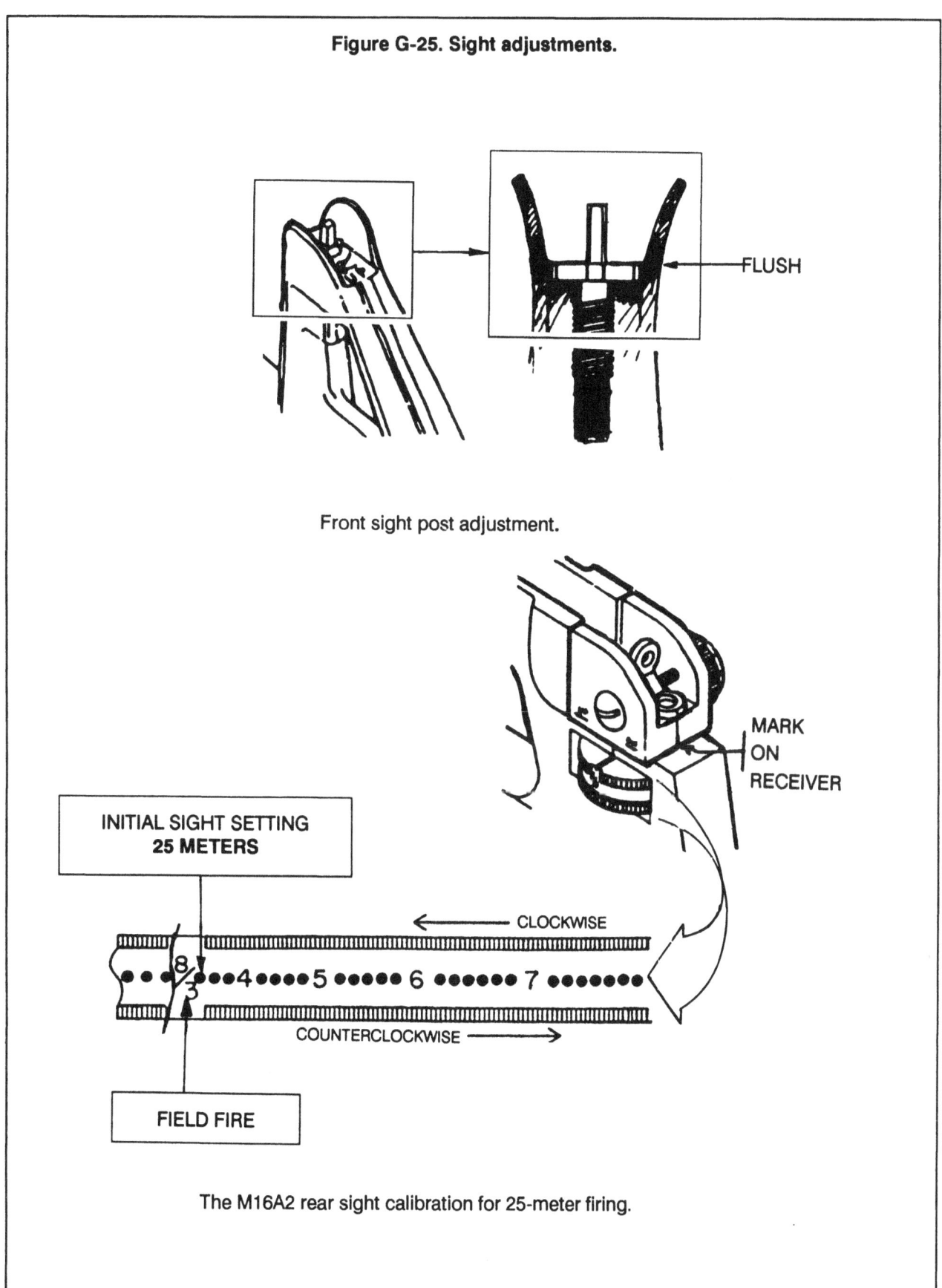

Front sight post adjustment.

The M16A2 rear sight calibration for 25-meter firing.

With the rear elevation knob calibrated and applying the fundamentals, the soldier consistently aims target center of mass as shown in Figure G-26.

The soldier fires two separate, three-round shot groups and numbers them as shown in Figure G-27.

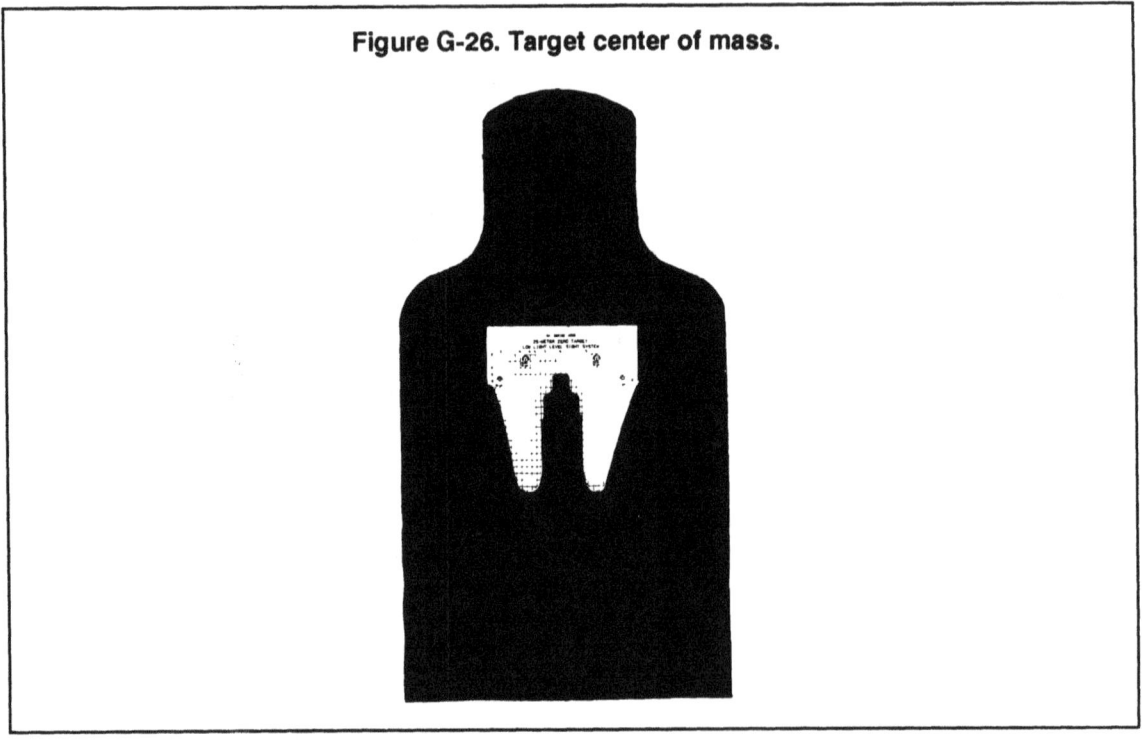

Figure G-26. Target center of mass.

Based on the shot groups in Figure G-27 and the information provided on the target, the soldier computes the changes as shown in Figure G-28 to move the group down and to the left.

Once the changes are made, the soldier fires two more three-round groups to confirm that the adjustments have moved the strike of the bullet into target center of mass (Figure G-29).

RECORDING OF ZERO, M16A2

The concept for zeroing on the M16A2 rifle is the same as for the M16A1 rifle. The conduct of fire on the 25-meter range follows the previous examples but uses the M16A2 (300 meters) zero target.

Section III. DOWNRANGE FEEDBACK

Downrange feedback provides precise knowledge of what happens to bullets at range. It provides for an effective transition between 25-meter firing and firing on the field fire range. Having precise knowledge of where all bullets are hitting or missing the target, the poor firer (with instructor/trainer assistance) can improve his performance and the good firer can bring his shots to target center. Firers develop the knowledge and skills required to perform with confidence on the field fire range, where only hit-or-miss information is available.

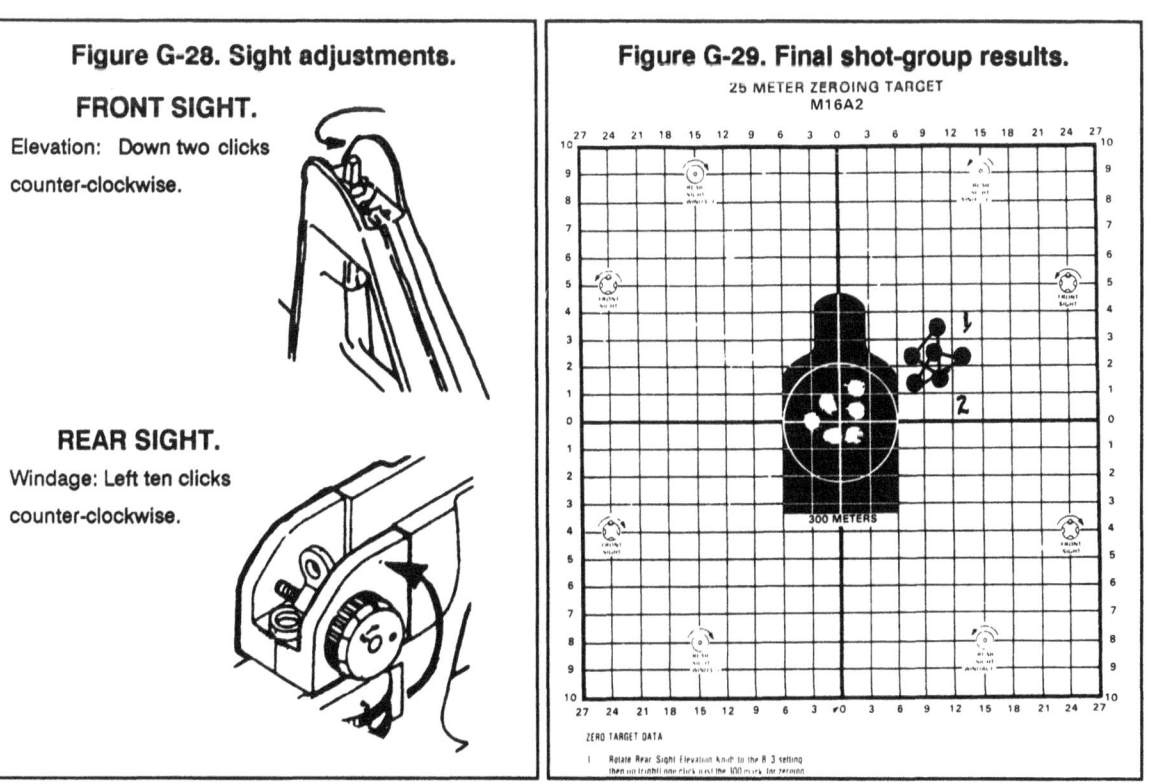

Figure G-27. Initial shot group results.

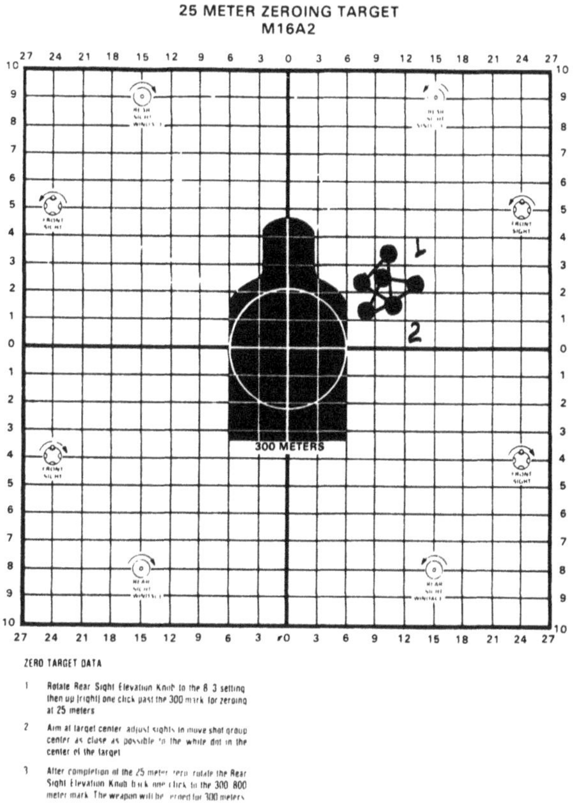

Figure G-28. Sight adjustments.

FRONT SIGHT.
Elevation: Down two clicks counter-clockwise.

REAR SIGHT.
Windage: Left ten clicks counter-clockwise.

Figure G-29. Final shot-group results.

CONDUCT OF DOWNRANGE FEEDBACK

During IET basic rifle marksmanship, downrange feedback is conducted with paper targets at 75 meters, 175 meters, and 300 meters. Shot groups are fired progressively at the 75-meter target, then the 175-meter and 300-meter range targets. Half of the bullets are fired from the supported fighting position and the other half from the prone unsupported position. After each group is fired, soldiers move downrange to mark their targets. Based on this feedback, soldiers receive a critique from their instructor/trainer/coach, and apply any needed sight changes or aiming adjustments. Army training centers conduct modified versions of downrange feedback based on the availability of the KD range or modified field fire range.

The downrange feedback exercise must be conducted within the constraints of time, ammunition, and available ranges. If 30 rounds of ammunition are available for training, firing 3-round shot groups 10 times is preferable over firing 5-round shot groups 6 times. Once the soldier understands the concept for adjusting the aiming point to compensate for the effects of wind and gravity, he is ready to apply his knowledge on the field-fire range.

To confirm zero, soldiers are placed in firing orders and issued one magazine of 6 to 10 rounds. Firers confirm their zero by firing two 3-round or 5-round shot groups at the 175- or 300-meter feedback targets. The firing line is then cleared, and firers move downrange to inspect their targets, to review their adjusted aiming performance, or to make sight adjustments. Each firer repairs his target by placing target pasters over the holes (black on black, white on white) and then moves back to the firing line.

Firing at 75-meter targets. Feedback can be provided after each round, each 3-round shot group, or each 5-round shot group on the 75-meter feedback targets. Soldiers fire from the supported and prone unsupported positions. The firing line is then cleared, and each firer and instructor/trainer/coach move downrange to inspect targets. Feedback consists of a critique of performance, adjustments to point of aim for gravity or wind effects, and evaluation of shot placement. Target spotters (markers) (Figure G-30) are placed in the bullet holes so hits can be viewed from the firing line.

Firing at 175-meter targets. Firers engage the 175-meter target using the same downrange procedure as the 75-meter target. The 175-meter target is engaged from the supported and prone unsupported positions.

Firing at 300-meter targets. The 300-meter target is engaged using the same downrange procedure as the 75-meter target. Firers use the supported and prone unsupported positions.

Target marking. When the initial shot group is fired, target spotters (markers) should be placed in each bullet hole, placing white spotters on the silhouette and black spotters off the silhouette.

This procedure ensures that the firer can see his performance when he returns to the firing line, that the tower operator can direct instructors/trainers to soldiers having the greatest problems, and that instructors/trainers can quickly assess firing problems (Figure G-30). Soldiers are motivated to fire better since their peers can observe their performance. On the second and subsequent trips to the target, the target spotters (markers) should be moved and placed in the holes of the new shot group. The old holes must be pasted, using black pasters on black and white pasters on white. Failure

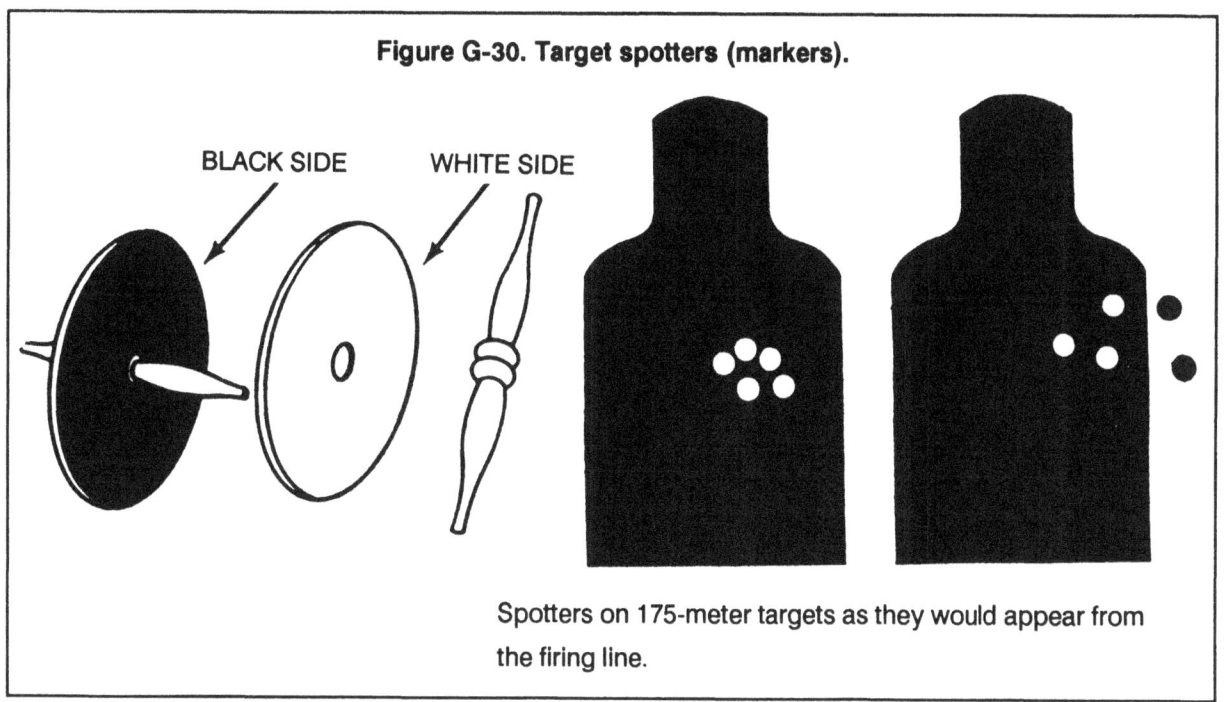

Figure G-30. Target spotters (markers).

Spotters on 175-meter targets as they would appear from the firing line.

to paste all bullet holes makes training ineffective since the soldier cannot clearly identify all bullets holes of his last shot group.

MODIFIED FIELD FIRE RANGE

A modified field fire range can be used for downrange feedback. To conduct downrange feedback, minor changes must be made to a standard field firing range (Figure G-31). Target frames, like those used on the 25-meter range, are placed at 75 meters, 175 meters, and 300 meters.

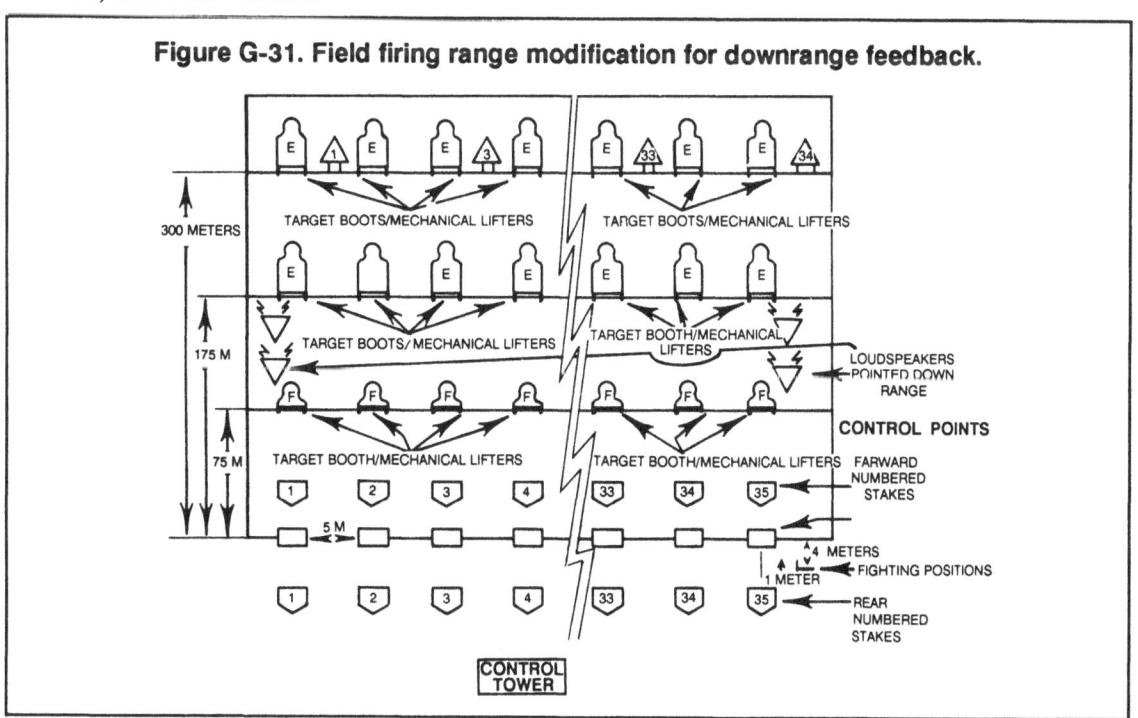

Figure G-31. Field firing range modification for downrange feedback.

Loudspeakers are placed downrange to allow firers at the 175-meter and 300-meter lines to hear the tower operator's instructions as firers move to the targets. After placing target spotters in the bullet holes, soldiers wait for an evaluation of their performance. The tower operator is also in a position to assess the effects of wind, which should be told to firers.

With these changes and the use of the downrange feedback targets, an effective training period can be conducted. A tower operator and one instructor/trainer for each five firing points are needed to operate the range.

RECORD OF PERFORMANCE

During the conduct of downrange feedback, a record of performance should be kept to facilitate, critique, and perform an after-action review.

As soldiers complete each phase and achieve the performance standard for that range, they should receive a critique. Instructors/trainers must ensure soldiers do not progress to a greater range until they become proficient at closer ranges. For example, the soldier who is having problems firing a 6-inch group at 75 meters should not be expected to fire an 11-inch group at 175 meters — progressing to greater ranges would only frustrate the soldier.

The DA Form 5239-R (75-, 175-, and 300-Meter Downrange Feedback Scorecard) is an effective way of recording target hit-and-miss information as the soldier completes each phase of firing (Figure G-32). (See Appendix H for a blank reproducible copy of this form.)

DOWNRANGE FEEDBACK TARGETS

The M16A1/M16A2 75-meter feedback target is the size of an actual F-type silhouette target with 6 inches cut from the bottom to allow for the target-holding mechanism and skip plate. When viewed from a distance of 75 meters, the target looks similar to a standard F-type silhouette on the field-fire range. However, there are important differences.

The target and surrounding space are large enough to capture all bullets fired. The firer moves to the target and actually locates each bullet hole. Information similar to that on the zero target has been overprinted to assist in applying sight adjustments. The 12-cm circle provides a standard equivalent to the 4-cm requirement on the zero range. An X is placed in the bottom portion of the circle to show the firer where he must aim on this target so his bullets will hit target center of mass when his rifle is zeroed. This target can be used at any range to capture bullet strike; however, the overprinted material is valid only at 75 meters.

The M16A1/M16A2 175-meter and 300-meter feedback targets are designed for use at 175 or 300 meters and have the same features as the 75-meter target (Figure G-33). Both targets can be used to confirm weapon zero or to refine the zero obtained on the 25-meter range. The zero obtained on these targets are more valid than the zero obtained on the 25-meter range. For example, when engaging this actual-size target at 175 or 300 meters from a supported position, it is a closer approximation to the actual task than the scaled target at 25 meters. By allowing for the adjusted aiming point and for wind, this exercise can contribute to a refinement of the zero.

Figure G-32. Example of completed DA Form 5239-R (75-, 175-, and 300-Meter Downrange Feedback Scorecard).

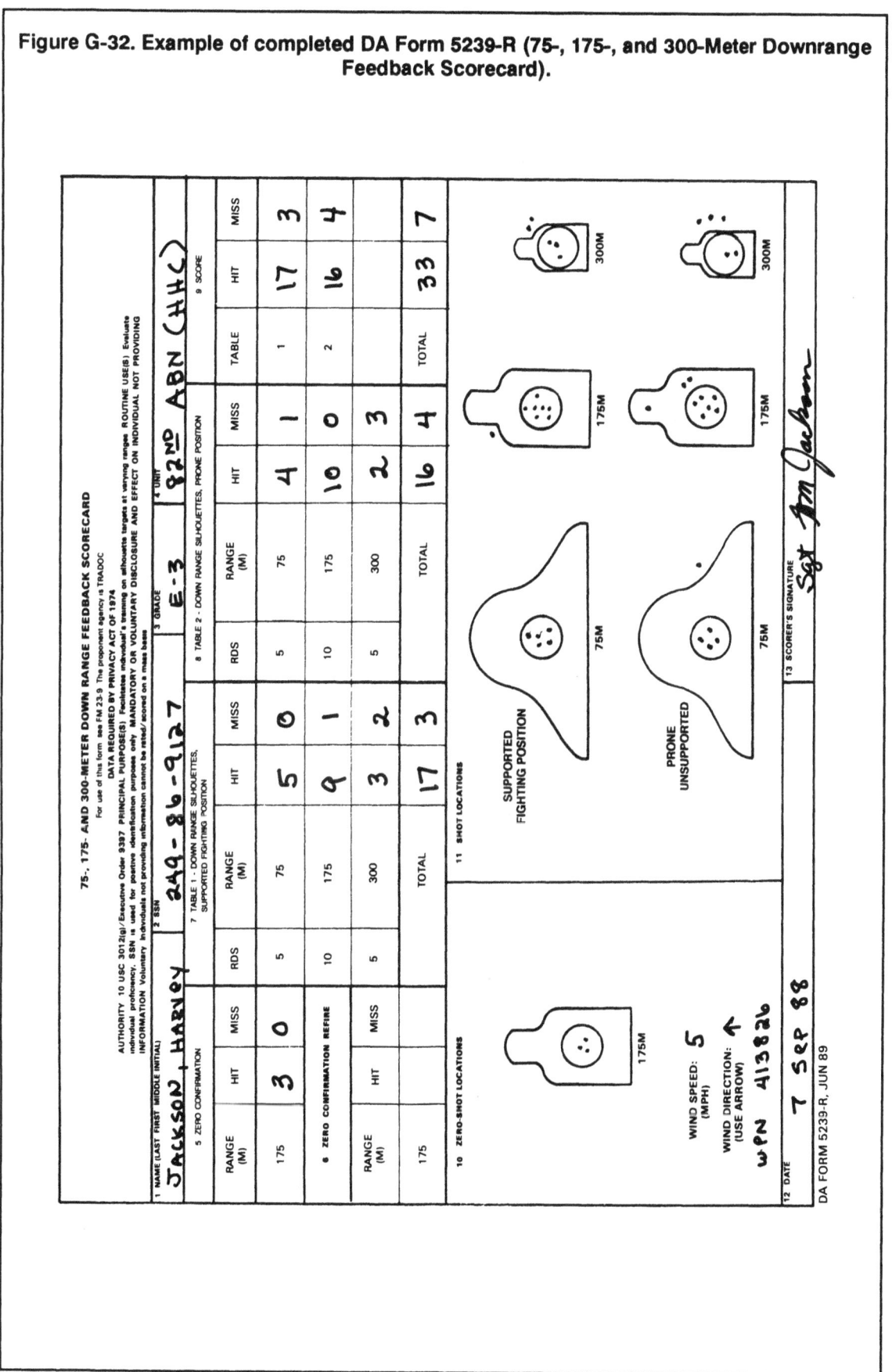

FM 23-9

Figure G-33. Feedback targets.

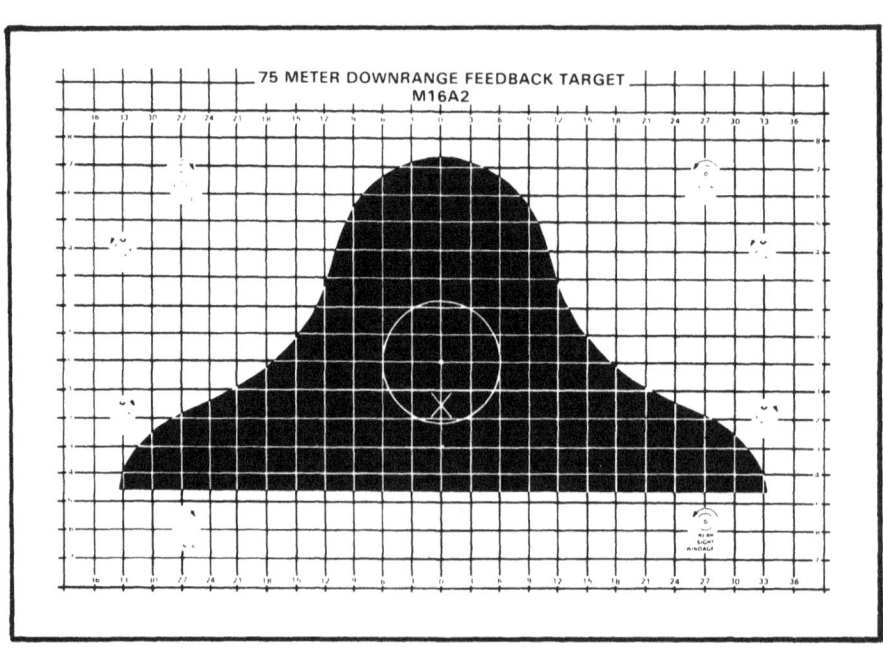

M16A1 75-meter feedback target (NSN 6920-01-169-6921).

M16A2 75-meter feedback target (NSN 6920-01-253-4006)

Figure G-33. Feedback targets (continued).

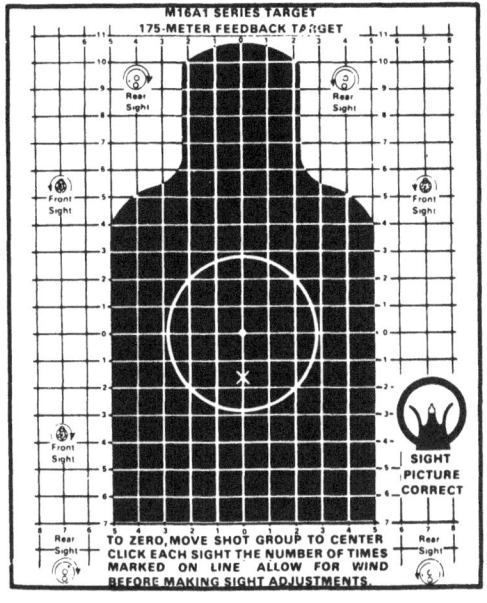

M16A1 175-meter feedback target
(NSN 6920-01-167-1395).

M16A2 175-meter feedback target
(NSN 6920-01-253-4008)

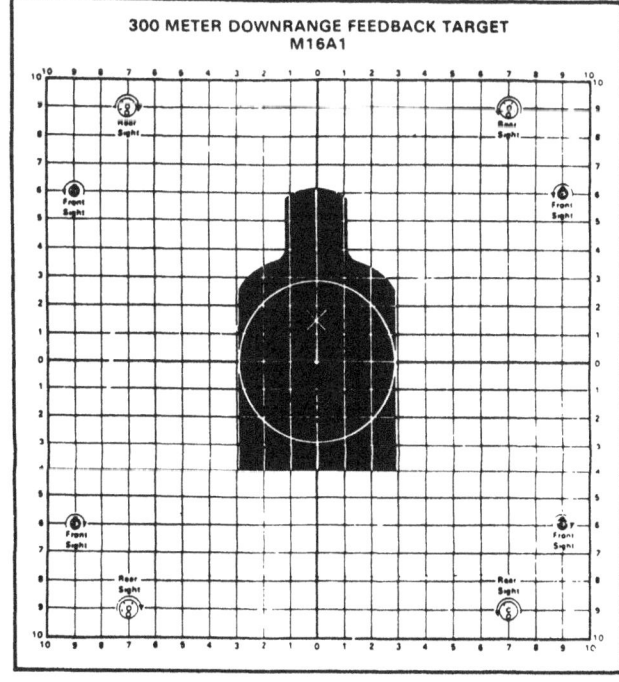

M16A1 300-meter feedback target
(NSN 6920-01-253-4009).

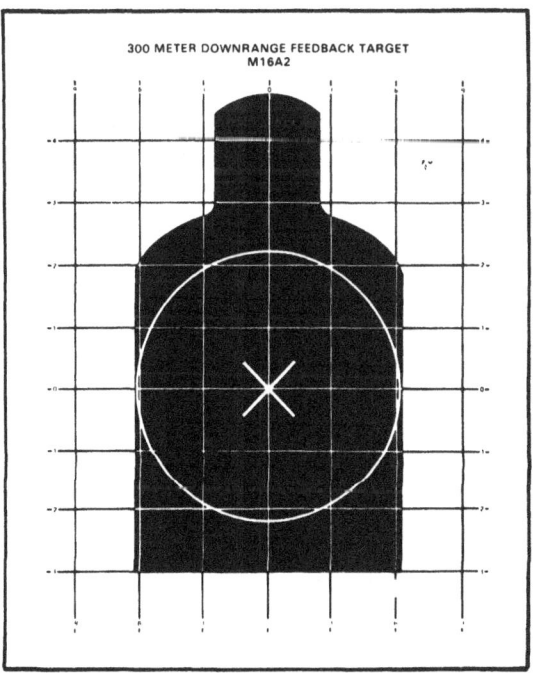

M16A2 300-meter feedback target
(NSN 6920-01-253-4007).

Downrange feedback training should include detailed explanations of the targets: how the 12-cm circle on the 75-meter target and 28-cm circle on the 175-meter equate to the 4-cm zeroing standard on the 25-meter range; the small difference that should be experienced between point of aim and bullet strike; how the grid can be used for making sight adjustments; and how the targets should be marked. Training should also include how wind can affect bullets and should emphasize the fundamentals of firing.

The information provided to the firer and instructor/trainer should be visualized for the targets shown in Figure G-34A. These two firing performances provide the same information back to the firing line — the target was hit one time and missed twice. When bullet locations can be seen, the firing problems are quite different. The firer on the left is failing to apply fundamentals, and the firer on the right has a zeroing problem. The targets in Figure G-34B indicate that both firers are turning in an excellent performance by hitting with all three bullets. However, when the actual bullet strikes can be seen, it is obvious that the firer on the left needs assistance.

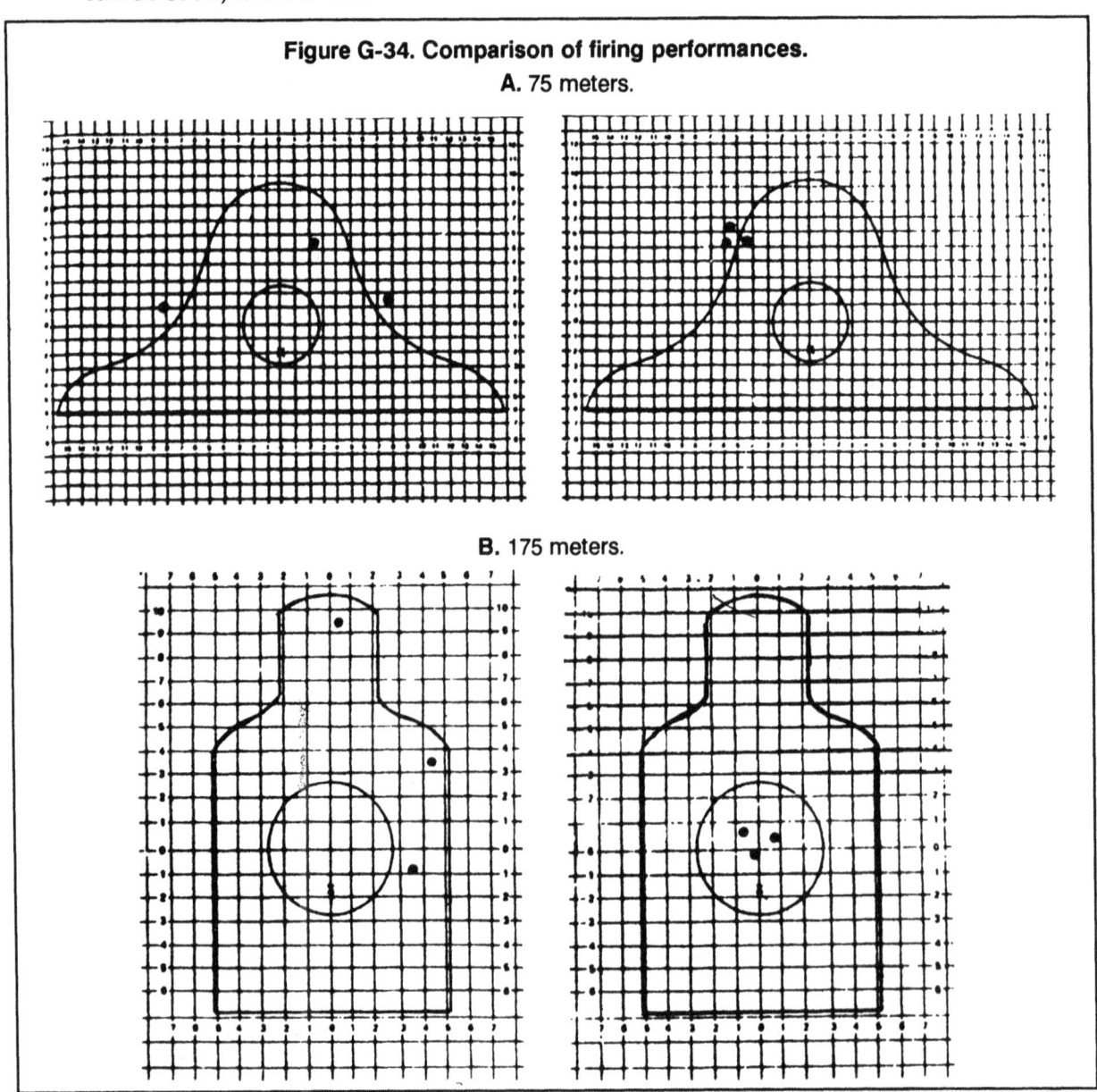

Figure G-34. Comparison of firing performances.
A. 75 meters.
B. 175 meters.

Section IV. KNOWN-DISTANCE RANGE

The known-distance range is used for testing and marksmanship training. The firing task on a known-distance range is an intermediate step toward the firing task of a combat soldier. Program changes (pop-up targets) and ranges have provided a much better simulation of combat requirements. The soldier is provided information concerning the precise hit-or-miss location of every bullet fired. KD firing is conducted with a single, clearly visible target at a known distance, and the soldier can establish a position that provides a natural point of aim on that single target. While this is good for its intended purpose, overuse often results in KD training being more competition-oriented than combat-oriented.

CONDUCT OF KNOWN-DISTANCE RANGE

The proper use of the known-distance range contributes to the unit rifle marksmanship program. Effective training can be conducted on the KD range (Figure G-35) using standard silhouette target facings; however, the downrange feedback targets are

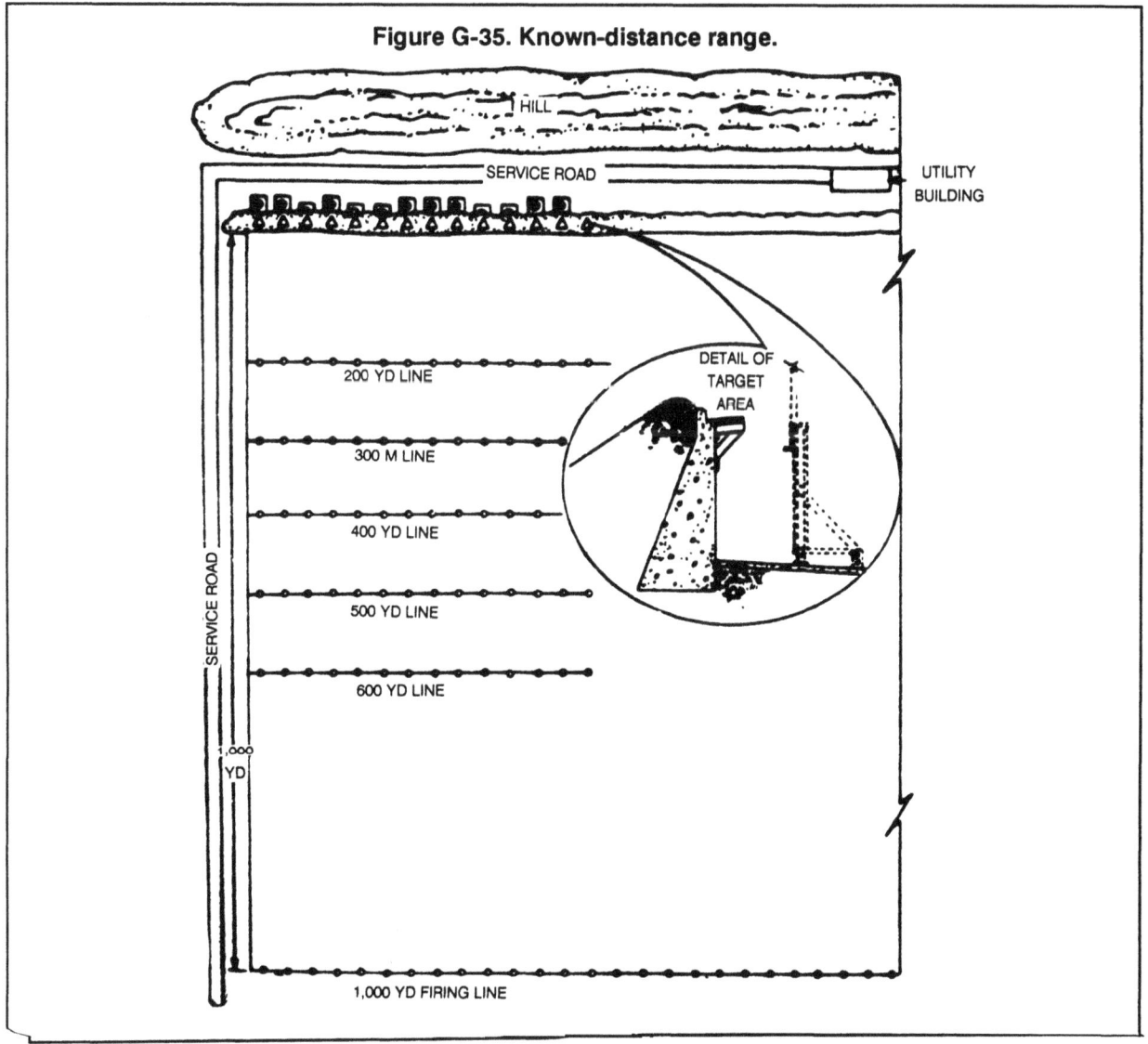

Figure G-35. Known-distance range.

recommended for the KD range. The F-type silhouette 75-meter feedback target can be engaged from the 100-yard line; the E-type silhouette 175-meter target from the 200-yard line; and the 300-meter target from the 300-yard line.

The KD range does not require the soldier to detect targets, to estimate range to targets, to scan a sector of fire, to respond to surprise targets, to respond to short exposure targets, to engage multiple targets, and so on.

NOTE: FM 25-7 and Appendix D provide information on pit crew operations, construction of target carriers for KD ranges, and ordering the sliding combination target frame (NSN 6920-049-9579).

An advantage of a KD range is the ability to see precisely where each bullet hits. Therefore, to benefit from this training, procedures must be established so soldiers can clearly see the results of each firing, whether a group, single shot, or 10-round exercise. This can be accomplished by single spotters alone at the closer ranges, using larger spotters as the range increases, supplementing the spotters with a marking disc, or using scopes and binoculars at the extended ranges.

RANGE ORGANIZATION

Known-distance ranges are organized and operated differently than other ranges. Personnel are organized into firing line safety and pit crews. The firing line safety crew operates under orders from the chief range officer. The pit crew operates under orders from the pit officer.

Chief Range Officer. The chief range officer ensures the safety of all personnel and proper operation of his range. He may appoint a tower operator to issue fire commands or perform those duties himself. The chief range officer performs the following:

- Gives the range safety briefings to all soldiers and support personnel before organizing firing orders and pit crews. Local regulations may require an appointed safety officer to perform this duty.

- Is responsible for ammunition details.

- Organizes soldiers into firing orders for testing. If a pit detail is provided, the chief range officer may use as many firing points as he chooses to conduct his range. When a pit detail is not provided, the chief range officer must assign four soldiers for each firing point. For example, if a unit has 100 personnel, the chief range officer should use 25 firing points.

- Numbers the firing orders. Firing orders 1 and 2 remain at the firing line, while firing orders 3 and 4 perform pit crew duties. Firing order 1 fires first and firing order 2 performs as coaches.

- Informs the pit officer when refires are to be conducted.

- Informs the pit officer when pit crew changeovers are executed. Pit crew changeovers occur when firing orders 1 and 2 have completed the known-distance course or the alternate course, including initial refires. Refires are conducted when that firing table is completed or during alibi firings.

Pit Officer. The pit officer is the main link in the range chain of command and operates under the orders of the chief range officer. He may be an NCO, depending on local restrictions. The pit officer performs the following:

- Maintains radio or telephone contact with the chief range officer during operation of the range.
- Serves as the timekeeper for target exposures.
- Is responsible to the chief range officer for the safety and conduct of the pit crew.
- Prepares and operates targets, scoring, and alibi target exposures.
- Conducts a pit safety briefing before firing and each time a pit crew changeover occurs.
- Briefs the pit crew to ensure each firing point pair understands their duties. The pit crew raises, lowers, disks, and repairs its respective targets upon command from the pit officer. All targets must be raised or lowered at the same time on order of the pit officer. The crew scores its assigned target only upon command from the pit officer.
- Announces alibi and refire targets. Alibi and refire targets are raised and lowered on order from the pit officer. All other targets remain lowered.
- Collects the scorecards from the pit crew at the end of each firing order.
- Verifies an signs the scorecards.
- Announces individual scores if requested by the chief range officer.

Firing Line Assistant Instructors/Trainers. Firing line assistant instructors/trainers perform as safety personnel and confirm allowable alibis during record fire.

Firing Line Safety Personnel. Firing line safety personnel are responsible for precise timing of individual alibi firers. They inform the chief range officer of rounds fired after the time allowed for alibis. The chief range officer informs the pit officer of rounds fired after the allotted time, which are scored as misses.

All Personnel. All personnel complete scorecards. The pit officer maintains scorecards by firing order and firing point number in the pits.

CONDUCT OF TRAINING

Before engaging targets, soldiers are issued one magazine of 6 to 10 rounds and confirm or refine their zero at the 300-yard line by firing two 3-round or 5-round shot groups. Targets are spotted by the pit crews after each group.

NOTES: 1. M16A1. The unmarked aperture (short-range) is used on the M16A1 for refinement of zero at 300 yards. For target engagement beyond the 300-yard line, the long-range aperture (L) is used.
    2. M16A2. The unmarked aperture is used for both refinement of zero and target engagement at all distances on the KD range.

When the zero is confirmed, the firing line is moved to the 100-yard line. Soldiers are issued one magazine of 6 to 10 rounds and engage the (75-meter feedback target) F-type silhouette target from either the prone unsupported position or prone supported position (if fighting positions are not available). After each group has been fired, the target is withdrawn and the rounds are spotted by the pit detail (white on black and black on white). The holes from earlier rounds fired are pasted (black on black and white on white). The instructors/trainers critique the firers after each round.

When all firers have expended 6 to 10 rounds or when proficiency has been demonstrated, the firing line is moved to the 200-yard line. The conduct of fire from the 200- and 300-yard lines is the same procedure as described for the 100-yard line.

If the pit crew has been drawn from the firing unit, the exercise is conducted twice. After completion, the firers on the firing line become the pit crew, and the firers performing pit detail complete the same firing exercise.

Section V. KNOWN-DISTANCE ALTERNATE COURSE

The known-distance alternate course gives soldiers the chance to engage targets at range with time constraints and feedback. The effects of wind and gravity are demonstrated while firing on the course. Before firing the course, all soldiers confirm the zero of their assigned rifles at 300 yards with six sighter rounds. Training/sustainment ammunition is used for sighter rounds if a zeroing exercise is not conducted on the day of record fire. The six sighter rounds are fired in the prone supported position from the 300-yard line before qualification — sighter rounds do not count for score. The known-distance alternate course is a 12-hour course of instruction as follows:

25-meter zeroing: 4 hours

Record fire: 8 hours

CONDUCT OF KNOWN-DISTANCE ALTERNATE COURSE

The chief range officer or tower operator gives fire commands verbatim as written herein to conduct the known-distance alternate course.

TABLE 1: Prone supported position, 300 yards, 2 magazines of 10 rounds each, 60 seconds for each exercise, E-type silhouette.

> FIRERS, ASSUME A PRONE SUPPORTED POSITION.
> COACHES, ISSUE THE FIRER ONE MAGAZINE OF TEN ROUNDS.
> THE FIRING LINE IS NO LONGER CLEAR.
> LOCK, ONE MAGAZINE OF TEN ROUNDS, LOAD. (Pause.)
> IS THE LINE READY? (Pause to observe the firing line to ensure all soldiers are in position and are ready. If not, try to get them in position. If all soldiers are ready, continue the commands.)
> READY ON THE RIGHT?
> READY ON THE LEFT?
> THE FIRING LINE IS READY.
> FIRERS, WATCH YOUR LANE.

At this time the pit officer issues the command TARGETS UP. When the targets are in the fully raised position, the pit officer starts his stopwatch. The raising of the targets is the command to commence fire. The chief range officer may also command COMMENCE FIRE. When the allotted time is completed, the pit officer commands TARGETS DOWN to the pit crew. The chief range officer commands CEASE FIRE, CEASE FIRE, CEASE FIRE when he observes the targets being lowered. The chief range officer can also keep time with the pit officer; however, the pit officer is the official timekeeper.

ARE THERE ANY ALIBIS?

Allowable alibis are allotted six seconds for each unfired round. An allowable alibi is a malfunction of the rifle or ammunition, which is not associated with firer error. Rounds not expended during the allotted time do not constitute an alibi and are counted as misses. Only one alibi for each table is authorized. If a rifle continues to malfunction, it is removed from the firing line so the armorer can inspect or repair it.

The chief range officer repeats the fire commands for alibi firers. Cross-fires are not allowable alibis for the cross-firing soldier. The recipient of the cross-fired rounds refire the table. The cross-firer is awarded misses for those shots on the wrong target. The cross-firer may refire the course. Regardless of his total hits during refire, he can be rated only as a marksman with a score of 26. The recipient is not penalized.

The firing line safety personnel "time" their firers. When alibi firings are completed and on order from the pit officer, all targets are scored, disked, and raised by the pit crew. Firers receive feedback regarding shot group size and location. The pit officer lowers the targets when all soldiers have observed their shot groups. The disks are removed and targets repaired for the next firing table. The chief range officer informs the pit officer if he elects to have all personnel fire at each yard line before moving to the next firing line. The firing line safety crew clears all rifles after completing each firing table and before changing yards lines. The tower operator commands:

CLEAR ALL WEAPONS.
CLEAR ON THE RIGHT?
CLEAR ON THE LEFT?
THE FIRING LINE IS CLEAR. (Commands are repeated for second 10-round exercise.)

TABLE 2: Prone unsupported position, 200 yards, 1 magazine of 10 rounds, 60 seconds, E-type silhouette.

TABLE 3: Prone unsupported position, 100 yards, 1 magazine of 10 rounds, 60 seconds, F-type silhouette.

Fire commands for Tables 2 and 3 are the same as for Table 1, including alibi and refire procedures. The pit officer has the targets changed from E-type to F-type silhouettes while the soldiers are moving from the 200-yard line to the 100-yard line. The pit officer informs the chief range officer when the target changeover is completed.

EQUIPMENT

E-type silhouette facing, paper, NSN 6920-00-600-6874	2 for each firer
F-type silhouette facing, paper, NSN 6920-00-610-9086	1 for each firer
Pasters, black, NSN 6920-00-165-6354	As required
Pasters, buff, NSN 6920-00-172-3572	As required
Disk, spotter with spindle, NSN 6920-00-713-8255	20 for each lane used
Communication set with loudspeakers	2 sets
Field telephone or wireless radio	2 each
Ammunition, zeroing	6 rounds for each firer
Ammunition, record firing	40 rounds for each firer
Magazines, ammunition	4 for each firer
Paste, wheat	Optional
Tacker, target with staples	As required
Stopwatch	1 each
Scorecard	1 for each firer

RECORD OF PERFORMANCE

The known-distance alternate course is fired IAW DA Form 5789-R (Record Firing Scorecard—Known Distance Course) (Figure G-36).(See Appendix H for a blank reproducible copy of this form.)

QUALIFICATION STANDARDS

The pit officer scores the targets. The pit crew responds promptly to commands from the pit officer and informs the pit officer of cross-fires. The pit crew accurately counts hits and misses. A hit is any bullet hole that is either completely within or touches some part of the silhouette facing. If a bullet hole does not touch some part of the silhouette facing, it is counted as a miss. Richochets are counted as hits or misses. The pit crew clearly completes scorecards and quickly repairs targets. Qualification standards for the known-distance alternate course are as follows:

- **Expert:** Hits 38 to 40 targets
- **Marksman:** Hits 26 to 32 targets
- **Sharpshooter**: Hits 33 to 37 targets
- **Unqualified:** Hits 25 targets and below

Figure G-36. Example of completed DA Form 5789-R (Record Firing Scorecard—Known-Distance Course)(front).

RECORD FIRING SCORECARD • KNOWN DISTANCE COURSE

For use of this form see back The proponent agency is TRADOC

DATA REQUIRED BY PRIVACY ACT OF 1974

AUTHORITY 10 USC 30129g) Executive Order 9397 **PRINCIPAL PURPOSE(S)** Records individual s performance on record fire range **ROUTINE USE(S)** Evaluation of individual s proficiency and basis for determination of award of proficiency badge. SSN is used for positive identification purposes only **MANDATORY OR VOLUNTARY DISCLOSURE AND EFFECT ON INDIVIDUAL NOT PROVIDING INFORMATION** Voluntary Individuals not providing information cannot be rated/scored on a mass basis

1 NAME (LAST FIRST MIDDLE INITIAL)	2 SSN	3 GRADE	4 UNIT	5 FIRING POINT AND ORDER
RIDLEY, GEORGE S.	420-05-2112	SFC	B BTRY 1/105 ARTY	22 / 3

6 TABLE 1 PRONE SUPPORTED

ROUND	RANGE 300	HIT	MISS
1	E SIL	X	
2			X
3		X	
4		X	
5		X	
6		X	
7			X
8		X	
9		X	
10		X	
11		X	
12		X	
13		X	
14			X
15		X	
16		X	
17		X	
18		X	
19		X	
20		X	
TIME 120 SEC	TOTAL	17	3

7 TABLE 2 PRONE UNSUPPORTED

ROUND	RANGE 200	HIT	MISS
1	E-SIL	X	
2		X	
3		X	
4		X	
5		X	
6			X
7		X	
8		X	
9		X	
10		X	
TIME 60 SEC	TOTAL	9	1

8 TABLE 3-PRONE UNSUPPORTED

ROUND	RANGE 100	HIT	MISS
1	F-SIL	X	
2		X	
3		X	
4		X	
5		X	
6		X	
7		X	
8		X	
9		X	
10		X	
TIME 60 SEC	TOTAL	10	

9 SCORE

[ ]	38-40 EXPERT
[X]	33-37 SHARPSHOOTER
[ ]	26-32 MARKSMAN
[ ]	25-BELOW UNQUALIFIED

11 REMARKS

NIGHT FIRE EXERCISE

DATE	HIT	MISS	GO	NO GO
8/5/88	8	2	X	

NBC FIRE EXERCISE

DATE	HIT	MISS	GO	NO GO
8/6/88	12	8	X	

WPN # 413821

10 LIGHT CLEAR

DIRECTION (12, 9, 3, 6)

WIND (12, 9, 3, 6) 5 MPH

ZERO 8-5-88	D-2	M16A2	FRONT
ELEV	8/3	12	◄
WIND	L-3	12	◄

*FIRER ISSUED 40 ROUNDS THE ROUNDS WILL BE PRELOADED IN FOUR 10-ROUND MAGAZINES - TWO FOR TABLE 1, ONE FOR EACH REMAINING TABLE ALL ROUNDS WILL BE FIRED WITH THE M16A1 SHORT RANGE SIGHT

12 DATE SIGNED	13 DATE SIGNED
8 Aug 88	8 AUG 88
14 SCORER S SIGNATURE	15 OFFICER'S SIGNATURE
Cris L. Hubbard SSG.	Douglas Woods CPT. AR.

DA FORM 5789-R, JUN 89

Figure G-36. Example of completed DA Form 5789-R (Record Firing Scorecard—Known-Distance Course)(back)(continued).

This scorecard is used to score Known Distance Course record fire qualification when the Known Distance Range is used. This course is used only when the standard record fire course is not available.

NOTE: If zeroing/grouping exercises are not performed on the day of record fire, 6 rounds of training/sustainment ammunition are fired from the 300 yard/meter line for confirmation of zero prior to conducting the Qualification Course.

CONDUCT OF FIRE

For Table 1, the firer is given two 10-round magazines to engage an E-silhouette at 300 yards within 120 seconds in the prone supported position. Table 2 is fired with a 10-round magazine at an E-silhouette at 200 yards within 60 seconds in the prone unsupported position. Table 3 is fired with a 10-round magazine at an F-silhouette at 100 yards within 60 seconds in the prone unsupported position.

SCORING

Scoring is conducted in the pits, with the results provided after each table. One point is awarded for each round hitting the target. A hit is scored for any bullethole that is within or touches some part of the silhouette facing.

Section VI. FIELD FIRE RANGE

When soldiers receive adequate training before field firing, they gain confidence in their ability to hit targets. When hit-and-miss information is available, soldiers develop techniques for observing their sector of fire, for engaging targets rapidly, and for engaging multiple targets—all preparing them for practice on the record fire course. Field firing should be scheduled to follow downrange feedback or known-distance firing, or other suitable training that assures soldiers they have developed adequate firing skills.

CONDUCT OF FIELD FIRE RANGE

The field fire range usually consists of 35 points—each point consisting of an F-type silhouette at 75 meters, and E-type silhouettes at 175 meters and 300 meters. When a bullet hits the target, the vibration activates a mechanism that causes the target to fall, simulating a kill (Figures G-37 and G-38).

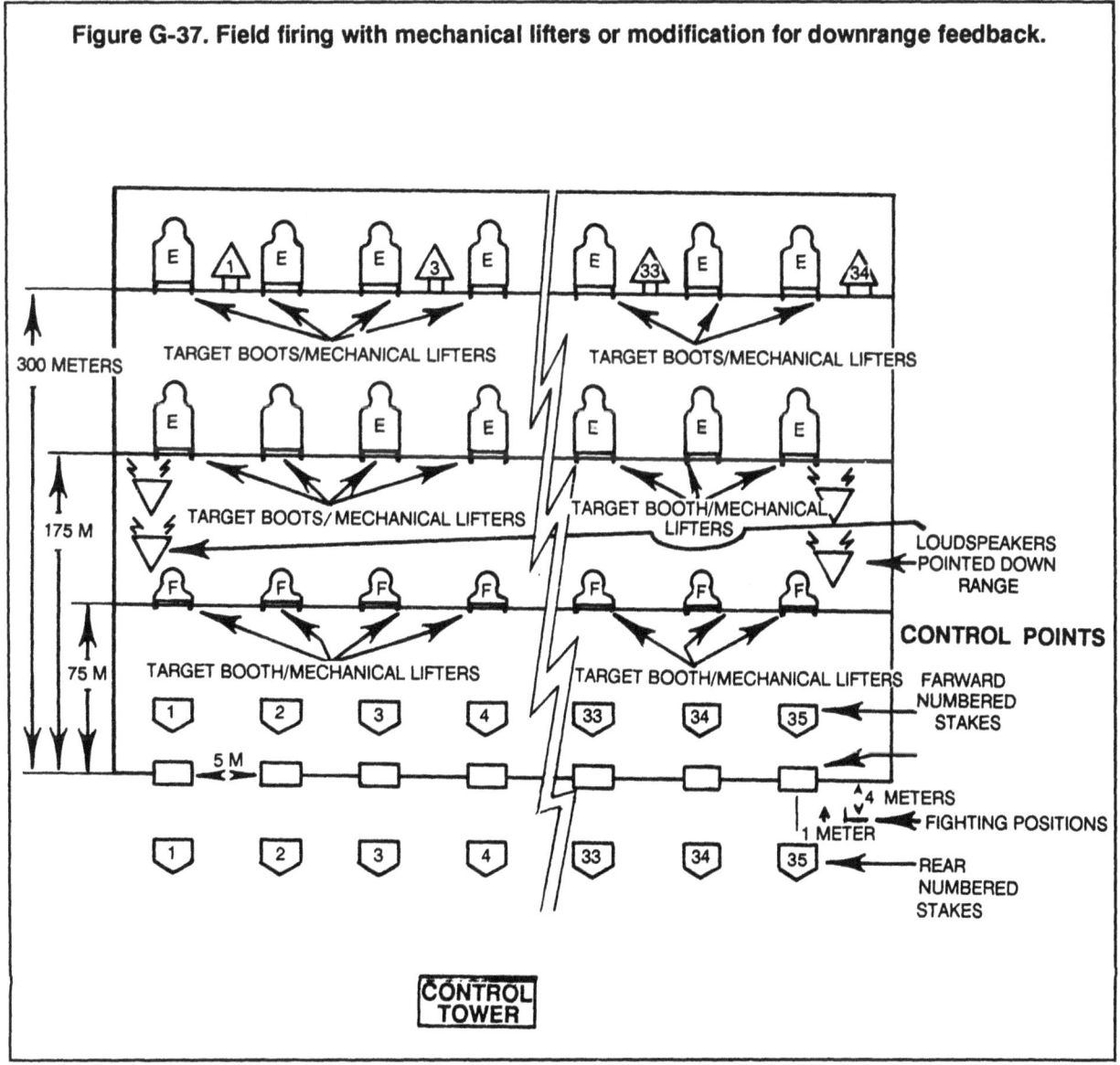

Figure G-37. Field firing with mechanical lifters or modification for downrange feedback.

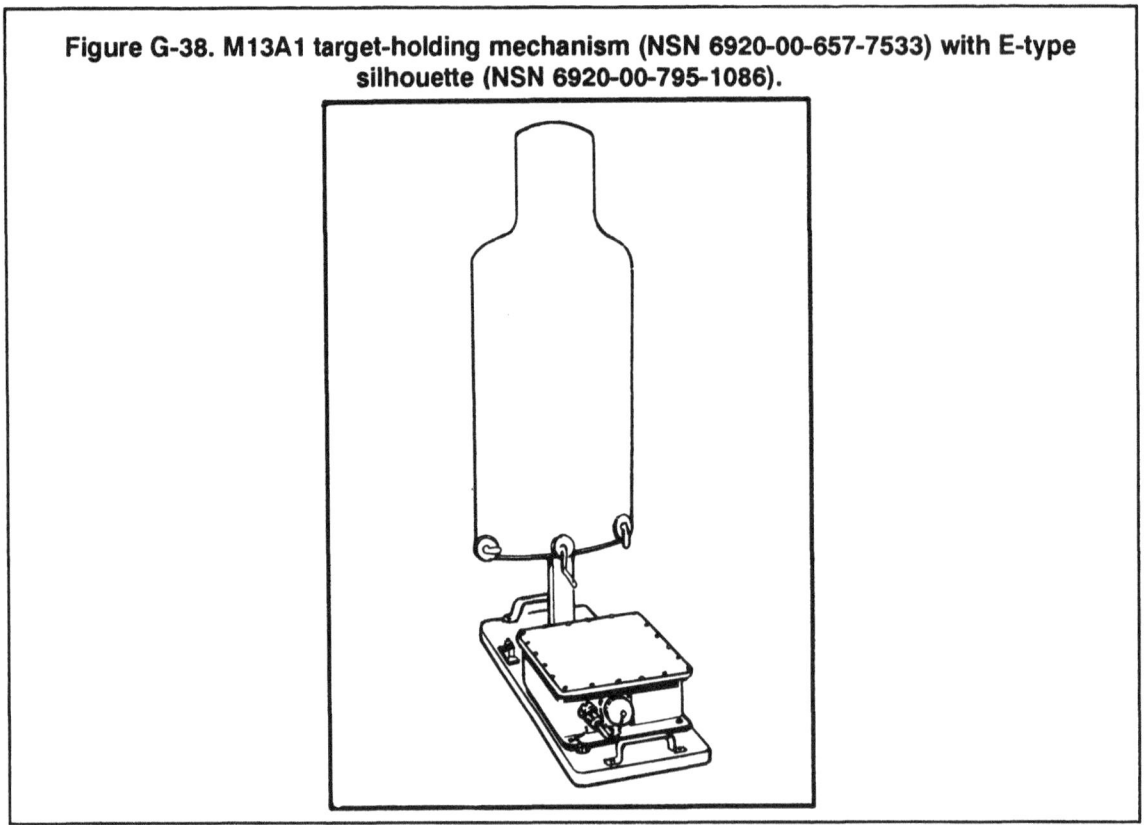

Figure G-38. M13A1 target-holding mechanism (NSN 6920-00-657-7533) with E-type silhouette (NSN 6920-00-795-1086).

RANGE CHECKS

Field fire ranges must be checked for proper operation of targets, mechanisms, and hit recording equipment. Soldiers must learn to fire while receiving only hit-and-miss information; however, when incorrect information is provided (for example, when the target is hit and does not go down), the learning process is disrupted. Therefore, range mechanisms must be in proper working condition, the target must be correctly and securely attached to the mechanism, and the holes on the target should be small enough to allow bullets to vibrate the target to activate the mechanism.

CONDUCT OF TRAINING

The field fire course should be fired from the supported fighting position and prone unsupported position. After these are mastered, other firing positions can be considered for training. Initial training should be with single exposed targets and increased time for engagement. As skills improve, multiple targets with shorter exposure times can be engaged.

Soldiers who miss most targets should be removed from the firing line for remedial training if their problem cannot be corrected. When a soldier fires a 300-meter target 10 times and misses it 10 times, it is obvious that he is not learning but instead is losing confidence in his ability. The typical soldier should hit the 300-meter target at least 7 out of 10 times.

Peer coaches should assist soldiers in observing the strike of rounds and in identifying firing problems. If the target is missed and the coach cannot observe the bullet strike, the coach should instruct the soldier to aim lower for the next shot, expecting to see the strike of the bullet in the ground. With this information, the coach can instruct the soldier on where to aim or how to adjust his point of aim to hit the target.

Live-fire training can be organized in several ways. A unit is divided into two or more firing orders based on the number of personnel to be trained. The first order is the firer, the second order is the coach, and (if required), the third order is scorers. At the conclusion of each exercise, positions rotate until all orders are complete.

Standard field-fire scenarios have been developed to provide several target exposures. Although they are recommended for initial training, any variety of more challenging target sequences can be developed by local commanders. Ammunition is allocated based on one round for each target exposure.

RECORD OF PERFORMANCE

During live fire, the soldier's hit-and-miss performance is recorded to facilitate the instructor/trainer's critiques or to indicate where more training is needed. A master score chart, indicating each soldier's scores for each exercise, encourages a competitive spirit within a unit. It also aids in determining which soldiers require close supervision or remedial instruction. Two methods used to record firing performance are manually marked scorecards and automated computer printouts.

Manual Recording. When manual recording is used, the unit provides soldiers for recording information on either DA Form 3601-R (Single Target—Field Firing Scorecard) (Figure G-39) or DA Form 5241-R (Single and Multiple Targets—Field Firing Scorecard) (Figure G-40). (See Appendix H for blank reproducible copies of these forms.)

NOTE: Peer coaches should not be used as scorers; their duty is to observe the firer.

Automated Recording. When field firing exercises are conducted on the new family of automated field fire ranges, a computer printout is provided for each firing order. At the conclusion of each firing order, the range NCOIC completes the printout and ensures the soldier identification is matched with each firing point(Figure G-41). He adds the soldier's name or roster number to the top of each lane/firing point data column. Based on a one-round allocation for each target exposure, data should be collected on hits, misses, no-fires, and repeated shots to assist the instructor/trainer in assessing firing proficiency.

FM 23-9

Figure G-39. Example of completed DA Form 3601-R (Single Target — Field Firing Scorecard).

SINGLE TARGET FIELD FIRING SCORECARD

For use of this form see FM 23-9. The proponent agency is TRADOC

DATA REQUIRED BY PRIVACY ACT OF 1974

AUTHORITY 10 USC 3012(g)/Executive Order 9397 PRINCIPAL PURPOSE(S): Facilitates individual's transition to distant target and provides feedback ROUTINE USE(S): Evaluate individual proficiency. SSN is used for positive identification purposes only. MANDATORY OR VOLUNTARY DISCLOSURE AND EFFECT ON INDIVIDUAL NOT PROVIDING INFORMATION: Voluntary individuals not providing information cannot be rated/scored on a mass base

1. NAME (LAST, FIRST, MIDDLE INITIAL)	2. SSN	3. GRADE	4. UNIT	5. ROSTER NO
THOMAS, BILL T.	357-50-8120	PVT2	B1/31	409

6. INTRODUCTION

TABLE 1 - SUPPORTED FIGHTING POSITION

RD	RANGE (M)	TIME (SEC)	HIT	MISS
1	75	6	X	
2	75	6	X	
3	75	6	X	
4	75	6	X	
5	75	6	X	
6	75	8	X	
7	175	8		X
8	175	8	XX	
9	175	8	XX	
10	175	8	XX	
11	175	8	XX	
12	175	8		X
13	300	10		X
14	300	10		
15	300	10	XX	
16	300	10	XX	
17	300	10		
18	300	10		X
		TOTAL	14	4

7. TABLE 2 - SUPPORTING FIGHTING POSITION

RD	RANGE (M)	TIME (SEC)	HIT	MISS
1	75	6	X	
2	175	8	XX	
3	300	10		
4	175	8		X
5	75	6	X	
6	300	10	XX	
7	300	6		X
8	75	8	XX	
9	175	8	XX	
10	175	7		X
11	300	8	X	
12	175	8	XX	
13	75	6	XX	
14	300	10		
15	175	8	X	
16	75	6		
17	300	10		X
18	75	8	X	
		TOTAL	14	4

8. TABLE 3 - PRONE POSITION

RD	RANGE (M)	TIME (SEC)	HIT	MISS
1	75	7	X	
2	175	9		X
3	300	11		X
4	175	9	X	
5	75	7	XX	
6	300	11		X
7	300	11	XX	
8	75	7	XX	
9	175	9	XX	
10	175	11		
11	300	11	XX	
12	175	9	XX	
13	75	7		X
14	300	11		
15	175	9	X	
16	75	7		X
17	300	11	XX	
18	75	7		
		TOTAL	12	6

9. SCORE

TABLE	HIT	MISS
2	14	4
3	12	6
TOTAL	26	10

10. REMARKS

11. DATE SIGNED	12. SCORER'S SIGNATURE
2 April 1989	SPC Frank Calero

DA FORM 3601-R, JUN 89.

FM 23-9

Figure G-40. Example of completed DA Form 5241-R (Single and Multiple Targets— Field Firing Scorecard).

SINGLE AND MULTIPLE TARGETS
FIELD FIRING SCORECARD

For use of this form see FM 23-9. The proponent agency is TRADOC

AUTHORITY: 10 USC 3012/(g)/Executive Order 9397 PRINCIPAL PURPOSE(S): Facilitates individual's transition to distant target and provides feedback. ROUTINE USE(S): Evaluate individual proficiency. SSN is used for positive identification purposes only. MANDATORY OR VOLUNTARY DISCLOSURE AND EFFECT ON INDIVIDUAL NOT PROVIDING INFORMATION: Voluntary. Individual cannot be rated/scored on a mass basis. DATA REQUIRED BY PRIVACY ACT OF 1974

1 NAME (LAST FIRST MIDDLE INITIAL)	2 SSN	3 GRADE	4 UNIT	5 ROSTER NO
KING, LYNN O.	123-02-1111	PVT 2	A 1/31	413

6. INTRODUCTION

TABLE 1 – SUPPORTED FIGHTING POSITION

RD	RANGE (M)	TIME (SEC)	HIT	MISS
1	75	5	X	
2	175	7	X X	
3	75	11		X
4	300			X
5	75	9	X	
6	175			X
7	75	10	X	
8	300			X
9	175	11	X	
10	300			X
TOTAL			6	4

7. TABLE 2 – SUPPORTING FIGHTING POSITION

RD	RANGE (M)	TIME (SEC)	HIT	MISS
1	175	7	X	
2	75	10	X	
3	300			X
4	75	9	X X	
5	175		X X	
6	300	9	X X	
7	75		X X	
8	175		X	
9	175	11	X	
10	300			X
11	75		X	
12	175		X X	
13	175	11	X	
14	300		X	
15	75	5	X	
16	175	11	X X	
17	300		X X	
18	75	9	X X	
19	175		X	
20	75	10	X	
21	300			X
22	175	7	X	
TOTAL			18	4

8. TABLE 3 – PRONE POSITION

RD	RANGE (M)	TIME (SEC)	HIT	MISS
1	75	6	X	
2	175	8	X X	
3	75	13	X	
4	300			X
5	75	11	X X	
6	175		X X	
7	75	12	X X	
8	300		X	
9	175	13		X
10	300		X X	
11	75	11	X X	
12	175			X
13	175	8	X	
14	75	6	X X	
15	75	11	X X	
16	175			X
17	75	12	X X	
18	300		X X	
19	75	11	X X	
20	175			X X
21	175	13		
22	300			
TOTAL			16	6

9. SCORE

TABLE	HIT	MISS
2	18	4
3	16	6
TOTAL	34	10

10 REMARKS: CHECK ADJUSTED AIM @ 300M.

11 DATE SIGNED: 17 Mar 89

12 SCORER'S SIGNATURE: Sgt. J.P. Hogan

DA FORM 5241-R, JUN 89

G-37

FM 23-9

Figure G-41. Example of automated scoring printout.

RECORD FIRE SCORE REPORT

SCENARIO I.D.: RECFIRB
UNIT I.D.: D/CO 1/19 FIRING ORDER: 4
TABLE PERIOD: RECORD ELEVEN
8/JUL/1988 8:54

ID IRN 105 LANE 1 TABLE-1		ID IRN 106 LANE 1 TABLE-1		ID IRN 107 LANE 2 TABLE-1		ID IRN 108 LANE 3 TABLE-1		ID IRN 109 LANE 4 TABLE-1		ID IRN 110 LANE 5 TABLE-1		ID IRN 111 LANE 6 TABLE-1		ID IRN 112 LANE 7 TABLE-1	8
50A	1	50A	1	50A	1	50A	1	50A	1	50A	1	50A	1	50A	1
200	1	200	0	200	1	200	1	200	1	200	0	200	0	200	1
100	1	100	0	100	0	100	1	100	1	100	1	100	0	100	1
150	1	150	0	150	1	150	0	150	1	150	1	150	1	150	1
300	1	300	1	300	0	300	0	300	0	300	0	300	1	300	0
250	1	250	0	250	1	250	1	250	0	250	0	250	0	250	1
50B	0	50B	1	50B	1	50B	1	50B	0	50B	1	50B	0	50B	1
200	1	200	1	200	1	200	1	200	1	200	0	200	1	200	1
150	1	150	1	150	1	150	1	150	1	150	1	150	1	150	1
250	1	250	1	250	1	250	1	250	0	250	1	250	0	250	1
100	1	100	1	100	1	100	1	100	1	100	1	100	1	100	0
200	1	200	0	200	1	200	0	200	1	200	1	200	0	200	0
150	1	150	1	150	1	150	1	150	1	150	1	150	1	150	1
300	0	300	0	300	0	300	0	300	0	300	1	300	1	300	0
100	1	100	1	100	1	100	1	100	1	100	1	100	1	100	0
250	1	250	1	250	0	250	1	250	1	250	0	250	1	250	0
200	1	200	0	200	0	200	1	200	1	200	1	200	1	200	0
150	1	150	0	150	1	150	1	150	1	150	1	150	1	150	1
50A	1	50A	1	50A	1	50A	1	50A	1	50A	1	50A	1	50A	1
100	0	100	0	100	0	100	0	100	0	100	0	100	0	100	1
TOTAL	**17**	**TOTAL**	**11**	**TOTAL**	**14**	**TOTAL**	**15**	**TOTAL**	**15**	**TOTAL**	**13**	**TOTAL**	**13**	**TOTAL**	**13**
TABLE-2		TABLE-2		TABLE-2		TABLE-2		TABLE-2		TABLE-2		TABLE-2		TABLE-2	
100	0	100	1	100	1	100	1	100	1	100	1	100	1	100	1
250	1	250	0	250	1	250	1	250	0	250	0	250	1	250	1
150	1	150	0	150	1	150	0	150	1	150	0	150	1	150	1
50B	1	50B	0	50B	1	50B	1	50B	1	50B	1	50B	1	50B	1
200	0	200	0	200	1	200	1	200	0	200	0	200	1	200	0
150	1	150	0	150	1	150	1	150	1	150	1	150	1	150	1
200	1	200	1	200	1	200	1	200	1	200	1	200	0	200	1
50A	1	50A	0	50A	1	50A	1	50A	1	50A	1	50A	1	50A	1
150	1	150	0	150	1	150	1	150	0	150	0	150	0	150	1
100	1	100	1	100	1	100	1	100	1	100	1	100	1	100	0
150	1	150	0	150	1	150	1	150	1	150	1	150	0	150	0
300	0	300	0	300	0	300	0	300	1	300	1	300	0	300	0
100	1	100	1	100	1	100	1	100	1	100	1	100	1	100	1
200	1	200	0	200	1	200	1	200	1	200	1	200	0	200	1
150	1	150	1	150	1	150	1	150	1	150	1	150	1	150	1
250	1	250	1	250	1	250	0	250	1	250	1	250	0	250	1
100	1	100	1	100	1	100	1	100	1	100	1	100	1	100	1
150	1	150	0	150	1	150	1	150	1	150	0	150	0	150	1
200	0	200	1	200	0	200	0	200	0	200	1	200	0	200	0
100	1	100	1	100	0	100	1	100	1	100	1	100	0	100	0
TOTAL	**16**	**TOTAL**	**9**	**TOTAL**	**17**	**TOTAL**	**16**	**TOTAL**	**16**	**TOTAL**	**14**	**TOTAL**	**11**	**TOTAL**	**14**
SCORE	**33**	**SCORE**	**20**	**SCORE**	**31**	**SCORE**	**31**	**SCORE**	**31**	**SCORE**	**27**	**SCORE**	**24**	**SCORE**	**27**
SHARPSHOOTER		UNQUALIFIED		SHARPSHOOTER		SHARPSHOOTER		SHARPSHOOTER		MARKSMAN		MARKSMAN		MARKSMAN	

FM 23-9

Section VII. PRACTICE RECORD FIRE RANGE

Although the soldier receives a practice rating based on the number of target hits, practice record fire should also be considered a valuable training exercise. When practice record fire is correctly conducted, all soldiers gain valuable experience and become more confident in engaging combat targets.

CONDUCT OF PRACTICE RECORD FIRE RANGE

The unit receives an orientation on the conduct of practice record fire and exercise scenarios to include a review of the fundamentals, lanes and target detection, immediate-action drills, and practice record fire performance standards. After the orientation, soldiers are divided into firing orders: the first order is the firer, the second order is coaches/scorers, and the third order is in the ready area. As each order is completed, duties are rotated.

The standard practice record fire range is divided into 16 lanes (see Figure G-42). Each lane is 30 meters wide with one fighting position and seven targets in each lane. The E-type and F-type silhouette targets (attached to RETS or M31A1 target mechanisms) are used for record fire. Two targets are placed 10 meters apart at a range of 50 meters from the line of fighting positions. Subsequent targets are placed at 50-meter intervals out to 300 meters.

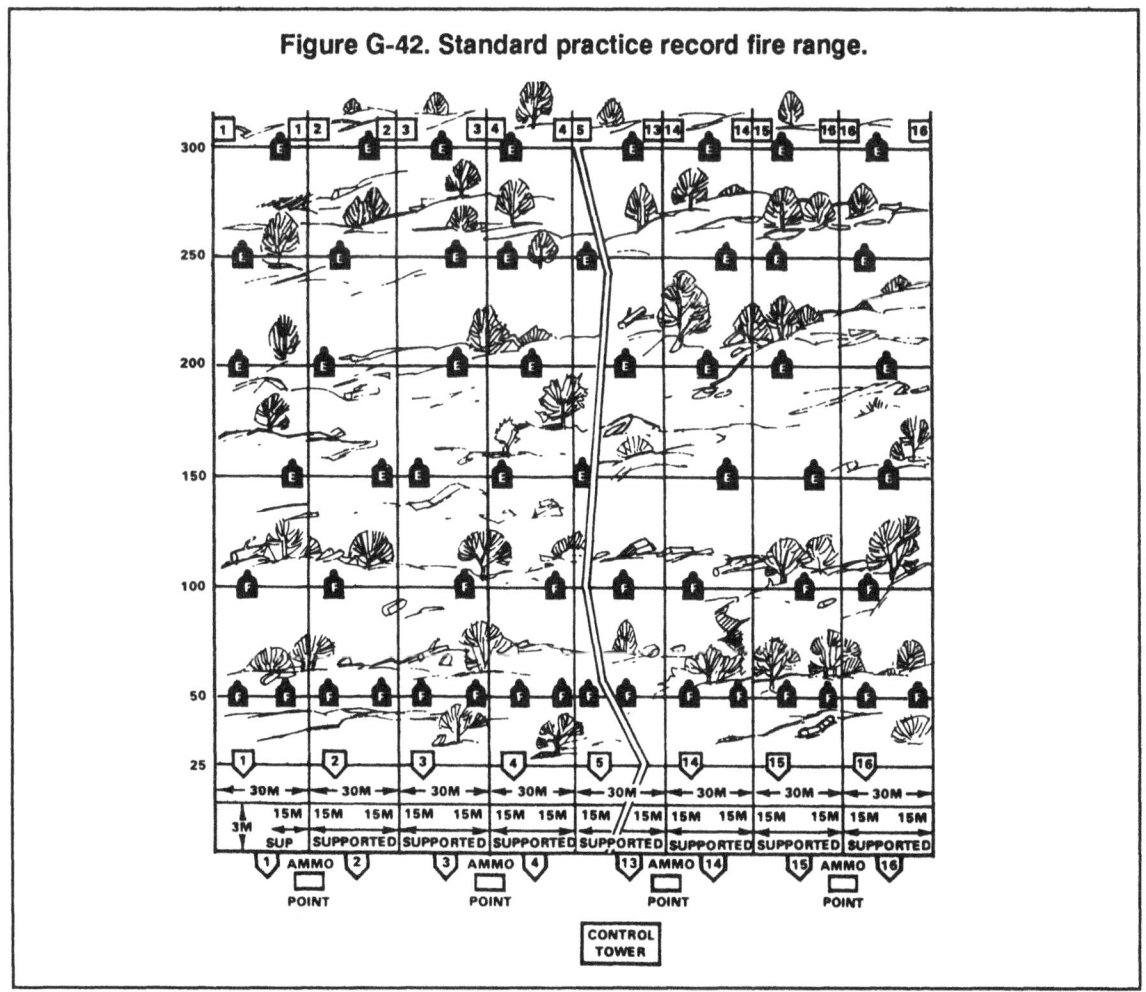

Figure G-42. Standard practice record fire range.

G-39

RANGE CHECKS

For exercises to provide effective practice and training, the operation of target mechanisms and recording equipment must be verified. A firer can lose confidence in his abilities due to targets that do not fall when hit or by hits that are not recorded. Also, accurate information cannot be provided for after-action reviews.

CONDUCT OF TRAINING

During practice record fire, soldiers fire at 40 single or multiple target exposures. They are issued 20 rounds of ammunition to be fired from the supported fighting position, and 20 rounds to be fired from the prone unsupported position. Based on the total number of hits achieved in each table, soldiers are critiqued on the practice record fire score.

Exposure times are three to seven seconds at ranges of 50 to 300 meters. Since it requires one to two seconds for the manually activated target mechanism to raise the target, timing begins when the target is fully exposed rather then when the target switch is activated by the tower operator. When practice record fire is conducted on the new family of automated record fire ranges, these factors are included in the computer program.

Alibi Firing. Alibi firing should be conducted at the end of each firing table and IAW the tower operator commands. Alibis are provided during practice record fire for three reasons: malfunction of the rifle, malfunction of the target mechanism, or faulty ammunition.

Uniform and Equipment. Soldiers do not need to wear full field equipment while firing the practice record fire course. Wearing helmets and LBE have little or no effect on performance; however, local commanders could require that they be worn. If so, the same equipment should be worn during the official record fire.

Range Training Areas. Three range/training areas are as follows:

Orientation Area. This area is located so firers cannot see the firing area. Practice record fire orientation includes conduct of fire, instructions on safety, and range operations (procedures in ready and retired areas).

Ready Area. This area is near the firing range and is located so firers cannot see targets on the range. The firer blackens the rifle sights, lubricates the rifle, and checks for defects that might cause malfunctions.

Retired Area. This area is about 100 meters behind the ready area. Soldiers completing practice record fire move to the retired area to clean their rifles and to be critiqued on their firing performance.

RECORD OF PERFORMANCE

Practice record fire is conducted IAW DA Form 3595-R (Record Firing Scorecard) (Figure G-43). (See Appendix H for a blank reproducible copy of DA Form 3595-R.)

QUALIFICATION STANDARDS

Accurate performance data are critical. The firer's scores are recorded in two ways: manually by using the practice record fire scorecard, or automated by using a computer printout provided on the automated range (see Section VI). Based on the data

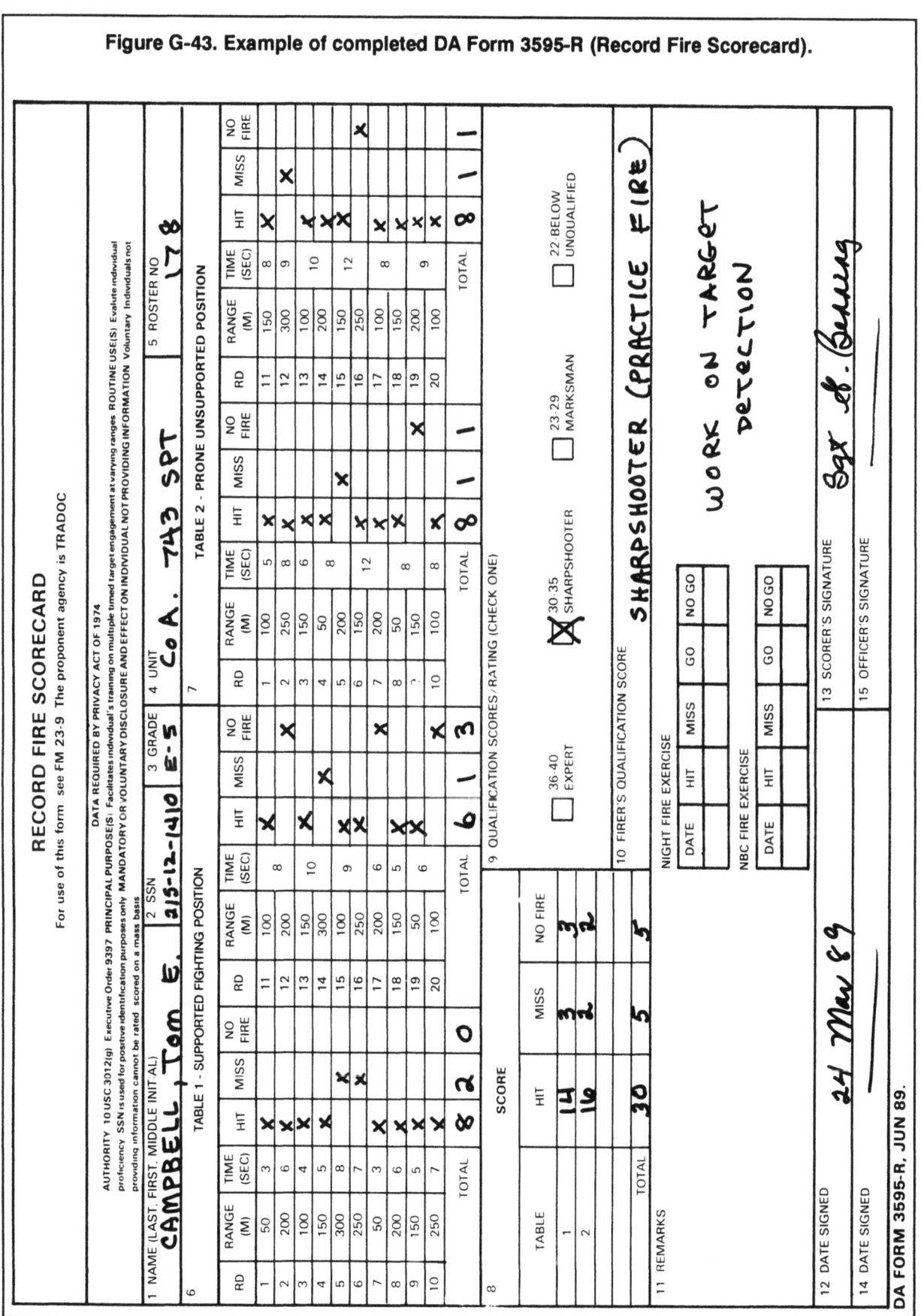

Figure G-43. Example of completed DA Form 3595-R (Record Fire Scorecard).

recorded, an after-action review can be performed by range and firing position to discuss firing performance.

A practice qualification rating is granted to soldiers as follows:

- **Marksman:** Hits 23 to 29 targets.
- **Sharpshooter:** Hits 30 to 35 targets.
- **Expert:** Hits 36 to 40 targets.

A firer who fails to qualify on his first try should refire the practice record fire range after his problem has been diagnosed and remedial training provided.

Section VIII. RECORD FIRE RANGE

The intent of record fire is to facilitate the commander's evaluation of several individual tasks and integrated marksmanship skill performance, and to provide unit readiness indicators. The qualification standards are specifically related to a prescribed procedure for the conduct of record fire. Individual performance must be evaluated IAW three components:

1. What test was used (standard, known-distance, or scaled).
2. How the test was administered.
3. What were the individual and unit performance distributions (23 to 40 or 26 to 40 for alternates), and at what target ranges.

CONDUCT OF RECORD FIRE RANGE

Since all soldiers must fire the record fire course at least once each year for qualification, the record fire course can provide excellent firing performance evaluations. It also provides excellent diagnostic information for instructors/trainers who are concerned with scheduling training to overcome the most serious firing weaknesses. The standard qualification course should be used for all soldiers. However, there are times when a qualification exercise must be conducted on an alternate course.

The following information concerning the development of the record fire course is provided to assist in understanding how standards were established.

The testing and development indicates that the soldier should hit at least 39 of 40 targets if he applies the marksmanship fundamentals correctly (assumes target mechanisms have been checked and are functional). This probability of hit (P_h) is provided as a guide concerning the capability of the typical rifle, ammunition, and soldier firing a standard course (Table G-1).

Table G-1. Probability of hits.		
Range (meters)	Ph	Number of Targets
50	1.0	05
100	1.0	09
150	1.0	10
200	.99	08
250	.95	05
300	.90	03

When the IET BRM POI, or an adequate unit training program, is conducted, it is expected that the following Ph will result (Table G-2):

Table G-2. Results from an adequate unit training program.

Range	Targets	Low Ph	Average Ph	High Ph
50	5	.80	.95	.98
100	9	.70	.90	.95
150	10	.65	.90	.95
200	8	.45	.70	.90
250	5	.35	.60	.85
300	3	.25	.50	.80
		23 hits	32 hits	37 hits

The new ratings represent a significant increase in the number of targets that are required to hit; and have been adjusted for alternate courses to provide a better correlation to performance on the standard course (see Table G-3).

Table G-3. New ratings.

Rating	Old	New Standard	New Alternates
Expert	28-40	36-40	38-40
Sharpshooter	24-27	30-35	33-37
Marksman	17-23	23-29	26-32
Unqualified	16-below	22-below	25-below

The standard record fire course was developed with the assumption that target systems would function. Therefore, a first objective is to ensure that all targets are functioning properly. When in doubt, a lane should be fired to ensure that a bullet strike will activate each target. Sometimes slapping a target with a cleaning rod can cause it to activate, but a bullet will not. When it is hot, the plastic targets may allow the fast moving 5.56-mm bullet to pass without causing sufficient vibration to activate the mechanism, resulting in a requirement to change targets more often, to use double targets, or to use different silhouettes for a positive indication of hits.

RANGE ORGANIZATION

The standard record firing course is the most realistic in that it presents the soldier with various surprise target situations that could be encountered in combat. Except where modification is necessary to install and maintain target mechanisms, the terrain is left intact. The number of targets hit by each soldier is totaled upon completion of record firing. Based on this score, marksmanship qualification ratings and appropriate badges are awarded. While competition between individuals and platoons is inevitable and encouraged, the goal should be to achieve the highest qualification rating on the first attempt.

Commanders should consider using disinterested or outside evaluators for the official record fire to assist in objective collection and analysis of data. Soldiers, squads, or platoons should not score themselves.

Commanders should be concerned about rating distributions. High distribution in the unqualified lowest category indicates skill erosion in the unit. A normal distribution for every 100 soldiers in an average unit should be at least 30 experts, 60 sharpshooters, and 10 marksmen. A well-trained unit will be higher.

The standard fire course designed to measure and provide indicators on the application or performance of several individual combat tasks/skills. When record fire is conducted correctly, most of the following tasks can be observed and objectively measured.

- Maintenance of weapons and magazines.
- Conduct of a serviceability check.
- Demonstration of an understanding of the rifle.
- Application of immediate action.
- Scanning of a designated area/sector and detect targets.
- Quick and consistent application of the four fundamentals of marksmanship.
- Engagement of targets from supported and unsupported positions.
- Knowledge of the effects of wind and gravity.
- Management of ammunition.
- Accurate battlesight zero of the rifle.

Commanders may designate what uniform and equipment will be worn during record fire. Firers should wear LBE and a helmet while firing the record course. When record firing is done for qualification, the soldier does not receive coaching or assistance. If a rifle malfunctions, the soldier applies immediate action and tries to clear the stoppage. Soldiers should prepare for training before live fire in the orientation and ready areas.

Orientation Area. This area is located close to the firing area so soldiers cannot see the firing area. Record fire orientation includes conduct of record fire, instructions on safety, range operations, ammunition handling, and scoring.

Ready Area. This area is located near the firing range so soldiers cannot see targets on the range. The soldier blackens the rifle sights, lubricates the rifle (if needed), and checks for defects in the rifle, magazines, and ammunition that might cause malfunctions. Defective magazines or ammunition should be exchanged before firing. The NCOIC supervises the activities of soldiers in the ready area. The unit armorer replaces damaged or broken parts discovered before firing. Soldiers should load ammunition into their own magazines.

NOTE: Replacing any element of the sight system changes the battlesight zero of the rifle. When replacing the part, the ordnance repairman informs the ready area NCO so the rifle can be zeroed again.

When the M16 pencil is used, everybody loses. Automatic scoring eliminates many potential problems, but most ranges must use manual scoring. A buzzer or whistle should be sounded when the targets are lowered at the end of their exposure time so the scorer knows when he should no longer give credit for a target hit. However, efforts to be fair also include giving the soldier the benefit of the doubt when warranted. The scorer should immediately report suspected malfunctioning targets.

CONDUCT OF TRAINING

The record fire course provides 40 bullets for the engagement of two 20-round exercises. Twenty single or multiple targets are engaged from the supported fighting position. Twenty targets are engaged from the prone unsupported position. Once firing begins, no cross-loading is allowed.

Credit for target hits should not be given when bullets are "saved" from difficult targets to be used on easier targets—for example, not firing a 300-meter target so an additional bullet can be fired at a 150-meter target. However, when double targets are exposed, the soldier should fire two bullets. If the first target is missed, he may fire at that same target with the second bullet.

While it is doctrinally sound to first engage the target what poses the greatest threat (normally assumed to be the closer target), no scoring distinction is made between near targets and far targets or the sequence in which they are engaged. Also, credit is not given if unused ammunition from one 20-round table is added to the magazine provided for the next table.

Soldiers who fail to qualify on the first attempt should be given appropriate remedial training and allowed to refire within a few days. When a soldier refires the course, he will be unqualified if he hits 22 targets or less and will be rated as a marksman if he hits 23 to 40. When automated scoring procedures are available that allow the performance of the soldier to be stored and retrieved before the malfunction, his performance is added to the score of his first attempt after weapons repair and refire. If a soldier's weapon becomes inoperable but his performance before the malfunction precludes qualification, he is considered unqualified and must refire.

Alibi firing is reserved for those soldiers who have encountered a malfunctioning target, ammunition, or rifle. A soldier will not be issued more than 20 rounds of ammunition for each table. If he fires all 20 rounds despite a target malfunction, he will not be issued anymore alibi rounds. Also, there are no alibis for soldier-induced weapon malfunctions or for targets missed during application of immediate action. The following are the procedures that must be strictly adhered to when a malfunction occurs.

NOTE: The ammunition procedures/allocation and alibi procedures for practice record fire and record fire are conducted the same. The only exception is that for practice record fire, coaching is authorized.

The NCOIC/scorer monitoring that lane must verify the target malfunction. The soldier continues to fire the exercise. On a computerized range, the tower operator confirms which target and how many malfunctions occurred.

The NCOIC verifies the malfunction. The soldier is permitted to fire at that target(s) with the exact number of rounds equal to the target malfunctions. For example, the soldier had two confirmed target malfunctions at 250 meters, although he may have had five rounds left from the overall exercise. The soldier would be given only two rounds to engage the two 250-meter target exposures, if repaired, or the next closer target. He would not be allowed to fire all remaining five rounds at the two 250-meter target exposures.

The soldier must apply immediate action and continue to fire the exercise. After firing, the soldier notifies the NCOIC to determine if the ammunition was bad or target malfunctioned.

Inoperable weapons are uncorrectable malfunctions such as a broken firing pin, jam caused by double feed which was not caused by the soldier, failure to extract due to broken extractor, or round in the bore. The soldier must have attempted to apply correct immediate action to eliminate the stoppages. If the stoppage is determined to be correctable — for example, the soldier did not apply correct immediate action — and as a result did not engage the required amount of targets, he is at fault.

Qualified weapons personnel/NCOIC must verify weapon malfunctions before the soldier can refire the course. Soldiers who erroneously claim a malfunction on the firing line are considered an unqualified and refire as a second-time firer.

On-site observation, detailed analysis and evaluation of individual results, and unit performance identify weaknesses. Training can then focus on combat tasks, skills, or other factors that address these weaknesses. Examples are: rifles that are not serviceable could be the cause of poor zeros or failures to fire and, therefore, failures to qualify. Some soldiers may not qualify because of a lack of understanding of immediate-action procedures or maintenance of the rifle and magazine. Soldiers who miss targets are not applying the four fundamentals or are not accurately zeroing the rifle. Soldiers who do not fire at exposed targets during qualification may indicate —

- Failure to scan the designated area.
- Lack of ability to detect targets.
- Lack of ability to shift from one target to another.
- Failure to manage ammunition.
- A stoppage.

RECORD OF PERFORMANCE

The record fire range is fired IAW DA Form 3595-R (Record Fire Scorecard). (Figure G-44). (See Appendix H for a blank reproducible copy of DA Form 3595-R.)

QUALIFICATION STANDARDS

To achieve the lowest possible individual qualification rating, a soldier must achieve a minimum score of 23 target hits on a standard record fire range.

A qualification rating is granted to soldiers who demonstrate performance as follows:

Expert: Hits 36 to 40 targets.
Sharpshooter: Hits 30 to 35 targets.
Marksman: Hits 23 to 29 targets.

A qualification rating is granted to soldiers on the alternate course as follows:
Expert: Hits 38 to 40 targets.
Sharpshooter: Hits 33 to 37 targets.
Marksman: Hits 26 to 32 targets.

Figure G-44. Example of a completed DA Form 3595-R (Record Fire Scorecard).

RECORD FIRE SCORECARD

For use of this form, see FM 23-9. The proponent agency is TRADOC

DATA REQUIRED BY PRIVACY ACT OF 1974
AUTHORITY 10 USC 3012(g); Executive Order 9397 PRINCIPAL PURPOSE(S) Facilitates individual's training on multiple timed target engagement at varying ranges ROUTINE USE(S) Evaluate individual proficiency. SSN is used for positive identification purposes only MANDATORY OR VOLUNTARY DISCLOSURE AND EFFECT ON INDIVIDUAL NOT PROVIDING INFORMATION Voluntary. Individuals not providing information cannot be rated / scored on a mass basis

1 NAME (LAST, FIRST, MIDDLE INITIAL)	2 SSN	3 GRADE	4 UNIT	5 ROSTER NO
HUBBARD, GARY L.	478-50-2142	SP4	824th MAINT CO	812

6   TABLE 1 - SUPPORTED FIGHTING POSITION

RD	RANGE (M)	TIME (SEC)	HIT	MISS	NO FIRE	RD	RANGE (M)	TIME (SEC)	HIT	MISS	NO FIRE
1	50	3	X			11	100	8	X		
2	200	6	X			12	200		X		
3	100	4				13	150	10	X		
4	150	5			X	14	300		X		
5	300	8		X		15	100	9	X		
6	250	7	X			16	250			X	
7	50	3	X			17	200	6	X		
8	200	6	X			18	150	5	X		
9	150	5	X			19	50		X		
10	250	7	X			20	100	6	X		
		TOTAL	8	1	1			TOTAL	9	1	

7   TABLE 2 - PRONE UNSUPPORTED POSITION

RD	RANGE (M)	TIME (SEC)	HIT	MISS	NO FIRE	RD	RANGE (M)	TIME (SEC)	HIT	MISS	NO FIRE
1	100	5	X			11	150	8	X		
2	250	8	X			12	300	9	X		
3	150	6	X			13	100	10	X		
4	50	8	X			14	200		X		
5	200		X			15	150	12	X		
6	150	12	X			16	250			X	
7	200	8	X			17	100	8	X		
8	50		X			18	150		X		
9	150	8	X			19	200	9	X		
10	100	8	X			20	100		X		
		TOTAL	10		1			TOTAL	9	1	

8 SCORE

TABLE	HIT	MISS	NO FIRE
1	17	2	1
2	19	1	
TOTAL	36	3	1

9 QUALIFICATION SCORES/RATING (CHECK ONE)

[X] 36-40 EXPERT   [ ] 30-35 SHARPSHOOTER   [ ] 23-29 MARKSMAN   [ ] 22-BELOW UNQUALIFIED

10 FIRER'S QUALIFICATION SCORE EXPERT

NIGHT FIRE EXERCISE

DATE	HIT	MISS	GO	NO GO
6/19/89	8	2	X	

NBC FIRE EXERCISE

DATE	HIT	MISS	GO	NO GO
6/18/89	12	8	X	

11 REMARKS

12 DATE SIGNED	17 June 89	13 SCORER'S SIGNATURE	/s/ Joey Carver SFC
14 DATE SIGNED	17 June 89	15 OFFICER'S SIGNATURE	/s/ Robert Franks Maj

DA FORM 3595-R, JUN 89

If the test is changed or not conducted as prescribed, the standards are irrelevant. Many performance indicators are not observed and give false results as to individual and unit readiness.

A soldier who fails to qualify on his first try must receive remedial training before firing again. However, whether or not the soldier exceeds the minimum qualification standard, the official rating is marksman. For example, if the soldier fires a 37 on his second firing, the qualification rating is recorded as 23.

Section IX. ALTERNATE QUALIFICATION COURSES

Units should conduct rifle qualification on a standard record fire range. Convenience and comfort should not be the prime consideration when choosing a range. The known-distance alternate course is used by all components of the Active Army, US Army Reserve, and Army National Guard when a standard record fire range is not available. The 25-meter alternate course is used when neither a standard record fire nor a known-distance range is available for rifle qualification. Units are permitted to use the 15-meter scaled alternate course only if a 25-meter range is not available.

The official records of personnel who are using an alternate rifle qualification course are noted to distinguish alternate qualification ratings from standard record fire course ratings. For example, official personnel records are annotated as follows:

JONES, John Q. 000-00-0000 Expert 36 (RF)
JONES, John Q. 000-00-0000 Expert 38 (KDAC)
JONES, John Q. 000-00-0000 Expert 38 (AC)

THE 25-/15-METER ALTERNATE COURSE

The 25-/15-meter alternate course provides units a way to test a soldier's rifle marksmanship proficiency. A soldier undergoing rifle qualification should first confirm the zero setting on his rifle before engaging the alternate course. The zero may be confirmed with the 25-meter battlesight zero procedure of six sighter rounds, which are fired in the prone supported position. Sighter rounds do not count for score. Training/sustainment ammunition is used for sighter rounds if a zeroing exercise is not conducted the day of record fire.

Firing at scaled silhouettes gives the soldier the chance to engage targets with time limits and feedback. Engaging targets at 25/15 meters precludes any training value received on target detection or the effects of wind and gravity, which is learned when firing at longer distances. Rifle qualification requirements are scheduled on the 25-/15-meter alternate course when a standard record fire or known-distance range is not available. The alternate course is an eight-hour course of instruction, as follows:

25-/15-meter zeroing 4 hours

Record fire 4 hours

EQUIPMENT

Frame, target (local manufacture)	1 for each lane
E-type silhouette (NSN 6920-00-071-4780)	1 for each lane
Target, zeroing, 25-/15-meter,	
25-meter (NSN 6920-01-167-1392)(M16A1/A2)	1 for each firer*
15-meter (NSN 6920-01-167-1394)(M16A2)	1 for each firer*
15-meter (NSN 6920-01-253-4005) (M16A1)	1 for each firer
Target, scaled, silhouette	
25-meter (NSN 6920-01-167-1398)	2 for each firer
15-meter (NSN 6920-01-167-1396)	2 for each firer
Pasters, black, (NSN 6920-00-165-6354)	As required
Pasters, buff, (NSN 6920-00-172-3572)	As required
Tacker, target with staples	As required
Ammunition, zeroing	18(6) rounds for each firer**
Ammunition, record firing	40 rounds for each firer
Magazines, ammunition	2 for each firer
Paste, wheat	Optional
Stopwatch	1 each
Scorecard	1 for each firer
Whistle, buzzer, or horn	1 each

* Also used if sighter rounds are fired.

** Six if sighter rounds are fired.

RANGE ORGANIZATION

The alternate course can be conducted on any 25-meter, 1,000-inch, or 15-meter (50-foot) indoor range where service ammunition can be fired. Becuase range facilities differ, so will range equipment for each unit conducting training. Target frames can be built locally. Target tackers or paste can be used to affix target sheets to target frames. Target sheets can be repaired with pasters or changed after each soldier completes each table. (Local supplies dictate target repair and replacement procedures.)

    The chief range officer ensures the safety of all personnel and the proper operation of his range. His duties include organizing personnel into firing orders, assigning numbers, managing ammunition, and assigning target details. He can assign NCOs to perform all these duties.

    The tower operator issues fire commands, or the chief range officer performs this duty.

FM 23-9

The safety officer gives required range safety briefings and organizes firing orders, or the chief range officer performs this duty. Local regulations can require appointing a safety officer.

Personnel in firing orders perform as firing line safety crew, coaches, and scorers.

CONDUCT OF TRAINING

Firers engage each of the 10 scaled silhouettes with one round from the first magazine. They perform a rapid magazine change and engage each scaled silhouette again with one round from the second magazine. Fire commands are given by the chief range officer or tower operator.

TABLE 1: Prone supported position, 2 magazines of 10 rounds each, 120 seconds.
FIRERS, ASSUME A PRONE SUPPORTED POSITION.
COACHES, ISSUE THE FIRER TWO MAGAZINES OF TEN ROUNDS EACH.
THE FIRING LINE IS NO LONGER CLEAR.
LOCK, ONE MAGAZINE OF TEN ROUNDS, LOAD. (Pause.)
LOAD YOUR SECOND MAGAZINE OF TEN ROUNDS AT YOUR OWN COMMAND.
IS THE LINE READY?

The chief range officer pauses to observe the firing line to ensure all soldiers are in position and ready to begin the engagement. If not, the firing line safety crew gets them in position and informs the chief range officer when all soldiers are ready. If no problems exist, the chief range officer continues with the fire commands.

READY ON THE RIGHT?
READY ON THE LEFT?
THE FIRING LINE IS READY.
FIRERS, WATCH YOUR LANE.

A whistle, buzzer, horn, or other loud audible signal is sounded to begin the exercise and sounded again to cease fire.

CEASE FIRE, CEASE FIRE, CEASE FIRE. (Given at the same time as the signal.)
ARE THERE ANY ALIBIS?

Allowable alibis are allotted six seconds for each unfired round. An allowable alibi is a malfunction of the rifle or ammunition — it is not associated with firer error. Rounds not expended during the allotted time do not constitute an alibi and are counted as misses. The firing line safety crew notes the number of alibi rounds to be fired and times the soldier. If a rifle continues to malfunction, the armorer removes it from the firing line for inspection and repair.

Cross-fires are not allowable alibis for the cross-firing soldier. The recipient of the cross-fired rounds refires the table. The cross-firer is awarded misses for those shots on the wrong target, and he may be allowed to refire the course. Regardless of his total hits during refire, he can be rated only as a marksman with a score of 26. The recipient is not penalized. If there are alibis, the chief range officer repeats the fire commands; otherwise, he continues the exercise.

CLEAR ALL WEAPONS.

CLEAR ON THE RIGHT?
CLEAR ON THE LEFT?
THE FIRING LINE IS CLEAR.
FIRERS AND COACHES MOVE DOWNRANGE, SCORE AND REPAIR OR REPLACE YOUR TARGET.

TABLE 2: Prone unsupported position, 2 magazines of 10 rounds each, 120 seconds.

NOTE: The fire commands and alibi procedures are the same as in Table 1.

QUALIFICATION STANDARDS

The chief range officer briefs all soldiers on the proper scoring procedures. The firing line safety crew —

- Perform as scorers.

- Inform the chief range officer of cross-fires.

- Inform the chief range officer of allowable alibis.

- Accurately count hits and misses. A hit is any bullet hole that is either completely within or touches some part of the scaled silhouette. If a bullet hole does not touch some part of the scaled silhouette, it is counted as a miss. Richochets are counted as hits or misses.

- Count only two hits for each silhouette for score in each table.

- Complete the scorecard.

- Assist the soldier with target repair.

- Total, sign, and return the completed scorecard to the chief range officer.

Qualification ratings for the alternate course are as follows:

- **Expert:** Hits 38 to 40 targets

- **Sharpshooter:** Hits 33 to 37 targets

- **Marksman:** Hits 26 to 32 targets

- **Unqualified:** Hits 25 and below

These courses are fired IAW DA Form 5790-R (Record Firing Scorecard—Scaled Target Alternate Course)(Figure G-45.)(See Appendix H for a blank reproducible copy of this form.)

FM 23-9

Figure G-45. Example of completed DA Form 5790-R (Record Firing Scorecard—Scaled Target Alternate Course)(front)(25 and 15 meters).

RECORD FIRING SCORECARD · SCALED TARGET ALTERNATE COURSE

For use of this form see back. The proponent agency is TRADOC

AUTHORITY 10 USC 30129(g); Executive Order 9397 **PRINCIPAL PURPOSE(S):** Records individual's performance on record fire range **ROUTINE USE(S):** Evaluation of individual's proficiency and basis for determination of award of proficiency badge. SSN is used for positive identification purposes only. **MANDATORY OR VOLUNTARY DISCLOSURE AND EFFECT ON INDIVIDUAL NOT PROVIDING INFORMATION** Voluntary Individuals not providing information cannot be rated scored on a mass basis

1 NAME (LAST FIRST MIDDLE INITIAL)	2 SSN	3 GRADE	4 UNIT	5 ROSTER NO
HARDY, LORI W.	255-10-881	2LT	A Co, 1st Of Avn Bn.	24

6 TABLE 1 - SUPPORTED FIGHTING/PRONE SUPPORTED			7 TABLE 2 - PRONE UNSUPPORTED			8 QUALIFICATION	9 REMARKS
TARGET	RANGE (M)	HIT (✓)	TARGET	RANGE (M)	HIT (✓)		
1	300	✓	1	300	✓	☐ 38-40 EXPERT	**NIGHT FIRE EXERCISE**
2	300	✓	2	300	✓		DATE | HIT | MISS | GO | NO GO
3	250	0	3	250	✓	☒ 33-37 SHARPSHOOTER	3-9-89 | 8 | 2 | | X
4	250	✓	4	250	✓		
5	200	✓	5	200	✓	☐ 26-32 MARKSMAN	**NBC FIRE EXERCISE**
6	200	✓	6	200	✓		DATE | HIT | MISS | GO | NO GO
7	200	✓	7	200	✓	☐ 25-BELOW UNQUALIFIED	3-10-89 | 12 | 4 | | X
8	200	0	8	150	✓		
9	150	✓	9	150	✓		
10	150	✓	10	150	00		
11	150	✓	11	100	✓		
12	150	✓	12	100	✓		
13	100	✓	13	100	✓		
14	100	✓	14	100	✓		
15	100	✓	15	100	✓		
16	100	✓	16	100	✓		
17	100	✓	17	100	✓		WPN # 526310
18	100	✓	18	100	✓		
19	50	✓	19	50	✓		
20	50	✓	20	50	✓		
TIME 120 SEC	HITS	18	TIME 120 SEC	HITS	18	TOTAL HITS	36

*FIRER ISSUED 40 ROUNDS TO ENGAGE 20 TARGETS — NO MORE THAN 2 RDS PER TARGET THE ROUNDS WILL BE PRELOADED IN 4. 10 ROUND MAGAZINES TWO PER TABLE ALL ROUNDS WILL BE FIRED WITH THE LONG RANGE SIGHT ON THE M16A1 RIFLE HITS ARE DENOTED BY A "✓".

12 DATE SIGNED	13 DATE SIGNED
3-9-89	3-10-89
14 SCORER'S SIGNATURE	15 OFFICER'S SIGNATURE
Laura Morley SGT.	Jeff B. Hoffman Cpt.

DA FORM 5790-R, JUN 89.

Figure G-45. Example of completed DA Form 5790-R (Record Firing Scorecard—Scaled Target Alternate Course)(back)(25 or 15 meters)(continued).

This scorecard will be used to score Alternate Course record fire qualification when the 25M (NSN 6920-01-167-1398) or 15M (NSN 6920-01-167-1396) scaled silhouette target is used. The Alternate Course will be used only when standard record fire and known Distance ranges are unavailable.

NOTE: If zeroing/grouping exercises are not performed on the day of record fire, 6 rounds of training/sustainment ammunition will be fired for 25 meter zero confirmation prior to conducting the Qualification Course

CONDUCT OF FIRE

The firer will be given two 10-round magazines to engage the 10 silhouettes on the target This includes 2 rounds for each silhouette from the foxhole supported position to be completed in 120 seconds, including the magazine change No more than two hits for each silhouette will be scored.

The firer will then be given 2 additional 10-round magazines to engage the 10 silhouettes on a second target sheet This includes 2 rounds for each silhouette from the prone unsupported position to be completed in 120 seconds, including the magazine change. No more than two hits for each target will be scored from the prone unsupported position

The prone supported position can be substituted for the foxhole position

SCORING

Award one hit for each round that is within or touches some part of the silhouette for a maximum of two hits for each silhouette on each target sheet.

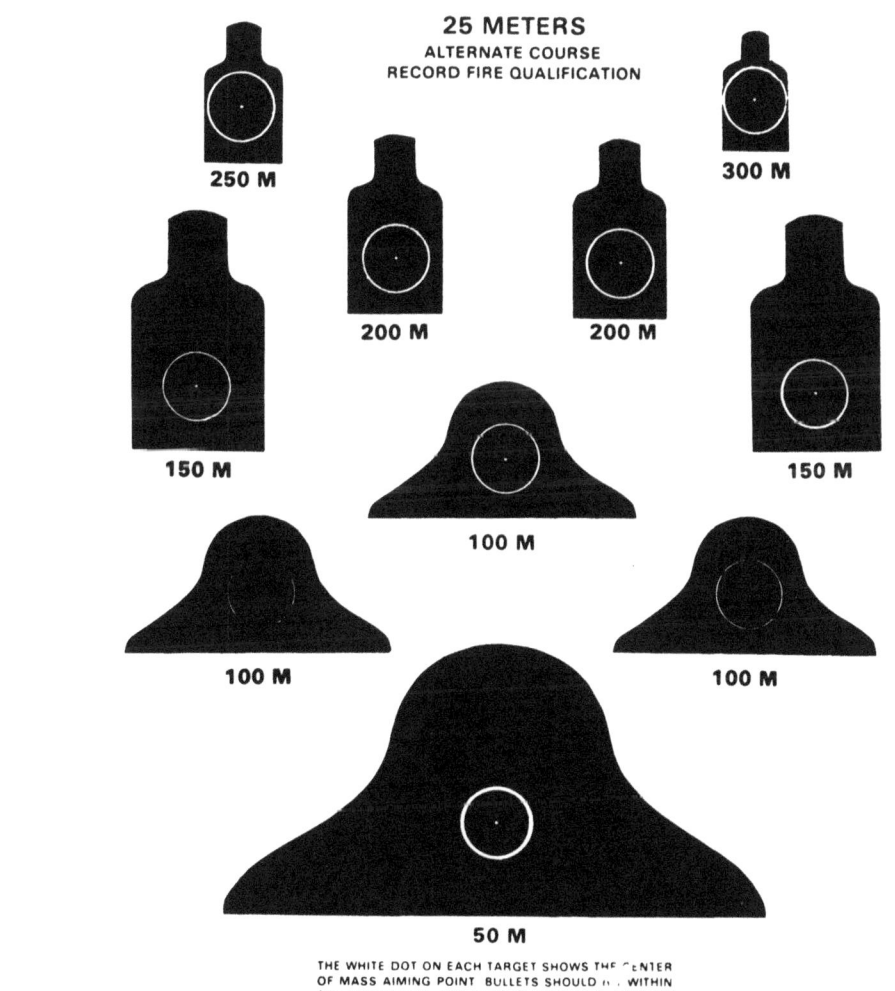

APPENDIX H

Reproducible Forms

This appendix provides a blank copy of DA Form 3009-R (Target Detection Exercise Answer Sheet, Periods 1, 2, and 8), DA Form 3010-R (Target Detection Exercise Answer Sheet, Period 3), DA Form 3011-R (Target Detection Exercise Answer Sheet, Period 5), DA Form 5791-R (Target Detection Exercise Answer Sheet, Period 7), DA Form 3014-R (Target Detection Exercise Answer Sheets Test No. 2 and No. 3, Period 9), DA Form 3601-R (Single Target Field Firing Scorecard), DA Form 3595-R (Record Fire Scorecard), DA Form 5241-R (Single and Multiple Targets—Field Firing Scorecard), DA Form 5239-R (75-, 175-, and 300-Meter Downrange Feedback Scorecard), DA Form 5789-R (Record Firing Scorecard—Known-Distance Course), and DA Form 5790-R (Record Firing Scorecard—Scaled Target Alternate Course). These forms are not available through normal supply channels. You may reproduce them locally on 8½ x 11 inch paper.

TARGET DETECTION EXERCISE
ANSWER SHEET
PERIODS 1, 2, AND 8

For use of this form, see FM 23-8 and FM 23-9; proponent agency is TRADOC.

NAME	(LAST)	(FIRST)	PLATOON	SQUAD	DATE

TRIAL NO.	PHASE NUMBER				WHERE (LETTER OF NEAREST LANDMARK)	RANGE (METERS)
	1	2	3	4		
1 -----						
2 -----						
3 -----						
4 -----						
5 -----						
6 -----						
7 -----						
8 -----						
9 -----						
10 -----						
11 -----						
12 -----						
13 -----						
14 -----						
15 -----						
16 -----						
TOTAL						

DA FORM 3009-R, 1 Nov 73   REPLACES DA FORM 3009-R, 1 JUN 65, WHICH IS OBSOLETE.

TARGET DETECTION EXERCISE
ANSWER SHEET
PERIOD 3

For use of this form, see FM 23-8 and FM 23-9; proponent agency is TRADOC.

NAME	PLATOON	SQUAD	DATE

TRIAL NO.	WHERE (LETTER OF NEAREST LANDMARK)	RANGE (METERS)
1		
2		
3		
4		
5		
6		
7		
8		
9		
10		

DA FORM 3010-R, 1 Nov 73   REPLACES DA FORM 3010-R, 1 JUN 65, WHICH IS OBSOLETE.

TARGET DECTION EXERCISE
ANSWER SHEET
PERIOD 5

For use of this form, see FM 23-8 and FM 23-9; the proponent agency is TRADOC.

OBSERVER'S NAME	(LAST)	(FIRST)	PLATOON

OBSERVATION POINT	DATE

TRIAL NO.	SOUND POSITION	TRIAL NO.	SOUND POSITION
1 --------		15 --------	
2 --------		16 --------	
3 --------		17 --------	
4 --------		18 --------	
5 --------		19 --------	
6 --------		20 --------	
7 --------		21 --------	
8 --------		22 --------	
9 --------		23 --------	
10 --------		24 --------	
11 --------		25 --------	
12 --------		26 --------	
13 --------		27 --------	
14 --------		28 --------	

TOTAL ---------- RIGHT _____ WRONG _____

DA FORM 3011-R, 1 Nov 73   REPLACES DA FORM 3011-R, 1 JUN 65, WHICH IS OBSOLETE.

TARGET DETECTION EXERCISE
ANSWER SHEETS TESTS NO. 2 AND 3
PERIOD 9

For use of this form, see FM 23-8 and FM 23-9; the proponent agency is TRADOC.

OBSERVER'S NAME	(LAST)	(FIRST)	PLATOON
OBSERVATION POINT			DATE

TRIAL NUMBER	NO. OF TARGETS PRESENTED	RIGHT	WRONG
1	1		
2	3		
3	4		
4	4		
5	2		
6	3		
7	3		
8	1		
9	2		
10	2		

TOTAL _____ RIGHT _____ WRONG _____

DA FORM 3014-R, JUN 89

SINGLE TARGET
FIELD FIRING SCORECARD

For use of this form, see FM 23-9. The proponent agency is TRADOC.

DATA REQUIRED BY PRIVACY ACT OF 1974

AUTHORITY: 10 USC 3012(g); Executive Order 9397. PRINCIPAL PURPOSE(S): Facilitates individual's transition to distant target and provides feedback. ROUTINE USE(S): Evaluate individual proficiency. SSN is used for positive identification purposes only. MANDATORY OR VOLUNTARY DISCLOSURE AND EFFECT ON INDIVIDUAL NOT PROVIDING INFORMATION: Voluntary, individuals not providing information cannot be rated/scored on a mass basis.

1. NAME (LAST, FIRST, MIDDLE INITIAL)	2. SSN	3. GRADE	4. UNIT	5. ROSTER NO

6. INTRODUCTION

TABLE 1 - SUPPORTED FIGHTING POSITION

RD	RANGE (M)	TIME (SEC)	HIT	MISS
1	75	6		
2	75	6		
3	75	6		
4	75	6		
5	75	6		
6	175	8		
7	175	8		
8	175	8		
9	175	8		
10	175	8		
11	175	8		
12	175	8		
13	300	10		
14	300	10		
15	300	10		
16	300	10		
17	300	10		
18	300	10		
TOTAL				

7. TABLE 2 - SUPPORTING FIGHTING POSITION

RD	RANGE (M)	TIME (SEC)	HIT	MISS
1	75	6		
2	175	8		
3	300	10		
4	175	8		
5	75	6		
6	300	10		
7	300	10		
8	75	6		
9	175	8		
10	175	7		
11	300	7		
12	175	8		
13	75	6		
14	300	10		
15	175	8		
16	75	6		
17	300	10		
18	75	8		
TOTAL				

8. TABLE 3 - PRONE POSITION

RD	RANGE (M)	TIME (SEC)	HIT	MISS
1	75	7		
2	175	9		
3	300	11		
4	175	9		
5	75	7		
6	300	11		
7	300	11		
8	75	7		
9	175	9		
10	175	7		
11	300	11		
12	175	9		
13	75	7		
14	300	11		
15	175	9		
16	75	7		
17	300	11		
18	75	7		
TOTAL				

9. SCORE

TABLE	HIT	MISS
2		
3		
TOTAL		

10. REMARKS

11. DATE SIGNED

12. SCORER'S SIGNATURE

DA FORM 3601-R, JUN 89

RECORD FIRE SCORECARD

For use of this form, see FM 23-9. The proponent agency is TRADOC

DATA REQUIRED BY PRIVACY ACT OF 1974

AUTHORITY: 10 USC 3012(g) Executive Order 9397 PRINCIPAL PURPOSE(S): Facilitates individual's training on multiple timed target engagement at varying ranges ROUTINE USE(S): Evalute individual proficiency. SSN is used for positive identification purposes only. MANDATORY OR VOLUNTARY DISCLOSURE AND EFFECT ON INDIVIDUAL NOT PROVIDING INFORMATION: Voluntary. Individuals not providing information cannot be rated, scored on a mass basis

1 NAME (LAST, FIRST, MIDDLE INITIAL)	2 SSN	3 GRADE	4 UNIT	5 ROSTER NO

6 TABLE 1 - SUPPORTED FIGHTING POSITION

RD	RANGE (M)	TIME (SEC)	HIT	MISS	NO FIRE	RD	RANGE (M)	TIME (SEC)	HIT	MISS	NO FIRE
1	50	3				11	100				
2	200	6				12	200	8			
3	100	4				13	150				
4	150	5				14	300	10			
5	300	8				15	100				
6	250	7				16	250	9			
7	50	3				17	200	6			
8	200	6				18	150	5			
9	150	5				19	50				
10	250	7				20	100	6			
							TOTAL				

7 TABLE 2 - PRONE UNSUPPORTED POSITION

RD	RANGE (M)	TIME (SEC)	HIT	MISS	NO FIRE	RD	RANGE (M)	TIME (SEC)	HIT	MISS	NO FIRE
1	100	5				11	150	8			
2	250	8				12	300	9			
3	150	6				13	100				
4	50					14	200	10			
5	200	8				15	150				
6	150					16	250	12			
7	200	12				17	100				
8	50					18	150	8			
9	150	8				19	200				
10	100	8				20	100	9			
							TOTAL				

SCORE

TABLE	HIT	MISS	NO FIRE
1			
2			
TOTAL			

9 QUALIFICATION SCORES RATING (CHECK ONE)

☐ 36-40 EXPERT ☐ 30-35 SHARPSHOOTER ☐ 23-29 MARKSMAN ☐ 22 BELOW UNQUALIFIED

10 FIRER'S QUALIFICATION SCORE

NIGHT FIRE EXERCISE

DATE	HIT	MISS	GO	NO GO

NBC FIRE EXERCISE

DATE	HIT	MISS	GO	NO GO

11 REMARKS

12 DATE SIGNED	13 SCORER'S SIGNATURE
14 DATE SIGNED	15 OFFICER'S SIGNATURE

DA FORM 3595-R, JUN 89

SINGLE AND MULTIPLE TARGETS
FIELD FIRING SCORECARD

For use of this form see FM 23-9. The proponent agency is TRADOC

AUTHORITY: 10 USC 3012/(g)/ Executive Order 9397 PRINCIPAL PURPOSE(S): Facilitates individual's transition to distant target and provides feedback. ROUTINE USE(S): Evaluate individual proficiency. SSN is used for positive identification purposes only. MANDATORY OR VOLUNTARY DISCLOSURE AND EFFECT ON INDIVIDUAL NOT PROVIDING INFORMATION: Voluntary. Individual not providing information cannot be rated/scored on a mass basis.

1 NAME (LAST FIRST MIDDLE INITIAL)	2 SSN	3 GRADE	4 UNIT	5 ROSTER NO

6. INTRODUCTION
TABLE 1 - SUPPORTED FIGHTING POSITION

RD	RANGE (M)	TIME (SEC)	HIT	MISS
1	75	5		
2	175	7		
3	75	11		
4	300			
5	75	9		
6	175			
7	75	10		
8	300			
9	175	11		
10	300			
TOTAL				

7. TABLE 2 - SUPPORTING FIGHTING POSITION

RD	RANGE (M)	TIME (SEC)	HIT	MISS
1	175	7		
2	75	10		
3	300			
4	75	9		
5	175	9		
6	300			
7	75	9		
8	175			
9	175	11		
10	300			
11	75	9		
12	175			
13	175	11		
14	300			
15	75	5		
16	175			
		9		
		10		
TOTAL				

8. TABLE 3 - PRONE POSITION

RD	RANGE (M)	TIME (SEC)	HIT	MISS
1	75	6		
2	175	8		
3	75	13		
4	300			
5	75	11		
6	175			
7	75	12		
8	300			
9	175	13		
10	300			
11	75	11		
12	175			
13	175	8		
14	75	6		
15	75			
16				
TOTAL				

9. SCORE

TABLE	HIT	MISS
2		
3		
TOTAL		

10 REMARKS

11 DATE SIGNED	12 SCORER'S SIGNATURE

DA FORM 5241-R, JUN 89

75-, 175- AND 300-METER DOWN RANGE FEEDBACK SCORECARD

For use of this form, see FM 23-9 The proponent agency is TRADOC

AUTHORITY 10 USC 3012(g)/Executive Order 9397 **PRINCIPAL PURPOSE(S):** Facilitates individual's training on silhouette targets at varying ranges **ROUTINE USE(S)** Evaluate individual proficiency. SSN is used for positive identification purposes only **MANDATORY OR VOLUNTARY DISCLOSURE AND EFFECT ON INDIVIDUAL NOT PROVIDING INFORMATION** Voluntary Individuals not providing information cannot be rated/scored on a mass basis

| 1 NAME (LAST, FIRST, MIDDLE INITIAL) | 2 SSN | 3 GRADE | 4 UNIT |

5. ZERO CONFIRMATION

RANGE (M)	RDS	HIT	MISS
175	5		

6. ZERO CONFIRMATION REFIRE

RANGE (M)	RDS	HIT	MISS
175	5		

7. TABLE 1 - DOWN RANGE SILHOUETTES, SUPPORTED FIGHTING POSITION

RANGE (M)	RDS	HIT	MISS
75	5		
175	10		
300	5		
TOTAL			

8. TABLE 2 - DOWN RANGE SILHOUETTES, PRONE POSITION

RANGE (M)	RDS	HIT	MISS
75	5		
175	10		
300	5		
TOTAL			

9. SCORE

TABLE	HIT	MISS
1		
2		
TOTAL		

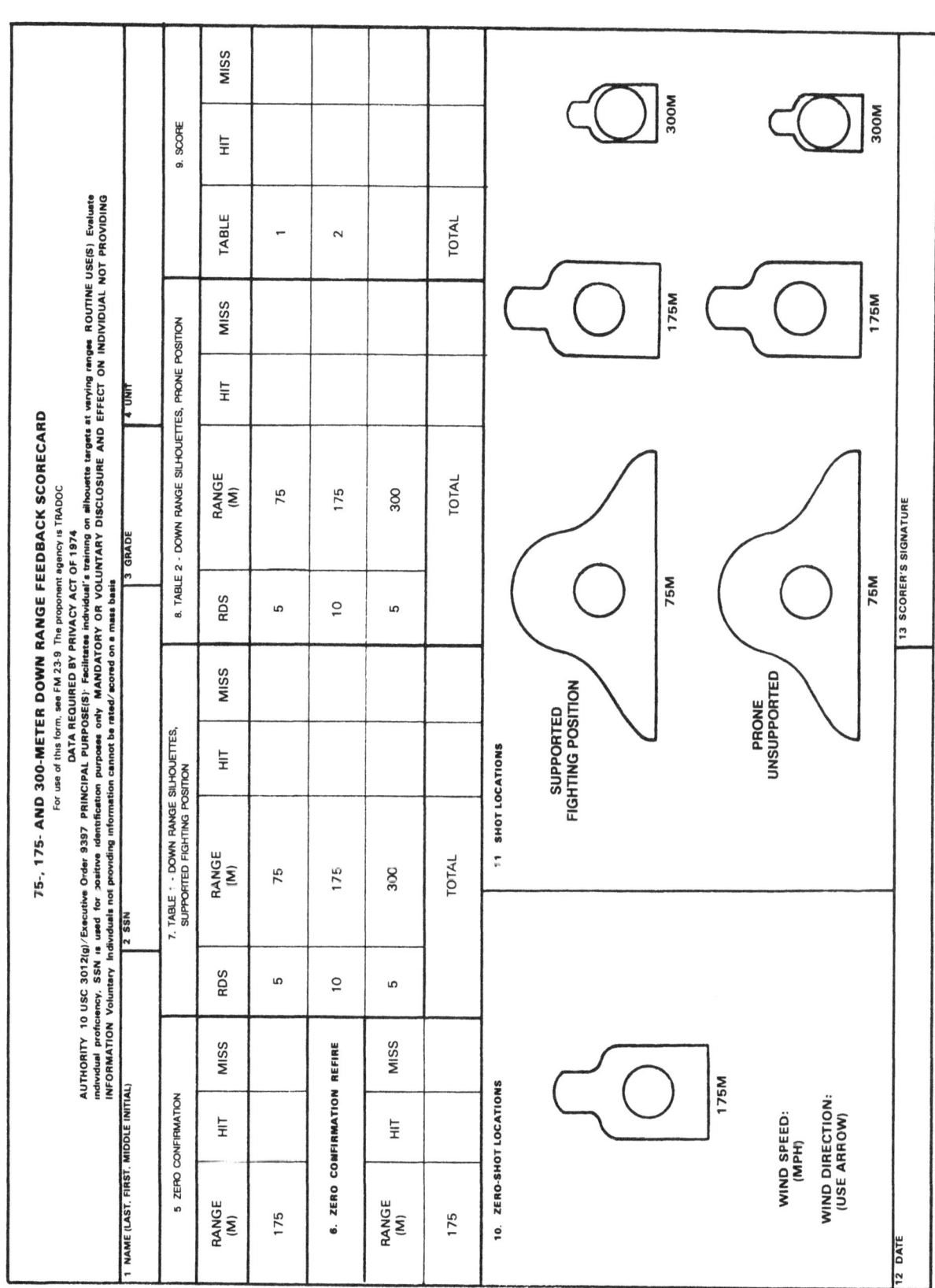

10. ZERO-SHOT LOCATIONS

11. SHOT LOCATIONS — SUPPORTED FIGHTING POSITION / PRONE UNSUPPORTED

WIND SPEED: (MPH)
WIND DIRECTION: (USE ARROW)

| 12 DATE | 13 SCORER'S SIGNATURE |

DA FORM 5239-R, JUN 89.

RECORD FIRING SCORECARD - KNOWN DISTANCE COURSE

For use of this form, see back. The proponent agency is TRADOC

DATA REQUIRED BY PRIVACY ACT OF 1974

AUTHORITY: 10 USC 30129(g) Executive Order 9397 **PRINCIPAL PURPOSE(S):** Records individual's performance on record fire range **ROUTINE USE(S):** Evaluation of individual's proficiency and basis for determination of award of proficiency badge. SSN is used for positive identification purposes only **MANDATORY OR VOLUNTARY DISCLOSURE AND EFFECT ON INDIVIDUAL NOT PROVIDING INFORMATION:** Voluntary Individuals not providing information cannot be rated/scored on a mass basis

1 NAME (LAST FIRST MIDDLE INITIAL)	2 SSN	3 GRADE	4 UNIT	5 FIRING POINT AND ORDER

6 TABLE 1 PRONE SUPPORTED				7 TABLE 2 PRONE UNSUPPORTED				8 TABLE 3 PRONE UNSUPPORTED				9 SCORE
ROUND	RANGE 300	HIT	MISS	ROUND	RANGE 200	HIT	MISS	ROUND	RANGE 100	HIT	MISS	
1				1				1				☐ 38-40 EXPERT
2				2				2				
3				3				3				☐ 33-37 SHARPSHOOTER
4				4				4				
5	E-SIL			5	E-SIL			5	F-SIL			☐ 26-32 MARKSMAN
6				6				6				
7				7				7				☐ 25-BELOW UNQUALIFIED
8				8				8				
9				9				9				
10				10				10				
11				TIME 60 SEC	TOTAL			TIME 60 SEC	TOTAL			
12												
13				11 REMARKS								
14				NIGHT FIRE EXERCISE								10 LIGHT
15				DATE	HIT	MISS	GO	NO GO				
16												DIRECTION WIND
17				NBC FIRE EXERCISE								
18				DATE	HIT	MISS	GO	NO GO				ZERO
19												ELEV
20												WIND MPH
TIME 120 SEC	TOTAL											

*FIRER ISSUED 40 ROUNDS. THE ROUNDS WILL BE PRELOADED IN FOUR 10-ROUND MAGAZINES - TWO FOR TABLE 1, ONE FOR EACH REMAINING TABLE. ALL ROUNDS WILL BE FIRED WITH THE M16A1 SHORT RANGE SIGHT

12 DATE SIGNED	13 DATE SIGNED
14 SCORER'S SIGNATURE	15 OFFICER'S SIGNATURE

DA FORM 5789-R, JUN 89.

This scorecard is used to score Known Distance Course record fire qualification when the Known Distance Range is used. This course is used only when the standard record fire course is not available.

NOTE: If zeroing/grouping exercises are not performed on the day of record fire, 6 rounds of training/sustainment ammunition are fired from the 300 yard/meter line for confirmation of zero prior to conducting the Qualification Course.

CONDUCT OF FIRE

For Table 1, the firer is given two 10-round magazines to engage an E-silhouette at 300 yards within 120 seconds in the prone supported position. Table 2 is fired with a 10-round magazine at an E-silhouette at 200 yards within 60 seconds in the prone unsupported position. Table 3 is fired with a 10-round magazine at an F-silhouette at 100 yards within 60 seconds in the prone unsupported position

SCORING

Scoring is conducted in the pits, with the results provided after each table. One point is awarded for each round hitting the target. A hit is scored for any bullethole that is within or touches some part of the silhouette facing.

RECORD FIRING SCORECARD - SCALED TARGET ALTERNATE COURSE

For use of this form see back. The proponent agency is TRADOC

DATA REQUIRED BY PRIVACY ACT OF 1974

AUTHORITY: 10 USC 3012(g); Executive Order 9397 **PRINCIPAL PURPOSE(S)**: Records individual's performance on record fire range **ROUTINE USE(S)**: Evaluation of individual's proficiency and basis for determination of award of proficiency badge. SSN is used for positive identification purposes only. **MANDATORY OR VOLUNTARY DISCLOSURE AND EFFECT ON INDIVIDUAL NOT PROVIDING INFORMATION**: Voluntary. Individuals not providing information cannot be rated/scored on a mass basis.

1. NAME (LAST, FIRST, MIDDLE INITIAL)	2. SSN	3. GRADE	4. UNIT	5. ROSTER NO

6. TABLE 1 - SUPPORTED FIGHTING/PRONE SUPPORTED

TARGET	RANGE (M)	HIT (✓)
1	300	
2	300	
3	250	
4	250	
5	200	
6	200	
7	200	
8	200	
9	150	
10	150	
11	150	
12	150	
13	100	
14	100	
15	100	
16	100	
17	100	
18	100	
19	50	
20	50	
TIME 120 SEC	HITS	

7. TABLE 2 - PRONE UNSUPPORTED

TARGET	RANGE (M)	HIT (✓)
1	300	
2	300	
3	250	
4	250	
5	200	
6	200	
7	200	
8	200	
9	150	
10	150	
11	150	
12	150	
13	100	
14	100	
15	100	
16	100	
17	100	
18	100	
19	50	
20	50	
TIME 120 SEC	HITS	

8. QUALIFICATION

- ☐ 38-40 EXPERT
- ☐ 33-37 SHARPSHOOTER
- ☐ 26-32 MARKSMAN
- ☐ 25-BELOW UNQUALIFIED

TOTAL HITS

9. REMARKS

NIGHT FIRE EXERCISE

DATE	HIT	MISS	GO	NO GO

NBC FIRE EXERCISE

DATE	HIT	MISS	GO	NO GO

*FIRER ISSUED 40 ROUNDS TO ENGAGE 20 TARGETS—NO MORE THAN 2 RDS PER TARGET. THE ROUNDS WILL BE PRELOADED IN 4, 10 ROUND MAGAZINES—TWO PER TABLE. ALL ROUNDS WILL BE FIRED WITH THE LONG RANGE SIGHT ON THE M16A1 RIFLE. HITS ARE DENOTED BY A "✓".

12. DATE SIGNED	13. DATE SIGNED
14. SCORER'S SIGNATURE	15. OFFICER'S SIGNATURE

DA FORM 5790-R, JUN 89

This scorecard will be used to score Alternate Course record fire qualification when the 25M (NSN 6920-01-167-1398) or 15M (NSN 6920-01-167-1396) scaled silhouette target is used. The Alternate Course will be used only when standard record fire and known Distance ranges are unavailable.

NOTE: If zeroing/grouping exercises are not performed on the day of record fire, 6 rounds of training/sustainment ammunition will be fired for 25 meter zero confirmation prior to conducting the Qualification Course

CONDUCT OF FIRE

The firer will be given two 10-round magazines to engage the 10 silhouettes on the target This includes 2 rounds for each silhouette from the foxhole supported position to be completed in 120 seconds, including the magazine change No more than two hits for each silhouette will be scored.

The firer will then be given 2 additional 10-round magazines to engage the 10 silhouettes on a second target sheet This includes 2 rounds for each silhouette from the prone unsupported position to be completed in 120 seconds, including the magazine change No more than two hits for each target will be scored from the prone unsupported position

The prone supported position can be substituted for the foxhole position

SCORING

Award one hit for each round that is within or touches some part of the silhouette for a maximum of two hits for each silhouette on each target sheet.

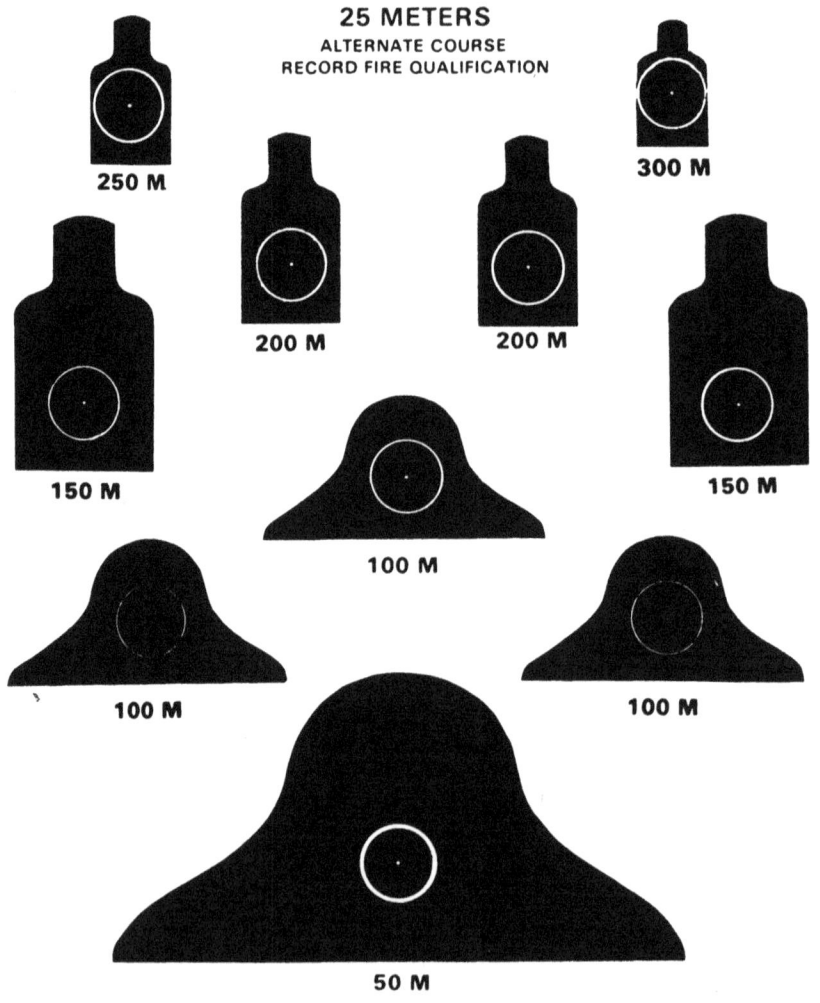

25 METERS
ALTERNATE COURSE
RECORD FIRE QUALIFICATION

THE WHITE DOT ON EACH TARGET SHOWS THE CENTER OF MASS AIMING POINT BULLETS SHOULD HIT WITHIN THE CIRCLE BUT ARE SCORED AS HITS IF THEY HIT ANY PART OF THE SILHOUETTE

**TARGET DETECTION EXERCISE
ANSWER SHEET
PERIOD 7**

For use of this form, see FM 23-8 and FM 23-9; proponent agency is TRADOC.

COMBINATION OF SOUND LOCALIZATION AND MULTIPLE MOVING TARGETS (OBSERVERS CHECK EACH OTHER'S ALINEMENT AND PLACE NUMBER OF TARGETS CORRECTLY ALINED IN SPACE OPPOSITE APPROPRIATE TRIAL NUMBER.)

NAME		PLATOON	SQUAD	DATE

TRIAL NO.	NO. CORRECT	TRIAL NO.	NO. CORRECT
1		12	
2		13	
3		14	
4		15	
5		16	
6		17	
7		18	
8		19	
9		20	
10		TOTAL CORRECT	
11			

TRIAL NO.	NO. CORRECT	TRIAL NO.	NO. CORRECT
1		12	
2		13	
3		14	
4		15	
5		16	
6		17	
7		18	
8		19	
9		20	
10		TOTAL CORRECT	
11			

DA FORM 5791-R, JUN 89

Glossary

Section I. ACRONYMS AND ABBREVIATIONS

AC – Alternate course
AMU – Army marksmanship unit
AR – Army regulation
ARM – advanced rifle marksmanship
ARTEP – Army training and evaluation program
ATC – US Army Training Command
BRM – basic rifle marksmanship
BT – basic training
C – centigrade
cm – centimeter
DA – Department of the Army
DS – direct support
F – Fahrenheit
FM – frequency modulation
fps – feet per second
FPF – final protective fire
FSN – Federal stock number
FTX – field training exercise
GS – general support
GTA – graphic training aid
Hz – hertz
IAW – in accordance with
IET – initial entry training

KD – known distance
kg – kilogram
km – kilometer
kmph – kilometers per hour
LBE – load-bearing equipment
LFX – live-fire exercise
LLLSS – low-light level sight system
LOI – letter of instruction
MAIT – maintenance assistance and instruction team
METL – mission-essential task list
m – meter
mg – milligram
MILES – multiple integrated laser engagement system
mm – millimeter
MOPP – mission-oriented protective posture
MOS – military occupational specialty
MOUT – military operations on urbanized terrain
mph – miles per hour
MPRC – multipurpose range complex
NBC – nuclear, biological, chemical
NCO – noncommissioned officer

NCOIC – noncommissioned officer in charge

NG – Army National Guard

NSN – national stock number

NVD – night vision device

OIC – officer in charge

OSUT – one station unit training

PH – probability of hit

POC – point of contact

POI – program(s) of instruction

PPA – plastic practice ammunition

PRI – preliminary rifle instruction

RATELO – radiotelephone operator

RETS – remote electronic target system

RFA – rimfire adapter

ROTC – Reserve Officers' Training Corps

SAAD – small-arms air defense

SAW – squad automatic weapon

SOP – standing operating procedure

SQT – skill qualification test

TASC – Training and Audiovisual Support Center

TEC – training extension course

TM – technical manual

TRADOC – Training and Doctrine Command

TVT – television tape

USAMU – US Army Marksmanship Unit

USAR – United States Army Reserve

Section II. DEFINITIONS

Active Army – All Regular Army (RA) forces in the Active Army.

Adjusted Aiming Point – An aiming point that allows for gravity, wind, target movement, zero changes, and MOPP firing.

Advanced Marksmanship – Normally refers to marksmanship skills taught during ARM.

Advanced Rifle Marksmanship (ARM) – Normally refers to the formal marksmanship instruction received by infantrymen upon completion of BRM during OSUT.

Aiming – A marksmanship fundamental; refers to the precise alignment of the rifle sights with the target.

Aiming Card – The M15A1 aiming card is a cardboard sleeve with a moveable insert. The rear sight aperture, front sight post, and target are pictured. This training device is used in conjunction with aiming instructions.

Aiming Point – A place on a target in which the rifle sights are aligned, normally the target center of mass.

Alibi Target — A target or additional target a soldier is allowed to engage during qualification firing when unable to complete a record fire scenario due to circumstances beyond his control; for example, a target mechanism, weapon, or ammunition malfunction.

Alternate Course (AC) — Alternatives to standard qualifications are discussed in Appendix G.

Ammunition Lot — A quantity of cartridges each of which is made by one manufacturer under uniform conditions and is expected to work in a uniform manner.

Ammunition Lot Number — Code number that identifies a particular quantity of ammunition from one manufacturer.

Aperture — The hole in the rear sight.

Armorer — One who services and makes repairs on small arms and performs similar duties to keep small arms ready for use.

Army Training and Evaluation Program (ARTEP) — A guide for the training and evaluation of critical unit combat missions — crew/squad through battalion/task force echelon.

Army Training Center (ATC) — Conducts OSUT and BRM. Locations are Fort Benning, GA; Fort Jackson, SC; Fort Dix, NJ; Fort Leonard Wood, MO; Fort McClellan, AL; Fort Knox, KY; Fort Sill, OK; and Fort Bliss, TX.

Artificial Illuminationa — Any light from a man-made source.

Assault Course — An area of ground used for training soldiers in attacking an enemy in close combat.

Automatic Fire — A firing mode that causes the weapon to continue firing as long as the trigger is held or until all ammunition has been expended.

Ball — The projectile; the bullet.

Ball Ammunition — General-purpose standard service ammunition with a solid core bullet.

Ball and Dummy — An exercise that substitutes a dummy round for a live round without the firer knowing it. An excellent exercise for identifying and correcting trigger jerk.

Ballistics — A science that deals with the motion and flight characteristics of projectiles.

Barrel Erosion — Wearing away of the surface of the bore due to the combined effects of gas washing, coring, and mechanical abrasion.

Basic Marksmanship — Fundamental marksmanship skills taught in BRM during IET and OSUT.

Basic Rifle Marksmanship (BRM) — The formal course of marksmanship instruction received by all soldiers.

Battlesight Zero — A sight setting that soldiers keep on their weapons. It provides the highest probability of hitting most high-priority combat targets with minimum adjustment to the aiming point, a 250-meter sight setting as on the M16A1 rifle, and a 300-meter sight setting as on the M16A2 rifle.

Blank Ammunition — A complete cartridge without the bullet used to simulate weapon firing.

Blank Firing Adapter (BFA) — A device that fits in the muzzle of the rifle; used only with blank ammunition.

Brass — An alloy of copper and zinc used to make cartridge cases and bullet jackets. Also, a common name for expended cases.

Breath Control — The third marksmanship fundamental; refers to the control of breathing to help keep the rifle steady during firing.

Bullet — The projectile or ball; the part that goes downrange. It may also be used to refer to the complete cartridge.

Bull's-Eye Target — Any target with a round black circle and scoring rings. Normally used in competitive marksmanship training.

Buttplate — Metal or rubber covering of the end of the stock on the rifle.

Cadre Coach — A trainer with expertise and knowledge exceeding that of the firer.

Caliber — Diameter of the bore; for example, the M16-series rifle bore is 5.56 mm (.223 inch).

Cartridge — A complete round of ammunition.

Center Of Mass — A point that is horizontally (left and right) and vertically (up and down) at the center of the target.

Chambering — The step in the cycle of operation that refers to fully seating the round in the chamber.

Chamber Plug — A range safety device that is a small plastic plug designed to fit into the chamber of the M16. A handle extends out the ejection port so safety personnel can see at a glance that the rifle is clear of ammunition.

Clock Method — Method of calling shots by referring to the figures on an imaginary clock dial assumed to have the target at its center. Also a method of determining the strength and direction of wind.

Coach — Any individual who assists firers on the firing line.

Coach-and-Pupil Method — Method of training in which pairs of pupils take turns practicing a procedure explained by the instructor/trainer.

Cocking — The step in the cycle of operation that refers to the rearward movement of the bolt riding over the hammer, resetting the weapon for subsequent firing.

Collective Firing Proficiency — Units delivering effective fire in a tactical setting. It requires individual skill plus command and control to engage all targets within an assigned sector.

Concurrent Training — Training that occurs at the same time other unit members are using the primary training facilities.

Cookoff — A round that fires as a result of a hot chamber without the trigger being pulled. It can occur any time until the weapon has cooled.

Crack and Thump — A method to determine the general direction and distance to an enemy firer who is shooting at you.

Cradle — A vise-like mechanism that holds a weapon in a secured position for test firing.

Cross Dominance — A soldier with a dominant hand and a dominant eye that are not the same; for example, a right-handed firer with a dominant left eye.

Cycle of Operation — The eight steps involved in firing a round: feeding, chambering, locking, firing, unlocking, extracting, ejecting, and cocking.

Cyclic Rate of Fire — The maximum rate at which a weapon will fire in the automatic mode.

Dime (Washer) Exercise — A dry-fire exercise used to practice trigger squeeze.

Downrange Feedback — Used to describe any training technique that provides precise knowledge of bullet strike (whether hit or miss).

Dry Fire — A technique used to simulate the firing of a live round with an empty weapon. Any application of the fundamentals of marksmanship without live ammunition may be referred to as dry fire.

Dry-Fire Moving Target Trainer — A small motorized scaled target device used to teach the engagement of moving personnel targets.

Dummy Ammunition — A cartridge without a primer or powder. Primarily used for ball-and-dummy exercises on the live-fire line.

Effective Wind — The average of all the varying winds encountered.

Ejection — The step in the cycle of operation that removes the expended cartridge from the weapon out of the ejection port.

Elevation Adjustment — Rotating the front sight post to cause the bullet to strike higher or lower on the target.

Expert — The highest qualification rating.

External Ballistics — What happens to the bullet between the time it leaves the rifle and the time it arrives at the target.

Extraction — The step in the cycle of operation that pulls the round from the chamber.

Eye Relief — The distance from the firing eye to the rear sight. Eye relief is a function of stock weld.

Feedback — Obtaining knowledge of performance.

Feedback Target — Targets designed for use at 75, 175, or 300 meters; includes an overprinted grid similar to a zero target.

Feeding — The step in the cycle of operation that is the forward movement of the bolt, stripping the top round from the magazine and moving it toward the chamber.

Field Firing — Training on the standard field firing range with target banks at 75, 175, and 300 meters.

Firing — The step in the cycle of operation that refers to pulling the trigger, releasing the hammer to strike the firing pin, which strikes the primer. The primer ignites and, in turn, ignites the powder charge within the cartridge case.

Firing Hand — The right hand of a right-handed firer. The left hand of a left-handed firer.

Firing Pin — Plunger in the bolt of a rifle that strikes the primer.

Fleeting Target — A moving target remains within observing or firing distance for such a short period that it affords little time for deliberate adjustment and fire against it.

Functioning — (See cycle of operation.)

Fundamentals of Rifle Marksmanship — The four essential elements needed to hit targets: steady position, aiming, breath control, and trigger squeeze.

Gravity — The natural pull of all objects to the center of the earth.

Grouping — A live-fire exercise with the objective of shooting tight shot groups.

Gun Bore Line — A reference line established by the linear extension of the bore axis of a gun.

Headspace — The distance between the face of the bolt (fully closed) and the face of a fully chambered cartridge.

Hold-off — (See adjusted aiming point.)

Horizontal Dispersion — The left-to-right displacement of bullets on a target.

Immediate Action — A procedure applied to rapidly reduce any rifle stoppage without determining its cause.

Individual Firing Proficiency — Individual firing skills; for example, an individual's performance on the record fire course.

Infantry Remoted Target System (IRETS) — (See RETS.)

Infrared Aiming Light — A unique night sighting system that uses infrared light to assist in the aiming process.

Initial Entry Training (IET) — Indicates the first training received by a new soldier, including the MOS-producing portion of his training such as one-station unit training (OSUT).

Initial Pressure — The application of about half of the total trigger pressure it takes to fire the rifle.

Instructor/Trainer Ratio — The number of soldiers for which each instructor/trainer is responsible.

Internal Ballistics — What happens to the bullet before it leaves the muzzle of the rifle.

Known Distance (KD) — Describes the older range complexes with large target frames behind a large berm and firing lines at 100-yard or 100-meter increments. (See FM 25-7.)

Laser — Light amplification by simulated emission of radiation.

Lead — Distance ahead of a moving target that a rifle must be aimed to hit the target.

Lead Rule — Provides a soldier guidance on how to adjust his aiming point to hit moving targets.

Line of Sight — A line between the rifle and the aiming point, extending from the firing eye through the center of the rear aperture, across the tip of the front sight post, and onto the target.

Location of Miss and Hit (LOMAH) — A projectile location system that provides immediate and precise information to the firer concerning bullet strike (hit or miss).

Locking — The step in the cycle of operation that is a counterclockwise rotation of the bolt, securing it into the barrel locking lugs.

Long-Range Sight — The aperture marked L on the M16A1 rifle equipped with standard sights; provides for a zero at 375 meters. The M16A1 rifle equipped with LLLSS has an aperture marked L, but it is a regular sight.

Low-Light Level Sight System (LLLSS) — A sighting system for low visibility firing that replaces the standard front and rear sights on the M16A1 rifle.

Marksman — The designation given to the lowest qualification rating.

Maximum Effective Range — The greatest distance at which a soldier may be expected to deliver a target hit.

Maximum Effective Rate of Fire — The highest rate of fire that can be maintained and still achieve target hits.

Maximum Range — The longest distance a projectile travels when fired from a weapon held at the optimum angle.

Minute of Angle — An angle that would cover 1 inch at a distance of 100 yards, 2 inches at 200 yards, and so on. Each click of sight adjustment on the M16A1 rifle with standard sights is equal to one minute of angle.

Multiple Integrated Laser Engagement System (MILES) — A tactical shooting device that uses a low-powered laser to activate detectors placed on people or vehicles.

Multipurpose Arcade Combat Simulator (MACS) — A part-task weapons trainer that is under development. The system consists of a light pen attached to the weapon, video monitor, and microcomputer.

Muzzle Velocity — The speed of a projectile as it leaves the muzzle of the weapon.

Natural Point of Aim — The direction the body/rifle combination is oriented while in a stable, relaxed firing position.

Natural Respiratory Pause — The temporary cessation of breathing between an exhale and inhale.

Night Firing — Firing performed under all conditions of limited visibility.

Nonfiring Hand — The opposite of the firing hand.

Optical Sight — Sight with lenses, prisms, or mirrors used in lieu of iron sights.

Paige Sighting Device — A device with a small-scaled target that fits into the muzzle of the weapon, allowing the soldier to practice aiming.

Pasters — Small white or black gum-backed paper used for covering bullet holes.

Peep Sight — The rear sight; a sight with a small aperture (hole).

Peer Coach — A soldier with shooting experience and knowledge equal to that of the firer he is coaching.

Pit — The target area behind the large berm of a KD range.

Plastic Ammunition — Ammunition with a plastic projectile, high-muzzle velocity (the light weight causes it to lose velocity rapidly with a maximum range of 500 meters or less) designed for use in close-in training areas; frangible bullet.

Point of Aim — The exact spot on a target the rifle sights are aligned with.

Point of Impact — The point that a bullet strikes; usually considered in relation to point of aim.

Pop, No Kick — A firing condition when the primer ignites and the powder charge does not. This normally results in lodging the bullet in the bore.

Pop-Up Target — A silhouette target that is activated remotely so it can suddenly appear and fall when struck by a bullet.

Practice Record — Firing conducted on a qualification course for practice.

Predetermined Fire — A technique of aligning the rifle during good visibility so the rifle can be aligned and fired on designated areas when they cannot be seen due to darkness, smoke, or fog.

Preparatory Marksmanship Training (PMT) — All marksmanship training that takes place before live fire.

Primer — A small explosive device in the center base of the cartridge case that is struck by the firing pin to fire the round.

Probability of Hit (PH) — Ranging from 0 to 1.0, it refers to the odds of a given round hitting the target at a given range.

Qualification Firing — Firing on any authorized course that results in meeting qualification requirements; may also be called record fire. (See record fire.)

Quick Fire — A technique of fire used to engage surprise targets at close ranges.

Range Card — Small chart on which ranges and directions to various targets and other important points in the area under fire are recorded.

Rapid Semiautomatic Fire — A firing procedure that results in an accurate shot being fired every one or two seconds.

Receiver — That portion of a firearm that holds the barrel and houses the bolt and firing mechanism.

Recoil — The rearward motion or kick of a gun upon firing.

Record Fire — Any course of fire used to determine if qualification standards are met. The standard record fire course consists of 40 target exposures at ranges between 50 and 300 meters. The standard course requires 23 hits to qualify as marksman, 30 for sharpshooter, and 36 for expert.

Reduced Range Ammunition — Ammunition that is designed to be a ballistic match with service ammunition to an appropriate range for training (may be less than maximum effective range) and a reduced maximum range.

Regular Rear Sight — The M16A1 rifle rear sight that is zeroed for 250 meters (the unmarked aperture on rifles with standard sights and the aperture marked L on rifles equipped with LLLSS).

Reinforcement Training — Training conducted that is over and above scheduled training.

Remedial Action — A procedure applied after immediate action has failed to correct a malfunction, which determines the cause of the malfunction.

Remedial Training — Additional training presented to soldiers who have demonstrated special shooting problems.

Remote Electronic Target System (RETS) — New range complexes. Some ranges include moving targets.

Reserve Components — Includes Army National Guard and Army Reserve forces.

Ricochet Fire — Fire in which the projectile glances from a surface after impact.

Riddle Sighting Device — A small magnetic device with a scaled target that attaches to the front sight assembly, allowing the soldier to practice aiming.

Rifle Cant — Any leaning of the rifle to the left or right from a vertical position during firing.

Rimfire Adapter — The caliber .22 rimfire adapter (M261) consists of a bolt and a magazine insert, which allows standard .22-caliber ammunition to be fired in the M16 rifle.

Round — May refer to a complete cartridge or to the bullet.

Scaled-Silhouette Target — Any target that is reduced in size. When it is observed from 25 meters, it looks the same size as though at a greater range.

Sector of Fire — An area assigned to an individual, weapon, or unit to be covered by fire.

Semiautomatic Fire — A mode of fire that allows one round to be fired each time the trigger is pulled.

Serviceability Checks — A technical inspection of the rifle to determine if it is safe to fire and in working condition. (May not ensure accuracy.)

Service Ammunition — Standard ammunition used by the military. Ammunition designed for combat.

Service Rifle — The primary rifle of a military force.

Service School — Branch schools such as the US Army Infantry School at Fort Benning, GA, and the Armor School at Fort Knox, KY.

Sharpshooter — The middle rating of qualification.

Shot Group — A number of shots fired using the same aiming point, which accounts for rifle, ammunition, and firer variability. Three shots are enough, but any number of rounds may be fired in a group.

Shot-Group Analysis — A procedure for analyzing the size of shot groups on a target to determine firer error.

Sight Alignment — Placing the center tip of the front sight post in the exact center of the rear aperture.

Sighter Rounds — Rounds fired that allow the bullet strike to be observed in relation to the aiming point.

Sight Picture — Placing correct sight alignment on a selected aiming point on a target.

Sight Radius — The distance from the front sight post to the rear sight aperture of a rifle.

Sighting Device (M16) — A small metal device with a tinted square of glass that is placed on the carrying handle, allowing a coach to see what the firer sees through the sights.

Silhouette Target — A target that represents the outline of a man.

Spotters — A round cardboard disk placed in bullet holes with a small wooden peg so that bullet strike can be observed from the firing line.

Squad Automatic Weapon (SAW) — A lightweight, one-man, 5.56-mm machine gun.

Starlight Scope — A weapon scope that amplifies ambient light so targets can be seen and effectively engaged during darkness. The AN/PVS-2 and AN/PVS-4 are used on the M16 rifle.

Steady Position — The first marksmanship fundamental, which refers to the establishment of a position that allows the weapon to be held still while it is being fired.

Stock Weld — The contact of the cheek with the stock of the weapon.

Supported Position — Any position that uses something other than the body to steady the weapon (artificial support).

Suppressive Fire — Any engagement that does not have a definite or visible target. Firing in the general direction of known or suspected enemy location.

Sustained Rate of Fire — Rate of fire that a weapon can continue to deliver for an indefinite period without overheating.

Terminal Ballistics — What happens to the bullet when it comes in contact with the target.

Tight Shot Group — A shot group with all bullet holes close together.

Tracer Ammunition — Ammunition with a substance at the rear of the bullet that ignites soon after firing. It burns brightly so the trajectory of the bullet can be seen.

Tracking — Engaging moving targets where the lead is established and maintained; moving with the target as the trigger is squeezed.

Train the Trainer — Describes any training that is designed to train marksmanship instructors or coaches.

Trainfire — A marksmanship program using pop-up targets in a realistic environment.

Trajectory — The flight path the bullet takes from the rifle to the target.

Trapping — A technique for engaging moving targets. The aiming point is established forward of the target. The rifle is held stationary and fired as the target approaches the aiming point.

Trigger Squeeze — The fourth fundamental; squeezing the trigger so that the movement of firing is a surprise, the lay of the weapon is not disturbed, and a large target hit can be expected.

Unit Marksmanship — All marksmanship training that is conducted by units.

Unlocking — The step in the cycle of operation that refers to the clockwise rotation of the bolt after firing, freeing the bolt from the barrel locking lugs.

Unsupported Position — Any position that requires the firer to hold the weapon steady using only his body (bone support).

Vertical Dispersion — The up-and-down displacement of bullets on a target.

Weaponeer — A training device that simulates the firing of the M16 rifle to provide performance feedback.

Windage Adjustment — Moving the rear sight aperture to cause the bullet to strike left or right on the target.

Wind Value — The effect the wind will have on the trajectory of the bullet.

Wobble Area — The natural movement of the weapon/sights on and around an aiming point when the weapon is being held in a steady position.

Zero Criterion — The standard or requirement for zeroing; 4-centimeter or smaller group at 25 meters.

Zeroing — Adjusting the rifle sights so bullets hit the aiming point at a given range.

Zero Target — A scaled-silhouette target with a superimposed grid for use at 25 meters.

INDEX

aiming card, C-2 (illus)
ammunition, 2-16
    bolt, M2, C-14
    care and handling, 2-17
    short-range training, C-14
    types and characteristics, 2-16 (illus)
automatic fire
    effectiveness, 4-8
    modifications, 4-9
    training, 4-20

ball-and-dummy exercise, C-7
ballistics
    bullet dispersion, F-17
    external, F-16
    internal, F-14
    terminal, F-17
blank firing attachment, C-4, C-5 (illus)

caliber .22 rimfire adapter, M261, C-12
characteristics, 2-2
    clearing, 2-1 (illus)
    M16A1 and M16A2, 2-2 (table)
coach checklist, 3-20
combat fire techniques
    automatic fire, 4-7
    MOPP firing, 4-14
    moving target engagement, 4-19
    quick fire, 4-11
    rapid semiautomatic fire, 4-3
    suppressive fire, 4-1

DA Form 3009-R (Target Detection Exercise Answer Sheet — Periods 1, 2, and 8), B-17
DA Form 3010-R (Target Detection Exercise Answer Sheet — Period 3), B-19
DA Form 3011-R (Target Detection Exercise Answer Sheet — Period 5), B-23
DA Form 3014-R (Target Detection Exercise Answer Sheets Tests No. 2 and 3 — Period 9), B-31
DA Form 3595-R (Record Fire Scorecard), G-41, G-47
DA Form 3601-R (Single Target — Field Firing Scorecard), G-36
DA Form 5239-R (75-, 175-, and 300-Meter Downragne Feedback Scorecard), G-21
DA Form 5241-R (Single and Multiple Targets — Field Firing Scorecard), G-37
DA Form 5789-R (Record Firing Scorecard — Known-Distance Course), G-31
DA Form 5790-R (Record Firing Scorecard — Scaled Target Alternate Course), G-52
DA Form 5791-R (Target Detection Answer Sheet — Period 7), B-27
destruction of materiel
    field-expedient, 2-18
    means, 2-18
dime (washer) exercise, C-8
downrange feedback, G-16
    training, 3-27, 4-18
dry fire
    concept of zeroing, 3-21
    conduct of, 3-18
    downrange feedback training, 3-27
    exercises, 4-11, 4-13, 4-18, 5-5
    field fire training, 3-27
    grouping, 3-21
    peer coaching, 3-18
    practice record fire, 3-27
    record fire, 3-27

elevation and windage rule, F-11
exercises,
    dry fire, 4-11, 4-13, 5-5
    live fire, 4-11, 4-14, 4-18, 5-5, A-15, G-1
    MOPP fire, 4-18
    target detection, B-1, B-15

field fire training, 3-27
firing positions
    advanced positions, 3-13 to 3-16 (illus)
    modified firing positions, 3-16
    MOUT firing positions, 3-16, 3-17 (illus)
    prone unsupported position, 3-13 (illus)
    supported fighting position, 3-12 (illus)
flag method, F-9
function, 2-1
    automatic fire mode, 2-8, 2-9 (illus)
    burst fire mode, 2-9 (illus)
    semiautomatic mode, 2-8
    steps, 2-3 to 2-8 (illus)

gravity, effects of
    aiming point, adjusted, F-4, F-6 (illus), F-7 (illus)
    M16A1 ammunition, F-1
    M16A2 ammunition, F-4
    on ammunition, F-1, F-2 (illus)
    trajectory, F-3 (illus)
grouping, 3-21
    range, G-1
    live fire exercises, 4-11, 4-14, 4-18, 5-5
    alternate qualification courses, G-48
    downrange feedback, G-16
    field fire range, G-33
    known-distance alternate course, G-28
    known-distance range, G-25
    grouping range, G-1
    practice record fire range, G-39
    record fire range, G-42
    zero range, G-10

location of miss and hit (LOMAH) system, C-12

M16 sighting device, C-3, C-4 (illus)

magazine changes, 4-4
malfunctions and corrections, 2-15 (illus)
    stoppage, 2-10
    major categories, 2-11 to 2-14
marksmanship fundamentals, 3-5, 4-15
    aiming, 3-7
    breath control, 3-10 (illus)
    steady position, 3-5, 3-6 (illus)
    trigger squeeze, 3-11
master trial sheet, B-8 (illus), B-10
mechanical training, 3-3
MOPP firing aiming modifications, 4-16
    effects, 4-14
    exercises, 4-18
    operation and function modifications, 4-17
moving target engagement
    fundamentals, 4-20
    lead requirements, 4-22
    multipurpose range complex train-up, 4-25
    single-lead rule, 4-21
    techniques, 4-19
multipurpose arcade combat
    simulator, C-13
multipurpose range complex train-up, 4-25
    night firing considerations, 5-1
    night vision, principles of, 5-2
    target engagement techniques, 5-4,
    training, 5-5

observation method, F-10

peer coaching, 3-18
pointing method, F-9
practice record fire, 3-27
precision firing
    ballistics, F-14
    elevation and windage rule, F-11
    gravity, effects of, F-1
    wind, effects of, F-7
program, rifle marksmanship
    basic, 3-1
    combat factors, 1-4
    fundamentals, 3-5
    instructor/trainer

selection, 3-1
duties, 3-2
qualification training, A-3
quick fire
effectiveness, 4-11
modifications, 4-13
training, 4-13

ranges
alternate qualification courses, G-48
field fire, G-33
grouping, G-1
known-distance, G-25
known-distance alternate course, G-28
operations checklist, D-6
personnel, D-2
practice record fire, G-39
procedures, D-3
record fire, G-42
target detection, B-9
zero, G-10
rapid magazine changes, 4-4, 4-5 (illus)
rapid semiautomatic fire, 4-3
effectiveness, 4-3
magazine changes, 4-4
modifications, 4-3
training, 4-7
rate of fire, 4-2
record fire, 3-27
Record Fire Scorecard (DA Form 3595-R) G-41, G-47
Record Firing Scorecard — Known-Distance Course (DA Form 5789-R), G-31
Record Firing Scorecard — Scaled Target Alternate Course (DA Form 5789-R), G-52
Riddle sighting device, C-2, C-3 (illus)
rifle holding device, C-6 (illus)

75-, 175-, and 300-Meter Downrange Feedback Scorecard (DA Form 5239-R), G-21
safety briefing, D-1
sights
standard (M16A1), 3-22, 3-24
standard (M16A2), 3-25
Single and Multiple Targets — Field Firing Scorecard (DA Form 5241-R), G-37
single-lead rule, 4-21
Single Target — Field Firing Scorecard (DA Form 3601-R), G-36
suppressive fire, 4-1

target box exercise, C-4, C-5 (illus)
target detection, B-1
exercises, B-1, B-15
MOPP firing, 4-15
range determination, B-4
ranges, B-7
target location, B-1
target marking, B-3
tests, B-14
training, conduct of, B-10
trials, conduct of, B-10
Target Detection Exercise Answer Sheet — Periods 1, 2, and 8 (DA Form 3009-R), B-17
Target Detection Exercise AnswerSheet — Period 3 (DA Form 3010-R), B-19
Target Detection Exercise Answer Sheet — Period 5 (DA Form 3011-R), B-23
Target Detection Exercise Answer Sheet— Period 7 (DA Form 5791-R), B-27
Target Detection Exercise Answer Sheets Tests No. 2 and 3 — Period 9 (DA Form 3014-R), B-31
target engagement techniques, 5-4
target marking, B-3
targets, scaled silhouette, E-1
slow-fire target, E-1, E-3 (illus), E-7(illus)
timed-fire target, E-2, E-5 (illus), E-7 (illus)
training, E-4
target trial card, B-10, B-11 (illus)
training aids and devices
caliber .22 rimfire adapter, M261, C-12
classifications, C-1
exercises, C-2
location of miss and hit (LOMAH) system, C-12

multipurpose arcade combat simulator, C-13
resources, C-1
selection of, C-8
target ordering numbers, C-10
Weaponeer, C-17

training, rifle marksmanship, 3-1
automatic fire, 4-10
basic program, 3-1
conduct of, 3-3
downrange feedback, 3-27
dry fire, 3-18
field fire, 3-27
firing positions, 3-12
night firing, 5-5
practice record fire, 3-27
quick fire, 4-13
rapid-fire, 4-7
record fiere, 3-27
target detection, B-10
training strategy, 1-1, 1-2 (illus)
initial training, 1-3
sustainment training, 1-3

training, year-round marksmanship, A-1
live-fire exercises, A-15
sample evaluation quide, A-16
tasks, A-4 to A-15
training the trainer, A-2
qualification training, A-3
unit sustainment, A-15

unit sustainment training, A-15, C-22

wind, effects of, F-7
aiming point, adjusted, F-10
measurement, F-9
flag method, F-9
observation method, F-10
pointing method, F-9
wind speed and direction, F-7

Weaponeer, C-17 (illus)
characteristics, C-15
description, C-16
equipment data, C-16
mobile configuration, C-19
mobile mounting stand, C-21 (illus)
mobile training unit, C-20 (illus)
use, C-18

zeroing
M16A1, 3-21, 3-22
M16A2, 3-25

www.ingramcontent.com/pod-product-compliance
Lightning Source LLC
Chambersburg PA
CBHW080533170426
43195CB00016B/2544